THE VERY EARLY UNIVERSE

THE VERY EARLY UNIVERSE

Proceedings of the Nuffield Workshop, Cambridge
21 June to 9 July, 1982

Edited by
G. W. GIBBONS and S. W. HAWKING
Department of Applied Mathematics and Theoretical Physics
University of Cambridge
and
S.T.C. SIKLOS
Newnham College, Cambridge

The right of the University of Cambridge to print and sell all manner of books was granted by Henry VIII in 1534. The University has printed and published continuously since 1584.

CAMBRIDGE UNIVERSITY PRESS
Cambridge
London New York New Rochelle
Melbourne Sydney

Published by the Press Syndicate of the University of Cambridge
The Pitt Building, Trumpington Street, Cambridge CB2 1RP
32 East 57th Street, New York, NY 10022, USA
10 Stamford Road, Oakleigh, Melbourne 3166, Australia

First published 1983
Reprinted 1984
First paperback edition 1985

Printed in Great Britain at the University Press, Cambridge

Library of Congress catalogue card Number: 83–7330

British Library Cataloguing in Publication Data

Gibbons, G. W.
The very early Universe

1. Cosmology
I. Title II. Hawking, S. W. III. Siklos, S.
523.1'01'84 QB981

ISBN 0 521 25349 7 hard covers
ISBN 0 521 31677 4 paperback

CONTENTS

INTRODUCTION

Ideas about cosmology have greatly changed in the last 60 years. At the start of this century it was accepted without question that the universe was essentially uniform and static apart from the local motions of stars and planets. Such a model suffered from problems with gravitational instability and with the second law of thermodynamics, but these were generally ignored. Most people preferred to believe that the universe had existed for ever because this avoided awkward questions about the initial data and about what happened before the beginning. This was still the accepted world picture when Einstein formulated his general theory of relavitity in 1915. He therefore introduced a cosmological constant term into his field equations in order to get a solution which represented a uniform static universe with a non-zero density of matter. In the context of the classical field equations, the cosmological constant was an adhoc addition but, nowadays, we believe that it should arise naturally from quantum field theory. The mystery is why it should be so small.

The static picture of the universe was abandoned in the 1920s when observations of nearby galaxies showed that they were receding from us at a rate proportional to their distance from us. People therefore adopted the next simplest model of the universe, one in which the universe was spatially uniform and isotropic but expanding. Under these assumptions, the Einstein field equations could be solved for the evolution of the model. It was found that, under reasonable assumptions, they always predicted a singularity of infinite density about ten thousand million years in the past. There were several attempts to avoid the conclusion that the universe had a beginning in time by dropping the assumptions of spatial uniformity and isotropy or by modifying the field equations to obtain a "Steady State" solution. However, these attempts failed or were shown to be inconsistent with observations. In fact all the observations to date are consistent with the "Standard Hot Big Bang" model in which the universe is assumed to have started off with infinite density and temperature and to have expanded in a spatially uniform and isotropic manner. We have direct evidence from the microwave background that this model is a good approximation at least back to the time at which the radiation decoupled from the matter at a red-shift of about 1,000. There is indirect evidence from the agreement between observations and predictions of the abundances of light elements that the model is good back to about one second after the Big Bang. However, the model does not explain why the universe was as it was at one second. It is simply assumed as an initial condition.

The aim of the Nuffield Workshop on "The Very Early Universe" was to try to account for the conditions of the standard model at one second in terms of physical processes in the very early stages. In particular one would like to explain the following features:

1) Why was the universe spatially uniform and isotropic over length scales much bigger than 10^{10}cm, the horizon size at 1 second

2) Why, despite the above uniformity, were there small irregularities which later developed into stars and galaxies

3) Why was the universe spatially flat to within one part in 10^{16} i.e. why was the universe expanding at almost exactly the critical rate required to avoid recollapse

4) Why was the ratio of baryons to photons about 10^{-9}

5) Did the universe contain significant numbers of exotic objects such as magnetic monopoles, strings, small black holes, axions, photini, gravitini or other -ini.

In addition to these questions which refer to the initial conditions for the standard model, there is also a puzzle about the cosmological constant: why is it so small that we have not been able to measure it whereas one might expect quantum effects to make it 10^{120} times larger.

The standard Hot Big Bang model and the observational evidence for it are reviewed in the articles by Wilczek and Rees. They describe what constraints can be placed on the present matter content of the universe and on the departures from spatial uniformity. They set the scene for the remaining articles in this volume which attempt to answer the questions posed above.

One possible answer would be to appeal to the Anthropic Principle: there exist very many universes with different initial conditions and values of the cosmological constant but only the small number of those that are like our universe are suitable for the development of intelligent beings to ask the question: "Why is the universe the way we see it?". However, most physicists would regard this as a counsel of despair. They would prefer to find an explanation in terms of a model with few or no adjustable parameters or initial conditions. One candidate, the Inflationary Model, received a lot of attention at the Workshop. It does not claim to explain the cosmological constant but it offers possible answers to the other questions although it seems that the irregularities it predicts in the universe may be too large, at least in the simpler versions of the model. About half the articles in this volume are directly related to the inflationary model. The remainder deal with a number of other problems and processes that are expected to occur in the very early universe.

The basic assumption in all the contributions to the Workshop is that the typical energies of particles increase as one goes back in time in the very early universe. When the energies are greater than the Planck energy, 10^{19} GeV, one would expect quantum fluctuations of the spacetime metric to be of order one and the classical concepts of space and time would probably break down. The ultimate description of the initial conditions of the universe must involve an understanding of this quantum gravity era. At the moment we do not have a complete theory of quantum gravity, but only some partial results. These are reviewed in the article by Hartle. However we expect that at energies lower than 10^{19} GeV, one can ignore quantum gravitational effects and treat the other quantum fields as evolving in a classical spacetime background. Following the success of the SU(2)xU(1) theory, which unifies the weak and the electromagnetic interactions at energies above 100 Gev, it is generally believed that these two interactions will be unified with the strong interactions in some Grand Unified Theory at energies above some Grand Unified Energy somewhere between 10^{15}GeV and 10^{19}GeV. These theories are reviewed in the article by Wilczek and are further discussed in the article by Ellis in the case that they are supersymmetric, that is, they possess a symmetry which relates particles of integral spin (bosons) with particles of half-integral spin (fermions). Such supersymmetric theories have a number of technical advantages because there are cancellations between fermion and boson closed loops. However no model has yet been proposed which is in complete agreement with experiments.

A general property of all Grand Unified Theories is that they allow interactions in which the number of baryons changes. It is therefore possible that the universe started out with equal numbers of baryons and anti-baryons and that the present observed excess of baryons over anti-baryons was generated by CP violating interactions in the T asymmetric environment of the early universe at the time that the energies had fallen to around the Grand Unified Energy. This might provide an answer to question 4 above though the ratios of baryons to photons produced are very model dependent so that one can hardly claim to predict the observed value of 10^{-9}. Another feature of Grand Unified Theories is that they must incorporate some form of symmetry breaking to explain why the weak and electromagnetic interactions are different from the strong interactions at energies below the Grand Unified Energy. In the very early stages there would have been complete symmetry but, as the universe expanded and the typical particle energies fell, there would have been some sort of phase transition in which the symmetry was broken. In the standard Hot Big Bang model, one would have expected the direction of symmetry breaking to be different in different regions because light would not have had time to travel from one region to another. At later times, when different regions came into causal contact with each other, the symmetry breaking would try to choose the same direction everywhere. However, there can be topological obstructions or "knots" which prevent the symmetry breaking from aligning itself everywhere. These topological obstructions or knots manifest themselves as massive monopoles, strings or domain walls. Monopoles are expected in any reasonable Grand Unified Theory but strings and domain walls occur only in certain theories. Such objects could have very important effects on the evolution of the universe and might be observable at the present time. They are discussed in the articles by Preskill, Shafi and Vilenkin.

In the standard Hot Big Bang model the number of monopoles

predicted would be much higher than is consistent with observation. This was one of the reasons that led Guth in 1981 to suggest the Inflationary Universe scenario. The idea was that as the universe expanded and particle energies fell below the Grand Unified Energy, the symmetry was not broken immediately but the universe supercooled in a metastable state of unbroken symmetry. As the thermal energy of the particles decreased, the energy momentum tensor would become dominated by the vacuum energy of the metastable state. The vacuum energy of the broken symmetry state that we live in today must be very small because otherwise it would act as an effective cosmological constant. The inflationary model does not attempt to explain why the present effective cosmological constant is very small but it uses the fact that it is, to predict that the vacuum energy of the metastable state of unbroken symmetry must have been high. This means that the vacuum energy would have caused the universe to expand exponentially in the very early stages. During this "inflationary" expansion any departures from uniformity or isotropy and any initial baryon number would have died away and the universe would have approached a state that was independent of the initial conditions. This is described in the articles by Boucher and Gibbons, and by Barrow.

In Guth's original proposal, the period of exponential expansion was supposed to end with the formation of "bubbles" of the broken symmetry state surrounded by the metastable state of unbroken symmetry. These bubbles would have expanded and collided as described in the article by Siklos and Wu and would have reheated the universe to a temperature near the Grand Unified Energy. However, the bubble collisions would have left the universe much more inhomogenous and anisotropic than is consistent with observation. This difficulty was removed by work by Linde, Albrecht and Steinhart, and Hawking and Moss who suggested that the symmetry might have been broken slowly in a spatially uniform manner. This "New Inflationary Scenario" is reviewed in the articles by Guth and Linde. It might provide answers to all the questions listed above with the exception of why the present cosmological constant is so small. In particular it predicts that the present density of the universe should be almost exactly the critical density that separates recollapse from indefinite expansion. It also predicts that the departures from exact uniformity and isotropy should have a scale free or almost scale free spectrum. It should be possible to test these predictions with observations in the fairly near future. The amplitude of the fluctuations differs somewhat in the articles by Guth, Hawking, Turner, Lukash and Novikov, but in all cases the predicted amplitude for the simpler Grand Unified Theories is too large to be consistent with observations of the isotropy of the microwave background radiation. One possible answer would be supersymmetric Grand Unified Theories. These are described in the articles by Ellis and Steinhardt.

The curvature of spacetime in the very early universe has important effects on the behaviour of the quantized matter fields. These are discussed in the articles by Allen, Ford, and Hu. Of particular interest is the effective potential of the scalar fields because this determines whether and in what manner symmetry breaking will occur. In order that the symmetry breaking should take place slowly enough for the new inflationary scenario to work, the potential has to be very flat. In the ordinary Grand Unified Theories, this requires fine tuning of the renormalized parameters but it may occur naturally in supersymmetric theories.

Monopoles and strings are not the only relics that may be left over from the very early universe. Other particles or objects might include massive neutrinos, gravitini, photini and other supersymmetric partners of normal particles, axions and low-mass black holes. Observational limits on their present abundance place important constraints on Grand Unified Theories and on conditions in the very early universe. These are described in the articles by Markov, Lazarides, Wilczek, Page, Sciama, and Khlopov and Polnarev. Finally, there is the problem of the cosmological constant, why it should be so small, and whether it would lead to instability. These issues are discussed in the articles by Dolgov, Perry, and Hull.

The results and successes of the Nuffield Workshop are summarized in the concluding article by Wilczek. The general opinion was that although we had not solved all the problems of the universe, we had made substantial progress and that the Inflationary Model gave grounds for hoping that we might be able to explain why the universe is the way it is. The Inflationary hypothesis has the great advantage that it makes definite predictions about the present density of the universe and about the spectrum of departures from spatial uniformity. It should be possible to test these against observations in the fairly near future and, either falsify the hypothesis, or strengthen it.

We would like to thank the Nuffield Foundation for the financial support which made this Workshop possible. We are grateful to all those who attended and made it such a success, and to those who contributed to these Proceedings. We would like to thank Judy Fella for doing much of the organisational work and Anne Ralph for typing a number of the articles. We hope this volume will be of some help to those interested in how the universe began.

S.W.H.

FOUNDATIONS AND WORKING PICTURES IN MICROPHYSICAL COSMOLOGY

F. Wilczek
Institute for Theoretical Physics
University of California, Santa Barbara
Santa Barbara, California 93106 USA

This is a brief, qualitative, but critical discussion of the main achievements and current prospects in the application of microphysics to cosmological problems.

I will divide the discussion into a description of the foundations, and the less secure or final working pictures. These are some speculations on the future of the subject. At the end there are some speculations on the ultimate future of the subject matter (scientific eschatology).

COSMOLOGICAL FOUNDATIONS

i) Recession of galaxies

Modern observational cosmology began with Hubble's discovery that the distant galaxies are receding from us, with velocities proportional to their distance:

$$v = Hr . \tag{1}$$

Almost sixty years later, much improved data and analysis has confirmed this formula, though the value of H has been revised downward by a factor 5 - 10 from Hubble's estimate. The modern value of H is

$$H = h\ 100 \text{km./sec. Mpc.} \tag{2}$$

where h is a numerical factor between 1/2 and 1. There has been lively debate recently concerning the value of h. It has become clear that many galaxies in our neighborhood are "falling" toward the center of the Virgo cluster; this effect must of course be separated from the general cosmological expansion. Programs of measurements are going forward and will be greatly aided by the space telescope; we can expect that the uncertainty in h will be much less than a factor two in the near future.

For very distant galaxies deviations from (1) are expected, very roughly speaking because the velocity of expansion was greater in the past, the galaxies having since been slowed by their mutual gravitation (of course the light we receive from distant galaxies presently reflects their past condition, because of the finite speed of light). The magnitude of these deviations would be a very important indicator of the overall density of the Universe, if they could be accurately measured. So far the great difficulty in defining suitable standard objects, whose physical properties such as size or luminosity are independent of the time in the history of the Universe at which they are observed, has limited the success of this program.

The form of (1) has two important implications. First, a general expansion of this form will leave the Universe homogeneous and isotropic if it starts out so. The distribution of galaxies is at best only roughly isotropic on very large scales, but the microwave background radiation (see below) is very accurately isotropic. It seems most probable that the Universe has been homogeneous and isotropic since very early in its history, with galaxies forming by gravitational condensation of small fluctuations. Accounting for both the approximate uniformity and the small deviations from it (and their evolution into galaxies) is a main goal of recent theoretical activity, to be discussed below.

Second, extrapolating the recession of galaxies backward in time we find that the galaxies must have emerged from a denser phase in the past. In fact extrapolating (1) backwards naively we would find the galaxies meeting at a single point of infinite density at a time H^{-1} ago.

ii) Microwave background

Pursuing the idea that the Universe was much denser -- and hotter, since matter cools on expansion -- in the past, we can try to reconstruct the early history of the Universe. In this regard it is very important to note that at high temperatures and densities thermal equilibrium may be achieved in a short time even for interactions which are quite feeble under ordinary conditions. The relevant time scale here is the rate of expansion of the Universe. If thermal equilibrium is achieved, the behavior of matter can be described using only microphysical laws and the principles of statistical mechanics. This is a profoundly important point, since it can obviate the need to impose initial conditions on our cosmologies.

At sufficiently high temperatures, of order leV., matter is ionized into electrically charged particles. These particles interact efficiently with photons, and equilibrium is rapidly established including a radiation gas of photons. As the Universe expands and the temperature falls, the ionization at some point drops precipitously. At this point there are few free charged particles, only neutral atoms, and the matter essentially becomes transparent to (decouples from) the ambient radiation gas. The radiation gas subsequently cools by expansion. Just such a gas, now cooled down to ∿2.7°K, was first deteced in the famous observations of Penzias and Wilson and subsequently confirmed by many other measurements.

A most remarkable feature of this microwave background radiation is its high degree of isotropy. The anisotropy is bounded above by 10^{-3} on a wide range of angular sizes. This is a strong indication for the overall uniformity of the Universe at $T \approx$ leV. when the photons originated. Fluctuations in the overall mass density would lead to position-dependent gravitational redshifts which would show up as an anisotropy. Moreover, since fluctuations tend if anything to grow with time by gravitational condensation, we are lead to think that the Universe, at least as far as overall mass density is concerned, must have been extremely homogeneous at earlier times. Thus the observed high degree of isotropy of the microwave background encourages us to take precisely homogeneous and isotropic cosmological models, the Friedmann models, very seriously as a starting point for modelling the actual Universe. It also sharpens the problem of galaxy formation, since we must manage to condense galaxies out of an apparently very uniform mass distribution.

The observed microwave background gives us a meaningful way of calibrating the density of matter in the Universe. By "meaningful" here I mean independent of the time at which it is observed, since of course the density is always decreasing in an expanding Universe. Neither number of photons nor the number of baryons in a comoving volume has changed significantly since the photons decoupled, so the ratio of number densities

$$n_B/n_\gamma \approx 10^{-11} \qquad (3)$$

is a relic from that time. Actually, insofar as baryon number is conserved and the photon density grows only slightly under ordinary conditions (through particle-antiparticle annihilations), n_B/n_γ is expected to be approximately constant since considerably earlier.

iii) Nuclear abundances

At still higher temperatures of order 1 MeV., we have another sort of ionization -- atomic nuclei decompose into their constituent protons and neutrons. Also roughly at these sorts of temperatures and above, by an apparently fortuitous coincidence, thermal equilibrium is established with respect to the weak interactions. In particular, reactions such as

$$\bar{\nu}_e + p \leftrightarrow e^+ + n \tag{4}$$

establish chemical equilibrium between protons and neutrons, so that the relative abundance of these particles, more protons than neutrons, is dictated by statistical mechanics.

As the Universe expands and cools, the protons and neutrons can start to combine into nuclei. The reaction

$$^2H + \gamma \rightarrow n + p \tag{5}$$

makes it difficult to get started (especially since there are so many photons around -- equation (3)!) -- this deuterium bottleneck is only passed for $T \lesssim 100$ keV. Once the deuterium bottleneck is passed, essentially all the available neutrons combine into deuterium and thence ultimately into ^{4}He. Heavier elements are not produced significantly in these conditions, because of the absence of stable nuclides at A = 5 or 8. (They are produced in stars, where there is longer to cook, by triple collisions).

At the crudest level, then, this picture tells us that material which has not undergone processing in stars should consist almost entirely of ^{1}H and ^{4}He. One can calculate much more precisely, with the result (under assumptions to be discussed further below)

$$^4He/(^1H + {}^4He) \approx .24 \quad \text{(by weight)} \tag{6}$$

This result seems to be confirmed, within the uncertainties, by observations. It is the most impressive quantitative result of Big Bang cosmology.

Much interest attaches also to the lighter nuclei ^2H and ^3He. In the cosmological context, these result mainly from incomplete burning since their equilibrium abundance at $T \lesssim 100$ keV. is negligibly small. Their final abundance depends sensitively on the density of baryons. A high density of baryons will make the burning efficient and leave us with no ^2H or ^3He. Observations indicate an abundance

$$^2H/^1H \approx 2 \times 10^{-5} \quad \text{(by weight)} \tag{7}$$

$$^3He/^1H \approx 9 \times 10^{-5} \quad \text{(by weight)} \tag{8}$$

It is difficult to see how ^2H can be produced significantly in a stellar environment, since it is weakly bound and therefore readily disintegrated or burned into more stable nuclides by collisions. Sharpening arguments of this kind, one can put significant bounds on the baryon density from the fact that so much ^2H is observed in interstellar material. This is of great interest in connection with the missing mass problem, as we shall see shortly.

There is one important ingredient in the successful calculation of ^4He abundance which deserves further notice. That is, the result of this calculation depends sensitively on the number of species of light neutrinos that exist.

This remarkable connection between cosmology and microphysics arises as follows. Recall that the fraction of ^4He produced depends directly on the density of neutrons relative to protons when they combine at $T \approx 100$ keV. Now weak interactions such as (4) are attempting to bring the n/p abundance in line with the equilibrium Boltzmann factor, which is very small. However at these temperatures the typical rate of the weak interactions is not quite enough to keep pace with the expansion of the Universe, so equilibrium is not maintained and n/p is larger than its equilibrium value. The faster the expansion of the Universe, the more important is this effect and therefore the larger is n/p at T = 100 keV. The more species of neutrinos (ν_e, ν_μ, ν_τ, ...?) there are, the faster is the rate of expansion of the Universe. (This is most easily seen by imagining everything running backwards in time. The larger the density for a given temperature, the faster the collapse). Thus the more species of neutrinos, the larger n/p and $^4He/H$. Another effect which works in the same direction is that a faster expansion gives neutrons less chance to decay before T = 100 keV. is obtained.

Detailed computations indicate that four neutrino species is at best marginally acceptable, while larger numbers are certainly ruled out. This remarkable conclusion will be tested in the forseeable future, when the width of the Z boson (which reflects $Z \to \nu\bar{\nu}$ as well as experimentally visible channels) is measured. It seems fair to mention that there is at present no good theory indicating why the number of neutrino species should be 3, 4, or any other number.

iv) Mass density

Although Big Bang cosmology *per se* does not predict the mass density of the Universe, it does correlate the density with other quantities, notably the age of the Universe. Also, the exciting new inflationary universe ideas do make a definite prediction for the mass density. Finally, ideas in particle physics suggest that the bulk of the mass density of the Universe may very well be in some exotic form such as massive neutrinos, axions, photinos, or gravitinos. For all these reasons measurements of the overall mass density of the Universe provide a very important foundation for cosmology.

One traditional method for observing mass indirectly is to observe luminosity and then to estimate the amount of mass needed to constitute the sources of the luminosity. The crucial parameter here is the mass-to-light ratio M/L (It is usually quoted as a multiple of $(M/L)_\odot$ for the sun). For typical stellar populations

$$M/L = 3 - 10\ (M/L)_\odot \qquad \text{(stars)} \tag{9}$$

A comparable amount of mass is believed to be in intergalactic gas clouds, which are observed through x-rays rather than visible light. Adding this to the stars, we find that the baryonic matter required by the theory of nucleosynthesis (i.e., of ^2H and ^3He) is pretty much accounted for. It would be very difficult to accomodate much more, for example a large population of stars which have exhausted their nuclear fuel and are no longer luminous.

It is appropriate and customary to parametrize the mass

density as a fraction of the critical density

$$\rho_c \equiv \frac{3H^2}{8\pi G} \simeq 10^{-29} \text{ gm./cm.}^3 \tag{10}$$

where H is Hubble's parameter and G is Newton's constant. If the observed density $\rho > \rho_c$ then the Universe will eventually recollapse, otherwise it will expand forever. In these terms, we have for the density in baryonic matter

$$\Omega_B \equiv \rho_B/\rho_C \approx .02 - .10 . \tag{11}$$

Another way to measure mass density is directly, through its gravitational effects. For a long time, this sort of measurement has given much larger values for ρ than the indirect measurements discussed above. The older measurements are based on applying the virial theorem to large clusters of galaxies which appear to be gravitationally bound. Such measurements consistently give values of ρ 10 to 50 times bigger than the indirect ones.

Recently the case for significant non-luminous mass has become even more compelling, and extended down to galactic scales, by measurements of rotation curves around galaxies. The principle of such measurements is extremely simple: for a body in circular orbit around a spherical mass distribution we have

$$v(r)^2 = \frac{GM(r)}{r} \tag{12}$$

where r is the distance of the body from the center, v(r) its speed, and M(r) the amount of mass contained within distance r from the center. Deviations from this simple geometry complicate the analysis; I will not discuss that here. The striking fact is that for each of the approximately 100 spiral galaxies for which measurements have been made, it is observed that v(r) is essentially constant for 5 kpc. $\lesssim r \lesssim$ 40 - 60 kpc. This is remarkable, since the visible light from these galaxies drops off rapidly for $r \gtrsim$ 10 kpc. If the mass distribution followed the luminosity, we would have the very different behavior $v(r) \propto r^{-1/2}$ for $r \gtrsim$ 10 kpc. These rotation curves show that the visible galaxies are surrounded by dark matter whose total mass $M(r) \propto r$, corresponding to a density $\rho(r) \propto 1/r^2$.

All this makes a very strong case for a non-luminous, non-baryonic material which forms the bulk of the mass of the Universe and clusters on galactic scales and above. It is a main problem of micro-physical cosmology to determine what this material is.

A $1/r^2$ fall of the density is characteristic of an isothermal gas sphere bound by gravitation. (For which the pressure $p \propto \rho$. We have for the mechanical and gravitational forces on a unit volume respectively $F = -\nabla p$ and $F = \rho\nabla\phi$, where the gravitational potential ϕ satisfies Poisson's equation $\nabla^2\phi = 4\pi G\rho$. The whole system is consistent, with the forces balancing, if $\rho \propto 1/r^2$.) It therefore requires no miracles and can be accounted for by fairly well understood dynamical mechanisms of condensation (violent relaxation), which do result in isothermal gas spheres.

An interesting numerical fact is that if we extend the linear growth of $M(r)$ around galaxies as far as we can (namely until the distributions around different galaxies meet) then the total average mass density is roughly the critical density ρ_c. It therefore seems not inconceivable that the average mass density of the Universe really is ρ_c, which as we shall see is suggested by inflationary universe ideas.

v) Age of the Universe

In Big Bang cosmology the age of the Universe, i.e. the time since the Big Bang, is correlated with the density of the Universe. As mentioned before, utterly naive extrapolation of the Hubble expansion backwards in time would put the Big Bang H^{-1} ago. This is too naive, since earlier on velocities were larger, having since been slowed by gravitation. The amount of deceleration due to gravity is of course dependent on the density. The result is that for critical density the age is

$$t_c = \frac{2}{3}H^{-1} = 6.7h^{-1} \times 10^9 \text{ years} \tag{13}$$

In any case, we should have $t \lesssim H^{-1}$. Ages longer than t_c would indicate an ever-expanding Universe, shorter ages would indicate eventual re-collapse.

Two methods have been used to estimate the age of the Universe from observations.

The first involves globular clusters. These clusters are

presumed to be at least as old as galaxies (whereas if they were produced after collapse of the galaxy into a disc, they would be concentrated in the plane of the disc). There are plausible arguments that galaxies formed quite early in the history of the Universe; if so, then the age of globular clusters would be a good indicator of the age of the Universe. In any case, globular cluster ages provide a lower bound on the age of the Universe. The age of globular clusters can be estimated from observations using the theory of stellar evolution. The more massive stars exhaust their hydrogen fuel first, and their temperature and luminosity then undergo characteristic changes (leaving the main sequence in the Hertzsprung-Russell diagram). The turn-off point can be observed and stellar structure theory used to calculate at what age the relevant stars, those which are just leaving the main sequence, exhaust their hydrogen. These very complex calculations lead to the estimates

$$t_{glob.} \simeq 10 - 15 \times 10^9 \text{ years.} \tag{14}$$

The other method uses very long-lived isotopes like ^{235}U and ^{238}U. One can estimate the ratio of abundances of these isotopes at formation (they are probably produced by rapid capture of neutrons, in supernova explosions). The present abundance ratio can be observed; given the lifetimes of the isotopes it is then readily calculated how long a decay time is necessary to reach the observed ratio from the ratio at production. This simple procedure is complicated by the fact that many sources of different ages can contribute to present abundances. Without entering further into details, I simply record the most recent careful estimates

$$8 \times 10^9 \text{ yrs.} \lesssim t_{nucl.} \lesssim 19 \times 10^9 \text{ yrs.} \tag{15}$$

These age measurements can probably not be much improved in the near future. They do not give a convincing determination of the density ρ, nor do they clearly decide whether the Universe will expand forever or eventually recollapse. However it is most impressive that the age of the Universe is not very different from H^{-1}, confirming a basic qualitative prediction of Big Bang cosmology.

This concludes the survey of the foundations of Big Bang cosmology. The main conclusions are that the Universe did emerge from a dense hot phase, with temperatures $\simeq 10^{11}$°K (= 10 MeV.), 10 - 20 x 10^9 years ago. Furthermore, the Universe was very homogeneous and isotropic, at least in overall mass density, at early times. Finally, there is a large amount of non-baryonic, non-luminous matter to be accounted for.

It should be emphasized that no very exotic physics has been invoked in all this. The highest energy scales considered were only tens of MeV., which is low by accelerator standards; and the nucleon densities were not much greater than in stars.

CRITIQUE OF CLASSICAL COSMOLOGY

Several important questions are left unanswered, and cannot be addressed, in the classical cosmological framework just outlined.

i) The relic quantity $n_B/n_\gamma \approx 10^{-11}$ is an unexplained small parameter.

ii) What is the missing mass?

iii) The approximate homogeneity and isotropy of the Universe, and in particular of the microwave background radiation, is not explained. This is particularly disturbing, because of the phenomenon of causal disconnection. This is the fact that when we look at the microwave background in different directions in the sky we can see regions which have never both been influenced by any single local physical event -- their past light-cones are disjoint (as can easily happen if time cannot be extended past the Big Bang). In fact, background radiation from points more than a few degrees apart in the sky was produced in causally disconnected regions. This makes it rather paradoxical that the microwave background is so isotropic on large scales, since no local physical effect could have correlated it.

iv) Likewise, the small deviations from isotropy, including the fluctuations which eventually lead to galaxies, are not explained. Naturally, this question cannot logically be addressed before iii). The spectrum of fluctuations is potentially very rich in phenomenological implications, since it can leave its imprint on the microwave background and also in principle allow us to simulate galaxy formation. A sub-question is the nature of the fluctuations: are they adiabatic -- the ratio n_B/ρ of baryon number density to total density constant while each fluctuates separately, or isothermal -- only n_B fluctuates?

v) Why is the Universe at anywhere near the critical density? This is especially puzzling, because the deviation from criticality, $(\rho-\rho_c)/\rho_c$, is a rapidly increasing function of time. For it to be -1/2 now, it had to be $\sim -10^{-3}$ when photons decoupled, for instance.

Microphysics has at least the beginnings of plausible answers to all these questions. But to address them one must extrapolate above the $T \approx 10$ MeV. of classical cosmology, using developments of the past decade in particle physics.

MICROPHYSICAL FOUNDATIONS

i) Asymptotic Freedom

Matter gets very complicated to describe at nuclear densities or temperatures around $T \approx 300$ MeV. Fortunately, there are strong indications that things get simple again at higher densities and temperatures. This is because of asymptotic freedom, which is the property of certain gauge theories that the effective interaction they describe becomes weak at high energy scales. Quantum chromodynamics (QCD), the modern theory of the strong interaction, is asymptotically free. It promises that at high temperatures, $T \gtrsim 1$ GeV., strongly interacting hadrons essentially dissolve into a weakly interacting plasma of quarks and gluons. This makes statistical mechanical calculations to determine the behavior of matter in the extreme conditions of the very early Universe very easy -- basically we have a gas, which becomes more nearly ideal as T increases. QCD is in very satisfactory shape experimentally and seems securely established.

ii) Electroweak unified theories

Gauge theories uniting the weak and electromagnetic interactions have had remarkable success in providing a renormalizable theory of the weak interactions, explaining universality at a fundamental level, and in predicting the nature of neutral current interactions quantitatively. Many facets of the theories have been experimentally confirmed, and the characteristic W and Z bosons should be observed in the near future.

It would be surprising if these theories needed fundamental revision below scales of $\sim 10^4$ GeV. At the same time I do not think these theories can be in any sense regarded as complete, because quark and lepton masses and mixing angles are at present incorporated in an extremely arbitrary and unsatisfying way.

An important idea which plays a key role in these unified theories is that of spontaneous symmetry breaking. The Lagrangian, and therefore the equations of motion, of the theory has a larger symmetry than its realization in the world. This happens because the lowest energy states of the theory have less than the maximal symmetry. The physical vacuum is such a state, and the small fluctuations around it -- i.e. ordinary particles -- all feel its influence (it determines, for instance, their physical mass).

The no-particle states, then, acquire a dynamical life of their own; their energetics must be carefully considered in determining the physical vacuum. Indeed in some sense the particle content of the world is just an epiphenomenon bubbling up over the vacuum. This theme will be considerably elaborated below.

iii) Grand unified theories

Since QCD, based on the gauge group SU(3), and the Glashow-Weinberg-Salam SU(2) x U(1) theory of electroweak interactions are mathematically very similar, it is possible to speculate that they both arise from a single larger gauge group by spontaneous symmetry breaking. Georgi and Glashow observed that not only does the group SU(5) neatly accomodate SU(3) x SU(2) x U(1), but moreover the observed quarks and leptons fit very nicely into SU(5) multiplets.

The observed inequality of strong, electromagnetic, and weak couplings at low energies (i.e., the three couplings in SU(3) x SU(2) x U(1)) must then be explained as a renormalization effect -- asymptotic freedom extrapolated many orders of magnitude in scale, until the strong interactions are not stronger than the others. This happens at a scale $M \approx 10^{15}$ GeV. Since three couplings are supposed to meet at this scale, there is a constraint, which can be interpreted as a prediction for the Weinberg angle in SU(2) x U(1). This prediction is remarkably successful -- which, it should be stressed, is the only clear quantitative success to date of grand unification.

Grand unified theories in general, and particularly SU(5), in their simplest form predict nucleon decay at detectable rates. Experimental data on this question is eagerly awaited.

Grand unified theories in their present form suffer from exactly the same incompleteness regarding masses as electroweak unified theories, and in addition require grotesque adjustments of parameters to yield two very disparate scales $\mu \approx 300$ GeV. and $M \approx 10^{15}$ GeV. for SU(2) x U(1) and SU(5) breaking respectively.

Whatever their ultimate fate, grand unified theories have liberated us -- we are now very accustomed to thinking about extraordinarily large energy scales and effects which are small or rare at laboratory scales because they are fundamentally mediated at these extraordinarily large scales. One example, mentioned above, is baryon number violation. Another is lepton number violation, as manifested in Majorana masses for neutrinos. Both of these have significant cosmological implications, to be discussed.

MICROPHYSICAL COSMOLOGY -- WORKING PICTURES

The questions mooted in the critique of classical cosmology are now addressed.

i) Baryon density

If, as suggested by grand unified theories, baryon number is not strictly conserved then n_B/n_γ is a relic not unlike the microwave background or the abundance of nuclei -- but from an earlier epoch. Grand unified theories suggest that at $T \gtrsim 10^{15}$ GeV. exchange of ultraheavy particles establishes a statistical equilibrium in which the relative abundance of baryons and antibaryons is fixed to be one. As the Universe cools these B-violating exchanges do not abruptly cease, although they can no longer enforce chemical equilibrium. Under these circumstances, it is possible that a net asymmetry between baryon and antibaryon density is established and "freezes in" as the Universe cools further and B-violation becomes exceedingly rare.

Note that a non-zero density of baryons is asymmetric with respect to the discrete transformations CP and CPT. Thus generation of net baryon density inevitably involves CP-violating interactions. CPT violation interactions are really not required (fortunately!) since the expansion of the Universe provides a CPT-asymmetric background driving force.

Quantitative calculations of the baryon-antibaryon asymmetry are frought with uncertainties. The relevant interactions might be mediated by vectors or scalars, with poorly determined masses, and involve their couplings including CP violation parameters in a complicated way. To get a number out of this morass is akin to pulling a rabbit out of a soup of amino acids, but people have attempted it semi-quantitatively. These model calculations show that the observed n_B/n_γ is neither ridiculously small nor ridiculously large to fit in this picture.

The qualitative conclusion, that n_B/n_γ is determined by microphysics and need not be posited as a separate initial condition, is certainly very attractive. It also suggests that fluctuations should be adiabatic -- i.e., n_B/n_γ is a constant determined by microphysics even as the numerator and denominator separately fluctuate.

ii) Missing mass

Particle physics provides candidates for the non-baryonic dark matter.

The most straightforward candidates are undoubtedly neutrinos. There is no doubt a cosmological background gas of neutrinos, quite analogous to the cosmological background gas of photons, left over from $T \simeq 1$ MeV. when the Universe became transparent to neutrinos. There are about $100\ \nu + \bar{\nu}$ of each species per cubic centimeter. If the mass of one of these neutrino species is as much as ~ 50 eV., it will supply the critical density ρ_c.

Grand unified theories suggest Majorana neutrino masses of the form

$$m_\nu \sim m_q^2/M_G \tag{15}$$

where m_q is a quark mass and M_G some grand unification scale (the scale at which lepton number is violated). With $m_q = 30$ GeV. and $M_G \approx 10^{12}$ GeV. (very favorable values) m_ν is still just barely getting interesting at 1 eV.

The theory of neutrino masses is primitive, and one looks to experiment for guidance. Neutrino oscillation experiments indicate that neutrino masses above ~ 1 eV. are possible only if the mixing angles are very small. $\beta\beta$ decay experiments put limits on a Majorana mass for ν_e of 5 - 10 eV. A positive result for the mass of the electron neutrino 14 eV. $\lesssim m_\nu \lesssim 60$ eV. has been reported from measurements of the endpoint spectrum in ^3H decay, but is difficult to reconcile with the other results. It is very important to press these experiments further, but for now I would bet against cosmologically interesting neutrino masses.

A more exotic candidate for missing mass, the axion, has gained considerable attention recently. I cannot review here the internal necessities of particle theory which suggest the introduction of this particle. It is expected to be very light, 10^{-5} eV, and to interact exceedingly weakly with ordinary matter. Cosmologically, axions are produced copiously and very cold.

Supersymmetric theories predict a raft of new particles, and both photinos and gravitinos (fermionic partners of the photon and graviton, respectively) have been suggested as candidates for the missing mass. It is too soon to evaluate these ideas, which are changing rapidly.

Different forms of missing mass have different cosmological implications, in particular for the theory of clustering and galaxy formation. Neutrinos are relativistic when they decouple and stream freely, wiping out all fluctuations below $\sim 10^{16} M_\odot$ (clusters of galaxies). Smaller structures would have to emerge by hydrodynamical fragmentation when these large structures collapse. Axions, on the contrary, are produced cold and can lead directly to clustering on all scales. Further study of these questions might enable us to decide among the different candidates for missing mass. Of course, laboratory detection of neutrino masses or of axions would be decisive.

iii) Inflationary Universe

Alan Guth made a brilliant contribution to cosmology by suggesting that the Universe went through a phase of exponentially rapid expansion.

If this suggestion is correct, it immediately explains why the density of the Universe is close to critical. Indeed critical density corresponds in Friedmann universes to zero spatial curvature, while overcritical density corresponds to positive curvature (closed universe) and undercritical density corresponds to negative curvature (open universe). If the Universe expands by an enormous factor any preexisting curvature gets scaled away, just as a balloon blown up to the size of the Earth would be rather flat locally. Thus, a crucial quantitative consequence of the inflationary Universe concept is simply $\rho = \rho_c$.

Inflation also removes the paradox of causal disconnection.

In an inflationary cosmology the presently observable Universe originated from a very much smaller region than in standard Friedmann cosmology, and the small region can be causally connected.

We might also guess that inflation would tend to iron out inhomogeneities, which would explain the approximate homogeneity and isotropy of the Universe. The extent to which the guess is true, and in particular whether the residual fluctuations are of the right size to seed galaxy formation, is a matter which requires more detailed models and is still being actively investigated.

iv) Mining the vacuum

Clearly the inflationary Universe idea is a very important one. But are there plausible circumstances in which rapid expansion occurs? Recall that the dynamical equation of cosmology are

$$(\dot{R}/R)^2 = \frac{8\pi G}{3}\rho \tag{16}$$

$$\dot{\rho} = -3(\rho + p)\dot{R}/R \tag{17}$$

where R is the scale factor, ρ the density, p the pressure, and curvature terms (inessential here) have been suppressed. These equations simply represent conservation of energy for a comoving test particle and the first law of thermodynamics for the cosmological fluid, respectively. For a locally Lorentz invariant no particle ("vacuum") state the expectation value of the energy-momentum tensor must take the form

$$\langle T_{\mu\nu}\rangle = \Lambda\, g_{\mu\nu} \tag{18}$$

which corresponds to $\rho = \Lambda$, $p = -\Lambda$. The unusual feature of the equation of state of the vacuum is that positive energy means negative pressure, and *vice versa*. Supposing $\Lambda > 0$, we find from (17) that ρ = const., and from (16) that R grows exponentially. It is remarkable and very significant, that despite the rapid expansion of space the energy density of the vacuum does not go down. This sounds like getting something for nothing; it can work because negative pressure feeds energy into an expanding system!

Summing up, the Universe can undergo an exponential expansion if the energy-momentum density comes predominantly from the background no-particle state. Our present vacuum does not support exponential expansion, so $\Lambda = 0$ for it accurately (thereby hangs a tale, which will be discussed shortly!). It is quite possible, and occurs in reasonable models of grand or even electroweak unification, that there are other metastable "vacua" with higher energy, $\Lambda > 0$. If the Universe gets hung up in one of these for a while, it will cool by ordinary expansion until the vacuum energy-momentum dominates that of matter and then begin exponential expansion.

Why should such a metastable vacuum ever have occurred? The "vacuum" state of the Universe, which in models amounts to the values of certain scalar order parameter fields, is determined not simply by minimizing the energy or even the free energy, but kinetically. At very high temperatures symmetries tend to be restored, so we might imagine starting with a symmetric "vacuum" state at very large T. As the temperature cools, some "vacuum" with less symmetry may come to have lower free energy, and if there is no potential barrier the Universe will move toward it. However if there are barriers the Universe can get hung up in a thermodynamically unfavorable vacuum for a long time, and exponential expansion as discussed above will occur.

An important question is how inflation ends, how we can get back to an ordinary Universe dominated by particle energy. The energy locked up in the vacuum is not lost by expansion, as has been mentioned, so raw energy at least is not a problem. However if the potential barriers are large the transition from false to true vacuum takes place by nucleation of small bubbles, and although these grow and there are more and more of them they cannot keep up with the exponential expansion of space. The result is a swiss-cheese universe nothing like our own.

Recently Linde, Steinhardt, and others have suggested that the situation could be saved if the potential barrier is not very high so that one doesn't get hung up too long, but has a nearly flat segment so that the Universe expands sufficiently while the energy-momentum is still vacuum dominated as the scalar field slowly rolls down the nearly flat potential. Hawking and Moss, and Lapedes and Mottola, considered the possibility that the potential barrier is very small, so that after sufficient expansion a very large bubble forms, which could be our whole Universe. Notice that if the volume energy density differs very little between two states, very large bubbles will be needed if the unfavorable surface energy is to be overcome. These special behaviors of the potential occur only in very limited regions of the general space of parameters describing possible potentials, for which there is no obvious independent motivation. This situation could possibly be improved in supersymmetric theories, which can give very flat potentials and/or near degeneracy of different vacua in a natural way.

v) The cosmological term

Our present vacuum is a broiling sea of quantum fluctuations and is picked dynamically at the cost of many broken symmetries -- why is it transparent to gravity? This is a dramatic way of asking why the cosmological term Λ is negligibly small for our present vacuum. The energy densities characteristic of grand unified symmetry breaking are $(10^{15}\ \text{GeV.})^4$, the energy densities characteristic of electroweak symmetry breaking are $(10^{3}\ \text{GeV.})^4$, and even the energy density characteristic of chiral symmetry breaking in QCD is $(10^{-1}\ \text{GeV.})^4$ -- whereas observational limits on the present vacuum energy density Λ are roughly $(10^{-12}\ \text{GeV.})^4$! Clearly there is something to be understood here, and until it is, it's difficult to have confidence in schemes exploiting vacuum energies to drive inflation.

I would like to briefly mention one idea in this regard, that I am now exploring. It is to do something for the Λ-parameter very similar to what the axion does for the θ-parameter in QCD, another otherwise mysteriously tiny quantity. The basic idea is to promote these parameters to dynamical variables, and then see if perhaps small values will be chosen dynamically. In the case of the Λ-parameter, it is no great trick to promote it to a dynamical variable -- we need only add a scalar field ϕ with coupling $\phi\sqrt{g}$ (and possibly others) into the Lagrangian. The hard part is to arrange that the potential for ϕ is minimized when ϕ itself is of correct magnitude to cancel the vacuum energy of ordinary matter. It might be possible to arrange that the potential for ϕ does respond directly to $\langle T^\mu_\mu\rangle$ in some special (scale-invariant) theories. However it seems to me now that a more promising approach is to have the potential for ϕ contain terms which depend explicitly on the curvature. If it is energetically very favorable for the curvature to vanish, then ϕ has the capability of cancelling any Λ-term that might otherwise exist, allowing this to happen.

VACUUM ESCHATOLOGY

Speaking of metastable vacua, one begins to get a little nervous -- might our vacuum be metastable? If we do live in a metastable vacuum, at some stage a bubble of stable vacuum will nucleate and engulf us, changing the values of fundamental physical constants and ending civilization as we know it. Since the bubble wall moves at the speed of light, we shall never know what hit us. For our own curiosity and the peace of mind of future generations, it is important to investigate this question.

There are two parts to this investigation. First, we must consider whether the allowed domain of parameters in our microphysical theory permits our vacuum to be metastable with a long lifetime ($\gtrsim 10^{10}$ years!) but not absolutely stable. Second, we must follow the dynamical evolution of the Universe from the symmetric state at very high temperatures down to the present, to see if it would get hung up in our present (wrong) vacuum.

R. MacKenzie, M. Turner and I investigated these questions in SU(5) superunified models, and found that it is perfectly possible that we do live in a metastable vacuum. R. Flores and M. Sher recently investigated this question in the context of electroweak unified theories. They found that if the top quark mass is very heavy we do in fact live in a metastable vacuum. Of course the prediction that we will be engulfed by a bubble of different vacuum has a peculiar epistemological status, since it can never be experimentally verified (the observer is vaporized). Nevertheless, if $m_t \gtrsim 125$ GeV., watch out!

REFERENCES

General references on classical cosmology: (Sciama 1975, Silk 1980, Weinberg 1972).
General reference on gauge theories, including early cosmological applications: (Mahapatra and Lai 1981).
References on inflation: (Guth 1981, this book).
Vacuum eschatology: (Turner and Wilczek 1982, Flores and Sher 1982).

Flores, R. and Sher, M. 1982, UC Irvine preprint.
Guth, A. 1981, Phys. Rev. D23, 347.
Mohapatra, R. and Lai, C. 1981, Gauge Theories of Fundamental Interactions Singapore: World Scientific.
Sciama, D. 1975, Modern Cosmology, Cambridge: Cambridge University Press.
Silk, J. 1980, The Big Bang, San Francisco: W.H. Freeman
Turner, M. and Wilczek, F. 1982, Nature 298, 633.
Weinberg, S. 1972, Gravitation and Cosmology, New York: Wiley.

WHAT THE ASTROPHYSICIST WANTS FROM THE VERY EARLY UNIVERSE

M.J. Rees
Institute of Astronomy, Madingley Road, Cambridge CB3 OHA
England
and
Institute for Advanced Study, Princeton, N.J. 08540, U.S.A.

Abstract. This article briefly summarises the astrophysical evidence on the scale of the universe, its content (baryonic and non-baryonic) and its overall isotropy. Reasons are advanced for believing that even at very early stages the Friedmann models may be applicable. The upper limits on fluctuation amplitudes on various scales are reviewed, and compared with the lower limits required by alternative scenarios for galaxy formation.

1 SOME "GLOBAL" REQUIREMENTS

1.1 The gross scale of the observed Universe

A prime aim of modern cosmologists is to understand the scale and homogeneity of our Universe. Quantifying this introduces vast numbers: there are $\sim 10^{80}$ baryons and $\sim 10^{89}$ thermal photons within the part of the Universe we can observe, and one would like to understand why so many powers of ten separate the "cosmic" scale from the Planck scale. A Universe must have a large space-time volume in order that stars and other entities of astrophysical interest can form and evolve within it. The "large numbers" involved in stellar physics can be expressed in terms of known fundamental ratios, and it may be interesting to relate the minimum size of "astrophysically acceptable" universes to these same quantities.

The basic properties of stars can be expressed in terms of the inverse "gravitational fine structure constant" $\alpha_G^{-1} = \hbar c/Gm_p^2 \simeq 1.7 \times 10^{38}$, m_p being the proton mass. The vastness of this number (or other equivalent ones related to it by factors like $\alpha_f = e^2/\hbar c$, or m_p/m_e) expresses the weakness of gravity on the microphysical scale. Familiar arguments then tell us that stars — whether they are "main sequence" stars (gravitationally confined fusion reactors) or white dwarfs — have masses of order $\alpha_G^{-3/2}\, m_p$. Moreover, straightforward physical arguments show that their characteristic *lifetimes* exceed microphysical timescales ($\sim \hbar/m_p c^2$)) by a factor of order α_G^{-1} (Dicke 1962). The Planck mass is

$m_{Planck} = (\hbar c/G)^{\frac{1}{2}} \simeq \alpha_G^{-\frac{1}{2}} m_p$; so a universe in which stars can form and evolve must persist for a time $t \gtrsim \alpha_G^{-3/2} t_{Planck}$. Unless the density is orders of magnitude below "critical", or unless the mass-energy is overwhelmingly in some non-baryonic form (and it would be hard to form stars at all if either of these were the case) then the total number of baryons must be $\gtrsim \alpha_G^{-2}$. The corresponding quantitative statement of the "flatness" constraint is then that, at some early time t_i when the Universe starts being Friedmannian, the ratio of curvature radius to horizon size must be $\gtrsim \alpha_G^{-3/4} (t_i/t_{Planck})^{-\frac{1}{2}}$. (This requirement is slightly eased if $p << \frac{1}{3} \rho c^2$ for the later stages of the expansion).

A related line of argument can be developed which does not depend on the properties of stars. A hot Friedmann cosmological model will not permit "small-scale" departures from thermodynamic equilibrium until it cools to the stage when the free electrons combine to form atomic hydrogen. This happens at a "decoupling temperature" T such that $kT \simeq (0.02)\alpha_f^2 m_e c^2$, where the numerical factor 0.02 arises because the much smaller phase-space volume for bound electrons delays the recombination until well below the temperature of 1 Rydberg (equivalent to 0.5 $\alpha_f^2 m_e c^2$). The time t_{rec} taken to cool to this temperature in a radiation-dominated universe is $\sim 10^3 \alpha_G^{-1} (m_p/m_e)^2 \alpha_f^{-4} t_{Planck}$.

The above considerations show - for what it is worth - that it is because α_G^{-1} is so large that a vast universe is a prerequisite for the existence of stars, cosmologists, and other manifestation of thermodynamic disequilibrium (see Carr & Rees (1979) for further discussion of these points).

1.2 Overall isotropy and "Friedmann" behaviour

The Universe is manifestly inhomogeneous on all scales up to $\sim$ 10 Mpc, and maybe even up to $\sim$ 100 Mpc. But these are still small compared to the Hubble radius ℓ_H, and compared to the depth that can be probed with large telescopes; studies of remote faint galaxies, and of quasars and radio sources, show that these luminous objects are genuinely more isotropic around us on the largest scales we can observe. Unless we adopt anti-Copernican attitudes, this of course implies a high level of homogeneity on scales comparable with the Hubble radius (and probably on all scales $\ell > 10^{-2} \ell_H$).

The Hubble law tells us that the kinematics of the Universe are at least roughly isotropic; but much the best evidence for isotropy within our particle horizon comes from the $\lesssim 10^{-4}$ limits on any large-scale angular variations in the microwave background. So the Robertson-Walker metric must be an excellent approximation at least since the epoch when the microwave background photons decoupled from the matter. This does not in itself give much clue about the time-dependence of the Robertson-Walker scale factor R(t) – in particular it does not tell us whether it is governed by the equations of a Friedmann model. Actually we have better evidence that R(t) follows a Friedmann law at $t \simeq 1$ sec than at the present era : as discussed further in § 2.5, primordial nucleosynthesis, particularly of ^{4}He, is very sensitive to the expansion rate at the proton/neutron "freeze out" temperature, for which $kT \simeq 1$ Mev; the gratifying accordance of the standard prediction gives us confidence that the expansion was isotropic and that the Friedmann equations (for a radiation-dominated expansion) are valid when $t \simeq 1$ second. Indeed, some authors have inverted the argument and used the measured He abundance, plus the Friedmann equations, to constrain such things as the number of neutrino species.

Recent ideas on baryon synthesis — if the grand unified theories (GUTs) on which they are based are borne out by future developments — might allow us to test whether the Friedmann models apply back at temperatures of $\sim 10^{15}$ Gev, corresponding to times $\sim 10^{-36}$ sec. The observed baryon-to-photon ratio, a measure of the fractional excess of baryons over their antiparticles at early times when $kT > m_p c^2$, is $\sim 10^{-9}$. (If this number were much smaller than $\sim \alpha_G^{\frac{1}{4}}$, the Universe would not be baryon-dominated when its age was of order a characteristic stellar lifetime.) The value of the net baryon excess can be computed, given a specific GUT (Kolb & Wolfram 1980); it involves a small parameter related to the CP-violation parameter in weak interactions. This work is not yet on the same footing as the nucleosynthesis of helium and deuterium (see Weinberg (1982) for a recent review and assessment): it is perhaps at the same level as nucleosynthesis was in the pioneering days of Gamow and Lemaitre. But if it could be firmed up it would represent an extraordinary triumph. The mixture of radiation and matter characterizing our universe would not be "ad hoc" but would be a consequence of the simplest initial conditions. Also, as well as vindicating a GUT, it would reassure us about extrapolating in one bound, based on a Friedmann model, right back to the threshold of classical cosmology – almost back to the Planck time. On a logarithmic scale, this is a bigger extrapolation from the nucleosynthesis era than is involved in going to that era from the present time. It would also place constraints on dissipative processes arising from viscosity, phase transitions, black hole evaporation etc., which might occur as the Universe cooled through the "desert" between 10^{15} and 100 GeV. Although these ideas are still speculative, the "prediction" of the photon/baryon ratio may turn out to offer one of the few empirical tests of GUTs.

1.3 The cosmological constant

The cosmological constant, or "Λ-term", regarded by most relativists as an ugly appendage to Einstein's field equations, has now been resuscitated in a more modern guise, and plays an important role in recent ideas involving symmetry breaking. In GUT schemes involving a phase transition at a temperature $\sim 10^{15}$ GeV, the differing energy densities of the two vacua would be equivalent to different values of the Λ-term. Λ is now very small, corresponding to a mass energy density of less than 10^{-30} gms/cm^3 (or, in other units, to 10^{-44} GeV^4). One would expect the *change* in the effective Λ associated with symmetry breaking to be of the order of the energy density at the grand unified temperature, i.e. 10^{60} GeV^4. Why things should be fine-tuned to a precision better than one part in 10^{100}, so that the *post-transition* Λ is so small, is still unexplained. But it is certainly a requirement of any astrophysically acceptable universe that it should pass through a stage where the self-gravity of the matter can overwhelm the repulsive effect of the cosmological constant.

2 THE CONTENT OF THE UNIVERSE

2.1 "Luminous" matter: galaxies and gas

It is convenient to introduce the "critical" density

$$\rho_c = (\frac{8}{3}\pi G t^2_{Hubble})^{-1}$$

and to define a density parameter $\Omega = \rho/\rho_c$. Since t_{Hubble} is still uncertain, in quoting numerical values I will introduce a quantity

$h = (10^{10}\ \mathrm{yrs}/t_{\mathrm{Hubble}})$; the experts advocate values of h in the range 0.5 - 1 (see Sandage & Tammann (1982) and Hanes (1982) for recent assessments from differing viewpoints.) The baryon density is then $n_b = 3 \times 10^{-6}$ $(\Omega_b h^2)\ \mathrm{cm}^{-3}$, where Ω_b is defined as the fraction of the critical density in baryonic form. The baryon-to-photon ratio is then

$$\mathcal{S}^{-1} = n_b/n_\gamma \simeq 3 \times 10^{-8} \left(\frac{T}{2.7^{\circ}K}\right)^{-3} (\Omega_b h^2), \qquad (1)$$

$\mathcal{S}$ being a measure of the entropy per baryon.

A direct lower limit on Ω_b of order 10^{-2} can be set from the observed mean density of baryons in conspicuously "luminous" form (visible galaxies, and intergalactic gas revealed by its X-ray emission), implying a baryon-to-photon ratio of $\gtrsim 3 \times 10^{-10}\ h^2$. This is the number which GUT models must attempt to explain.

There is no firm evidence for any antimatter in the Universe, apart from a small fraction of cosmic ray anti-particles which could have been produced in high-energy collisions. Strong constraints on the presence of antimatter in and around our Galaxy are set by the measured limits to the γ-ray background. Nevertheless, if one is strictly agnostic and free from theoretical preconceptions, one can certainly envisage that the Universe might possess matter-antimatter symmetry (i.e. that the overall net baryon number, and $\langle n_b/n_\gamma \rangle$, are zero) provided that the scale of the regions of each "sign" is at least as large as a cluster of galaxies.

2.2 Evidence for "unseen" mass

There are dynamical indications of some "unseen" mass, which may or may not be baryonic. The most convincing evidence comes from applications of the virial theorem to rich clusters of galaxies (e.g. the Coma cluster) and from the statistical technique known as the "cosmic virial theorem", whereby one analyses the deviations from Hubble-law motions induced in galaxies by their neighbours. These studies are still bedevilled by observational problems, but they broadly suggest that Ω is in the range 0.1 - 0.2: in other words, there is perhaps ten times as much non-luminous matter as there is in stars and detectable gas; ten times as much gravitating stuff is implicated in the relative motions of galaxies as in the internal dynamics of individual galaxies. The "unseen" mass must be in diffuse halos around galaxies, or must pervade clusters or groups of galaxies.

The evidence suggesting $\Omega = (0.1 - 0.2)$ comes from studying the dynamics of galaxy groups on scales $(1 - 2)h^{-1}$ Mpc. It is interesting to ask whether the data *permit* $\Omega = 1$, the value favoured by advocates of "inflationary" cosmology. The short answer is that this may be consistent with the data provided that $\frac{M}{L}$ continues to increase with lengthscale out to $\gtrsim 10\ h^{-1}$ Mpc, so that on scales where the virial theorem can be reliably applied the "unseen" mass is less "clumped" than the luminous mass. Much attention has been given to the velocity field in the local supercluster. Our infall velocity towards the Virgo Cluster, relative to the mean Hubble flow, would be an indicator of the mean $\frac{M}{L}$ within the supercluster *if* the total mass was distributed like the galaxies. However, if one drops this assumption, the results become very inconclusive (Hoffmann & Salpeter 1982).

Only 10% (and maybe as little as 1%) of the mass-energy of the Universe is in "known" form. All that can confidently be said about the "unseen" mass is that it is "dark": it has a much higher "mass-to-light ratio" than the ordinary luminous content of galaxies. Whereas the inner parts of typical galaxies have $\frac{M}{L}$ which is $\lesssim 10$ times the solar mass-to-light ratio, if the unseen mass contributes a density parameter Ω it must have $\langle M/L \rangle$ exceeding $2300\ \Omega h$ solar units.

2.3 Baryonic forms for the "unseen" mass

If galaxies and clusters were "assembled" from sub-units that condensed earlier, most of the initial baryons might have been incorporated in a pregalactic population of stars: these stars, or their remnants, could perhaps now have a high $\frac{M}{L}$ and contribute to the "unseen" mass. Ideally, one would like to be able to calculate what happens when a cloud of $10^6 - 10^8\ M_\odot$ condenses out soon after recombination – does it form one (or a few) supermassive objects, or does fragmentation proceed efficiently down to low-mass stars? Our poor understanding of what determines the mass spectrum of stars forming now in, for instance, the Orion nebula, gives us little confidence that we can calculate the nature of pregalactic stars, born in an environment very different from our (present-day) Galaxy.

Although we cannot predict what these pregalactic stars would be like, there are several constraints which, in combination, imply that if there are enough of them to provide the "unseen" mass, the individual masses must either be $\lesssim 0.1\ M_\odot$ or else in the range $10^3 - 10^6\ M_\odot$. Masses above $\gtrsim 0.1\ M_\odot$ would contribute too much background light unless they had all evolved and died, leaving dark remnants. But the remnants of "ordinary" massive stars of 10 - 100 $M_\odot$ would produce too much material in the form of heavy elements. Limits on the range 100 - 1000 $M_\odot$ are uncertain because only ^{4}He may be ejected, the "heavies" in the core collapsing into a black hole remnant. An uncertainty in the evolution of massive or supermassive stars is the amount of loss during H-burning; however the hypotheses that most mass goes into very massive objects (VMOs) of $10^3\ M_\odot$ is compatible with the nucleosynthesis constraints. A further consideration favouring these high masses is that VMOs are likely to terminate their evolution by a collapse which swallows most of the mass; if most of the material were ejected, "recycling" through several generations would be necessary in order to end up with most of the material in black holes rather than gas. Detailed discussions of pregalactic stars are given by Carr, Bond & Arnett (1983) and by Tarbet & Rowan-Robinson (1982).

The upper limit of $10^6\ M_\odot$ comes from noting that objects above this mass would have various detectable effects — e.g. dynamical friction or accretion — and can be ruled out as major constituents of galactic halos (Carr 1978). Higher masses could contribute to Ω if they were in intergalactic space, though systematic searches for gravitational lensing effects could soon reveal them, or allow us to exclude masses above $10^6\ M_\odot$. The probability of observing lensing along a line of sight to $z \simeq 1$ due to a population of compact objects is of order their contribution to Ω. The characteristic angular size of the image is, however, $\sim (M/10^6\ M_\odot)^{\frac{1}{2}}$ milliarc seconds

2.4 Non-baryonic unseen mass?

If neutrinos have negligible rest mass, the present density expected for relic neutrinos from the big bang is $n_\nu = 110\ (T_\gamma/2.7\text{K})^3\ \text{cm}^{-3}$ for each two-component species. This conclusion holds for non-zero masses, provided that $m_\nu c^2$ is far below the thermal energy ($\sim$ 5 MeV) at which neutrinos decoupled from other species and that the neutrinos are stable for the Hubble time. Comparison with the baryon density shows that neutrinos outnumber baryons by such a big factor ($\sim \mathcal{S}$) that they can be dynamically dominant over baryons even if their masses are only a few eV. In fact a single species of neutrino would yield a contribution to Ω of

$$\Omega_\nu = 0.01\ h^{-2}\ (m_\nu)_{ev} \tag{2}$$

so if h = 0.5, only 25 ev is sufficient to provide the critical density.

The entire range 100 h^2 eV - 3 GeV is incompatible with the hot big bang model (Gunn *et al.* 1978). (For $m_\nu \gtrsim 3$ GeV, the rest mass term in the Boltzmann factor would kill off most of the neutrinos before they decoupled; the number surviving would be $\lesssim n_b$). If any species of neutrino *were* discovered to have a mass in this excluded range, it would show that one cannot extrapolate the hot big bang back to kT $\gtrsim$ 5 MeV, and that most of the photons must have been generated at later times.

[Such arguments are familiarly expressed by saying that the hypothetical particles of non-zero rest mass would 'close the Universe by a large factor'. This loose phrase is in fact rather misleading. The geometry of the Universe is likely to have been laid down at very early stages by mechanisms which do not "know" what the dominant constituents will be after 10^{10} years. If the Universe were indeed 'flat' (e.g. for "inflationary" reasons), it would expand with $\Omega = 1$ (i.e. with a density of $(\frac{8}{3}\pi G\ t^2)^{-1}$) for all t. If the neutrinos were in the excluded mass range, or if there were, for instance, "too many" primordial monopoles, the observational incompatibility would be that the baryonic fraction of the total density would, for $t \simeq 10^{10}$ yrs, be $<< 10^{-2}$.]

Physicists have other particles "in reserve" - right handed neutrinos, photinos or gravitinos, for instance - which could (if they existed) have been in thermal equilibrium with other species at very early times, and therefore contribute to Ω in an analogous way (see Dolgov & Zeldovich (1981) for a review). The only difference would be that (n/n_γ) could be less than for neutrinos because the other "...inos" may have decoupled before muons (or even hadron pairs) annihilated — the latter annihilations would then boost the neutrinos but not the still more weakly coupled "...inos". For gravitinos, we may have $(n/n_\gamma) \lesssim 10^{-2}$.

Any such particles would be dynamically important not only for the expanding Universe as a whole but also for large bound systems such as clusters of galaxies. This is because they would now be moving slowly: if the Universe had cooled homogeneously, primordial neutrinos would now be moving at $\sim 200\ (m_\nu)^{-1}_{eV}$ km s^{-1}. They would be influenced even by the weak ($\sim 10^{-5}\ c^2$) gravitational potential fluctuations of galaxies and clusters. If the three (or more) types of neutrinos have different masses, then the heaviest will obviously be gravitationally dominant, since the numbers of each species should be the same.

It was conjectured more than a decade ago (Cowsik & McLelland 1972; Marx & Szalay 1972) that neutrinos could provide the "unseen" mass in galactic halos and clusters. In recent years astrophysicists have explored this possibility in some detail, and considered scenarios for galaxy formation in which neutrino clustering and diffusion play a key role. These scenarios have several appealing features, even though they lead to some new problems.

We may inhabit a universe where, on the largest scales, the baryons are merely a "tracer" for the distribution of a gravitationally-dominant neutrino sea. Neutrinos of mass $\sim$ 10 eV, gravitinos, monopoles or axions are just some of many candidates for the "unseen" mass in the Universe: as far as the astronomical evidence goes, the "unseen" mass could equally well be low mass stars, or black holes of up to at least 10^6 solar masses (which could be either primordial, or the remnants of a generation of very massive pregalactic stars). There is no lack of candidates for "unseen" mass, either baryonic or non-baryonic.

2.5 Primordial nucleosynthesis: need for non-baryonic matter if $\Omega \simeq 1$?

Some considerations based on primordial nucleosynthesis seem to favour the non-baryonic option, especially if the total density corresponds to $\Omega = 1$. The primordial nucleosynthesis depends on two things: the expansion timescale at 0.1 - 1 Mev and the baryon density (which is proportional to $\mathcal{S}^{-1} \propto \Omega_b h^2$. The predicted ^{4}He abundance is rather insensitive to the matter density: for $\Omega_b h^2 > 10^{-2}$ (corresponding to $\mathcal{S} \lesssim 3 \times 10^9$) the density of baryons is high enough to ensure that most of the neutrons which survive at "freeze-out" get incorporated in ^{4}He. However, the resulting ^{4}He abundance *is* more sensitive to the *expansion rate* when $kT \simeq 1$ Mev : if the Universe expanded somewhat faster, then the neutron/proton ratio would "freeze out" at a higher temperature, when neutrons were less disfavoured by the Boltzmann factor; the ^{4}He abundance would then be *higher*.

Schwartzman (1969) pointed out that the observed fractional abundance of ^{4}He can be used to place interesting constraints on the number of types of neutrino. This is because the expansion rate (at a given T) depends on the number of independent species: it is in fact proportional to $(\frac{11}{4} + \frac{7}{8} N_\nu)^{\frac{1}{2}}$. One of the main uncertainties in quantifying this line of argument is that we do not know the primordial ^{4}He abundance — all that we can say is that is must be *below* the *lowest reliably-determined* abundance in any astronomical object (since stars can make extra helium during the course of galactic evolution). Most astronomers (cf. Pagel 1982) would assess that the primordial ^{4}He must be $\lesssim 25\%$. This is consistent with $N_\nu = 3$ only for $\mathcal{S} > 2 \times 10^9$ (which corresponds to $\Omega_b h^2 < 0.02$ and permits $N_\nu = 4$ for $\mathcal{S} > 6 \times 10^9$ ($\Omega_b h^2 < 0.01$). Even lower densities would be needed, for a given N_ν and expansion rate, if the primordial fraction of ^{4}He were less than (say) 23% rather than 25%. However, *very* low values of Ω_b lead to another inconsistency; the abundance of ^{3}He and D, intermediate products in ^{4}He synthesis, exceed what is observed (Wagoner, Fowler & Hoyle 1967, Audouze 1982). These isotopes are both produced in the big bang; D can be burnt into ^{3}He in stars, but the primordial abundance of (^{3}He + D) is unlikely to have exceeded the presently-observed value of 8×10^{-5}. This constrains $\Omega_b h^2$ to be $\gtrsim 0.01$ (i.e. $\mathcal{S} < 3 \times 10^9$). On the other hand, if the observed D is a relic of the big bang (as is generally believed, owing to the failure of astrophysicists to come up with a plausible alternative way of making it), then $\Omega_b h^2$ must be *less* than about 0.025.

These arguments, based on primordial nucleosynthesis, suggest that $N_\nu \lesssim 4$, *and* that $\Omega_b h^2$ is in the range 0.01 - 0.025 (permitting a value of Ω_b no higher than 0.1, even for h = 0.5). Higher values of N_ν could be reconciled only with more contrived and inhomogeneous models. If the lepton number for ν_e and $\bar{\nu}_e$ were non-zero, then the neutron-proton equilibrium ratio would be shifted, thereby affecting ^{4}He production. It is possible in principle to compensate for *any* speed-up factor in this way and thereby permit a higher value of N_ν or a higher Ω_b (David & Reeves 1980). However, in order to make much difference, the neutrino lepton number must be of order the photon number - i.e. $\sim \mathcal{S}$ times larger than the baryon number. Moreover there is an impressive consistency in the predictions for the various light elements based on the simplest hypothesis ("standard" model with $N_\nu = 3$).

There are other possible complications and "escape clauses" (involving large-amplitude inhomogeneities in the baryon distribution, etc). But these considerations suggest that, unless one is to abandon the standard hot big bang model completely, the idea of non-baryonic "unseen" mass is very appealing, especially if (for other reasons) one favours an overall cosmological density as high as the critical value ($\Omega = 1$). Because the relevant parameter in primordial nucleosynthesis is $n_\gamma/n_b \propto \Omega_b h^2$, more precise comparison of models with observation must await a firmer value of the Hubble constant. If $h \simeq 1$ (corresponding to Hubble time of 10^{10} yrs) then the simplest inference would be that most of the "unseen" mass – both in the halos of individual galaxies and in clusters and groups – was non-baryonic; but if $h = \frac{1}{2}$ (corresponding to a Hubble time of 2×10^{10} yrs) the *lower* limit to $\Omega_b h^2$ set by the requirement not to overproduce D + ^{3}He implies that some unseen matter – maybe that in halos, if not in intergalactic space – is baryonic, though only enough to contribute $\Omega_b \lesssim 0.1$. It is remarkable that the simplest and least arbitrary form of hot big bang model can, for a suitable choice of Ω_b, account for D, ^{4}He and ^{7}Li (Yang *et al.* 1983).

3 THE ORIGIN OF GALAXIES : PRIMORDIAL FLUCTUATIONS

3.1 Galaxies : the basic units

The basic units delineating the Universe's large scale structure are, of course the *galaxies*. Much is now understood about their morphology and internal dynamics. But we still do not know *why* galaxies exist – why the most conspicuous large-scale features of the cosmic scene should be these gravitating aggregates of 10^{10} - 10^{12} stars, with dimensions 10^4 - 10^5 light years. Galaxies are about as standardised as stars. But, whereas we understand physically why stellar masses are $\sim \alpha_G^{-1}$ m_{Planck} we have no very convincing commensurate expression for galactic masses (but see Rees & Ostriker (1977) and Silk (1977) for tentative suggestions on this point). Even worse, we do not know whether the explanation we seek lies within the province of the astrophysicist or the cosmologist. Conceivably the right characteristic mass is somehow "imprinted" in the early Universe; alternatively, the galactic mass may be singled out by physical processes, just as stars in the stable mass range $\sim \alpha_G^{-1}$ m_{Planck} are the end-product of condensation from a broad mass-spectrum of inhomogeneities (in, e.g. the Orion nebula) without having to be favoured in the interstellar medium.

Despite our ignorance on this issue, two points are uncontroversial.

(1) The early Universe must have contained *some* inhomogeneities (despite its overall Friedmannian character); otherwise its baryon content would now, after $\sim 10^{10}$ yrs, still just comprise uniform neutral H and He.

(2) Whereas gas dynamical and dissipative effects must have been important in the formation of individual galaxies (so that the characteristics of galaxies are related only indirectly to the form of the primordial fluctuation spectrum), any inhomogeneities on much larger scales must be induced primarily by gravitation. The clustering properties of galaxies are the best evidence we have on this.

3.2 Inhomogeneities in the distribution of observable matter

There is now a good deal of quantitative evidence on how galaxies are clustered. Note, however, that we can only use this to infer the total density fluctuations insofar as galaxies are a good tracer for the overall mass distribution.

The work of Peebles and his associates (codified in Peebles (1980)) has led to a determination of the galaxian 2-point correlation function $\xi(r)$, defined so that the probability δP of finding a galaxy in a volume element δV a distance r from a known galaxy is $\delta P = n\delta V(1 + \xi(r))$, where n is the mean spatial density of galaxies. The result of this body of research is that $\xi(r)$ seems to obey a power law,

$$\xi(r) \simeq \left(\frac{r}{r_o}\right)^{-1.8}, \quad (0.1\ h^{-1} \lesssim r(\mathrm{Mpc}) \lesssim 10\ h^{-1}\ \mathrm{Mpc})$$

with

$$r_o \simeq 5 \times 10^{\pm 0.3}\ h^{-1}\ \mathrm{Mpc} \tag{3}$$

This form of correlation can be modelled by a "clustering hierarchy", and is consistent with the gravitational clustering which develops in an expanding universe of point masses initially set down at random.

There is now mounting evidence that $\xi(r)$ does not continue to fall off as steeply as (3) for $r > 10\ h^{-1}$ Mpc, but that there are conspicuous large-scale structures on scales up to $\gtrsim 50\ h^{-1}$ Mpc. The people who have scrutinised "superclusters" in the greatest detail claim that these large-scale structures are filamentary (Einasto, Joeveer & Saar 1980;

Figure 1. At the bottom of the diagram are shown two possible forms of the density perturbation spectrum at t_{rec} ($z \simeq 1000$). Entropy fluctuations at the left may have a spectrum extending down to $\sim 10^6 M_\odot$; objects in the mass range of 10^6–$10^8 M_o$ may condense out soon after t_{rec}, and galaxies and clusters then form later via hierarchical build-up of subunits. If the subunits generate energy, and give rise to "secondary" perturbations, then galaxies and clusters could form even if there were a negligibly small ($\ll 10^{-3}$) amplitude on the relevant scales at t_{rec} (as in the dashed spectrum). At the right is the contrasting case of adiabatic perturbations. The first scales to condense out would be on a scale of galaxy clusters or larger, because $(\Delta\rho/\rho)_{rec}$ would have been attenuated on smaller scales by pre-combination damping processes. If massive neutrinos play a key role in galaxy formation, the scenario resembles that shown on the right, though the characteristic mass below which attenuation occurs is then set by phase mixing of collisionless neutrinos rather than by photon viscosity.

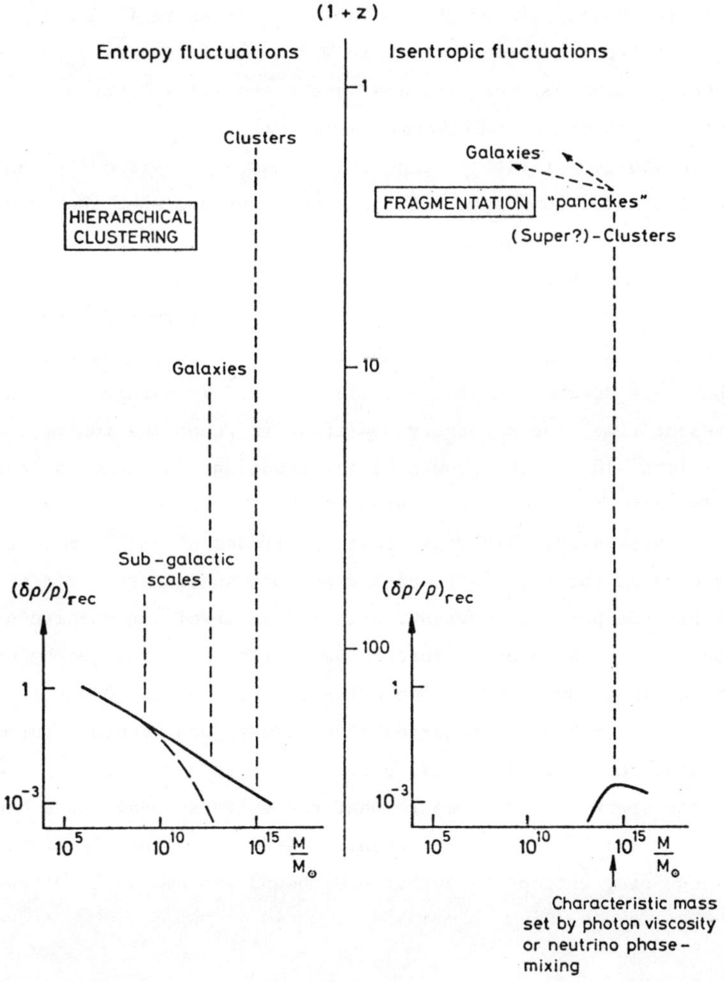

Davis *et al.* 1982; Oort 1982; Zeldovich, Einasto & Shandarin 1982). This is certainly the subjective impression that the data give, but only recently have quantitative statistical measures (e.g. those used in percolation theory) been applied which are sensitive to effects on scales much larger than the mean spacing between neighbouring galaxies (Einasto, Klypin & Shandarin 1982).

3.3 "Scenarios" for galaxy formation

The early Universe cannot have been completely smooth. Some initial irregularities must have given rise to bound systems, which then either themselves turned into galaxies, or triggered the formation of galaxies by an indirect route.

The two basic classes of perturbation are isothermal (or entropy) perturbations, in which the radiation pressure is unaltered; and adiabatic (isentropic) perturbations in which the ratio $\mathcal{S} = n_\gamma/n_b$ is unperturbed. Isothermal perturbations are essentially "frozen in" before recombination, and do not oscillate. A general perturbation whose oscillatory component is damped by viscosity can leave behind an isothermal component. After recombination, when the gravitational instability of the matter is opposed only by gradients in the gas pressure (less than radiation pressure by a factor $\sim 10^8\,\Omega^{-1}$), all scales $\gtrsim 10^6\,M_\odot$ grow, at least until bound systems form and generate enough energy to heat and reionise the gas. The linear growth is never faster than $\frac{\delta\rho}{\rho} \propto R$, where R is the scale factor. To obtain a bound system by the present time, the necessary amplitude at recombination must therefore be at least 10^{-3}. The growth of perturbations on the mass scales of galaxies is inhibited before recombination by the effects of radiation pressure and viscosity; this means that amplitudes of $\gtrsim 10^{-3}$ may be necessary even at the (earlier) epoch when such scales are first encompassed within the particle horizon. (In models involving massive neutrinos, when growth can occur before recombination, one can get by with perturbations whose amplitude on entering the horizon is somewhat smaller.)

The two basic "scenarios" for galaxy (and cluster) formation, are summarized in Figure 1. These pictures are very different: they contrast most starkly if we compare what the universe would have been like at, say, $R(t)/R(t_o) = \frac{1}{20}$ (a redshift of 20). In the hierarchical picture (requiring entropy perturbations) bound systems would already exist. Indeed, nearly all the initial mass could have been incorporated in pregalactic stars or massive objects which could have already augmented or distorted the microwave background spectrum. These pregalactic objects may have amplified perturbations on large scales, giving rise to galaxies by secondary effects. On the other hand, in the adiabatic picture essentially nothing of interest would have happened by $z \simeq 20$: density contrasts would then still be $\lesssim 1$, and the Universe would still consist of quiescent expanding neutral gas.

There is no reason to believe that a galactic (or cluster) mass-scale is specially imprinted into the early Universe: it is more natural to envisage a spectrum of fluctuation extending over all scales. One can then ask what we know about the fluctuations on other scales.

On larger scales than galaxies and clusters (i.e. $\ell > 10^{-2}\ \ell_H$) the microwave background isotropy sets upper limits on the fluctuation amplitude. These limits are sufficiently powerful to be worth discussing in some detail. Considered in conjunction with the *lower* limit on the amplitude of fluctuations on the relevant scale implied by the existence of galaxies, the microwave background data severely constrain the spectrum and character of the primordial fluctuations.

4 MICROWAVE BACKGROUND ISOTROPY, AND OTHER UPPER LIMITS ON PRIMORDIAL FLUCTUATIONS

4.1 The "cosmic photosphere"

The microwave background isotropy measurements (relative rather than absolute measurements) are already amazingly precise: effects at the level of one part in 10^4 would be detectable in a *total* cosmic background which is $\sim 10^2$ weaker than the contribution from the Earth itself. Recent reviews of the data have been given by Partridge (1982) and Wilkinson (1982). There are no confirmed anisotropies apart from the "dipole" anisotropy, due primarily to the motion of our Galaxy relative to the mean background. The upper limits are of order 10^{-4} on all angles from a few arc minutes up to 90° (quadrupole).

The microwave photons were last scattered as a "cosmic photosphere" whose redshift z_* and thickness depend on the thermal history of the matter. If the primordial plasma (re)combined as it cooled adiabatically below a temperature of a few thousand degrees, and there was no significant later reheating, then $z_* \simeq 10^3$, the effective thickness of the photosphere, determined by the width Δz_* of the function $d/dz\, e^{-\tau(z)}$, is only $\sim 0.1\, z_*$. This is because the reduction in the free electron density (which results in the "fog lifting" and the microwave photons becoming free to travel uninterruptedly) is rather sudden. If reheating caused the intergalactic medium to "fog up" again at a later epoch, with $\tau > 1$ at redshifts below 1000, then z_* would be smaller and $\Delta z_* \simeq z$. Three cases are illustrated in Figure 2. Note that reheating at $z < 10$ could never generate $\tau > 1$, even if the Universe contained enough ionized intergalactic gas to give $\Omega = 1$.

Figure 2. Optical depth $\tau(z)$ (dot-dash lines) of the Universe back to a redshift z, and 'visibility factors' $\tau e^{-\tau}$ (dotted lines) for two values of x_e, the fraction of the critical density in the form of ionized plasma (assumed independent of z for illustrative purposes). Even a small amount of reheating is sufficient to produce a last scattering surface that is more 'smeared out' in redshift than the standard model of recombination (labelled 'rec'). In some models for galaxy formation where energy (and maybe heavy elements) are generated at large redshifts, other kinds of opacity may be competitive with Thomson scattering. See Hogan, Kaiser & Rees (1982) for further discussion.

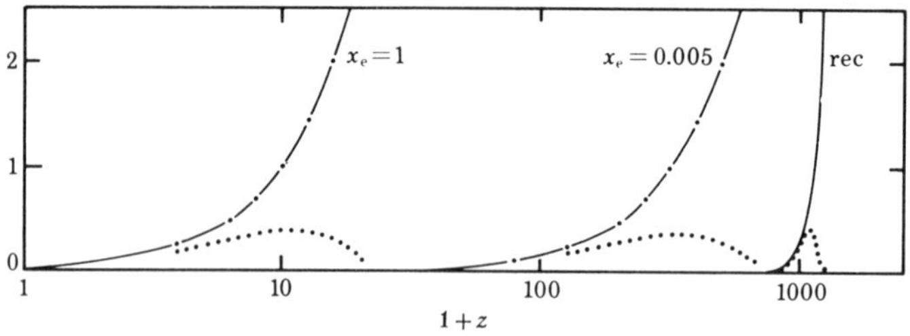

A Friedmann model with $\Omega_o < 1$ evolves very like a "flat" model at redshifts $> \Omega_o^{-1}$. Provided that $z_* > \Omega_o^{-1}$, which is so for almost all relevant models, the angle subtended by a given comoving region on the last scattering shell is essentially independent of z_*. A comoving length ℓ corresponds to an angle

$$\theta(\ell) \approx (\Omega_o/2)(\ell/\ell_H) \tag{4}$$

A region of present length 1 Mpc subtends an angle $\frac{1}{2}\Omega_o h$ arc minutes. A region that came within the horizon at z_* has present length $\ell_* \approx 2\ell_H \times (\Omega_o z_*)^{-\frac{1}{2}}$ and subtends an angle $\theta_* \approx (\Omega_o/z_*)^{\frac{1}{2}}$.

Another important scale if $\Omega_o < 1$, is the length corresponding to the intrinsic curvature of the spacelike hypersurfaces in the Robertson-Walker metric; this subtends an angle

$$\theta_c \simeq \Omega_o(1 - \Omega_o)^{-\frac{1}{2}} \tag{5}$$

The fluctuation amplitude on a given scale is conveniently characterised by

$$\varepsilon(\ell) = \begin{Bmatrix} \text{value of } \frac{\delta\rho}{\rho} \text{ when scale } \ell \text{ is first} \\ \text{encompassed within particle horizon} \end{Bmatrix}$$

If the growth were unaffected by pressure, this spectrum leads to $(\frac{\delta\rho}{\rho}) \propto \varepsilon(\ell) \times \ell^{-2}$ for all scales ℓ within the horizon at a given time - density contrasts, in the linear regime would grow $\propto R(t)$, and would eventually condense out as virialised systems with gravitational binding energy $\sim \varepsilon c^2$ per unit mass. The quantity ε is the "curvature fluctuation", which one would like to be able to calculate from first principles.

4.2 Scales $\ell_* \lesssim \ell \lesssim \ell_H$

Any fluctuation straddling the cosmic photosphere, whose co-moving length-scale exceeds ℓ_*, the horizon scale at z_*, but is less than ℓ_H would generate a temperature anisotropy ($\Delta T/T$) which is simply of order ε, as was first calculated by Sachs and Wolfe (1967). Similar fluctuations along the line of sight with redshifts less than z_* have smaller effects (Rees & Sciama 1968, Dyer 1976). If $z_* \simeq 1000$, then all fluctuations subtending angular scales > $(2\ \Omega_o^{\frac{1}{2}})$ degrees exceed the horizon scale; even if $z_* \simeq 10$ (an implausibly extreme lower limit, attained only if most of the critical density were in gas which had been reionized at z > 10), the horizon scale subtends $\lesssim 15^o$. Consequently the high isotropy of the background ($\frac{\Delta T}{T} \lesssim 10^{-4}$) implies that, if $\Omega_o \simeq 1$

$$\varepsilon \lesssim 10^{-4} \text{ for } \begin{cases} 1 > \ell/\ell_H > \frac{1}{3} & (z_* \simeq 10) \\ 1 > \ell/\ell_H > \frac{1}{30} & (z_* \simeq 10^3) \end{cases} \tag{6}$$

There seem to be no ambiguities in this argument; there are only minor changes if other values of Ω_o are taken. Fluctuations on scales $\sim \ell_H$ would cause large angle (quadrupole and octopole) effects; the smaller angular scale observations probe smaller values of ℓ (cf (3)).

4.3 Scales $\ell \gtrsim \ell_H$

If $\Omega_o \lesssim 1$ then our particle horizon will eventually grow to encompass an infinite amount of matter that we cannot yet observe. Inhomogeneities can induce gradients and shear across the region within our horizon, and may cause anisotropy in the microwave background.

Grishchuk & Zel'dovich (1978) found that for $\Omega_o = 1$ a very long wavelength density ripple in a spatially flat universe generates a quadrupole anisotropy

$$(\Delta T/T)_q \approx \varepsilon(\ell/\ell_H)^2 \approx (\frac{\Delta\rho}{\rho})_o$$

This may be understood in terms of photons becoming red-shifted or blue-shifted as they climb up or fall down the potential gradient associated with the perturbation. That there is no dipole term is because both we and the radiation are in free fall. If $\Omega_o = 1$ the limits on any quadrupole constrain $(\delta\rho/\rho)_o$ to be less than about 10^{-4} on all scales outside the horizon (subject to a "random phase" assumption); however, the curvature fluctuation amplitude ε may still diverge on scales larger than about $10^2 l_H$ and, since perturbations grow for all time if $\Omega_o = 1$, these perturbations may have $(\delta\rho/\rho) \gtrsim 1$ when they come within the horizon. Similar results can be derived for $\Omega \neq 1$ (Kaiser 1982).

The microwave isotropy thus allows us to probe perturbations on scales larger than those we can presently observe. However these limits, which are themselves somewhat model-dependent, become progressively less stringent (in terms of ε) on larger scales. We certainly cannot rule out a gross deviation from homogeneity (bubble wall? Minkowski space?) on scales $>> 10^2\ \ell_H$. There can only be "philosophical" reasons for extrapolating the observed homogeneity of the part of the Universe we can now study to a domain which (if the Universe is open) is infinitely larger.

4.4 Scales $\ell \lesssim \ell_*$

The scales to which (6) applies are larger than those on which we actually observe inhomogeneity in the distribution of galaxies. It would be of particular interest in constraining theories of galaxy formation if we could actually observe temperature fluctuations due to the same scales which now display clustering. The upper limits to $\Delta T/T$ on the relevant scales of a few arc minutes (cf (4)) are at the level of $\sim 2 \times 10^{-4}$. On these scales, however, the limits are harder to interpret.

On angles corresponding to scales $\ell > \ell_*$ we are basically just seeing "metric fluctuations" which have been unaffected by pressure gradients and have evolved acausally throughout the (post-inflationary?) Friedmannian expansion phase of the Universe. But if $\ell \lesssim \ell_*$ various complications arise:

(i) The Sachs-Wolfe contribution to $\Delta T/T$ may not be the dominant one. Doppler effects due to peculiar motions may be even more important; so also may effects due to the changes in the recombination and decoupling time in perturbed regions.

(ii) Pressure gradients and damping may have affected the perturbations on scales $< \ell_*$ at epochs before decoupling. This complicates the relationship between ε and $(\frac{\delta\rho}{\rho})_{z=z_*}$ for these scales.

(iii) On scales below $(\Delta z_*/z_*)\ell_*$ the temperature fluctuations are smeared out by the gradual nature of the decoupling — i.e. the fact that the last scattering surface is not sharp compared to the scale being studied. Furthermore, these scales would be attenuated by photon viscosity during recombination; this further complicates the relationship between $\Delta T/T$ and the value of $(\delta\rho/\rho)$ which survives after decoupling. (In the case of no reheating, these effects due to smearing and viscosity are important only for $\ell \lesssim \ell_H/300$.)

There is an extensive literature attempting to quantify the effects (i) - (iii) above: to relate ε, $\Delta T/T$ and the post-decoupling $(\delta\rho/\rho)$ on the scales relevant to cluster formation (see Press & Vishniac 1980; Peebles 1981, 1982a; Silk 1982 and references cited therein). This work is, however, almost all predicated on the assumption that there is no early reheating, and that the "cosmic photosphere" is determined by the recombination behaviour of the primeval H-He plasma. While this work offers a consistency check on scenarios which preclude early reheating, it is important to stress that the interpretation of $\Delta T/T$ isotropy limits on small angular scales is *very sensitive* to the value of z_* and Δz_*. Under more general assumptions, smearing effects can be very large for scales $\ell \simeq 10^{-2}\ \ell_H$; furthermore, any anisotropies actually observed on small angular scales may be of "secondary" origin, rather than being directly related to ε. (See Hogan, Kaiser & Rees (1982) for further discussion of this topic).

4.5 Limits to ε on subgalactic scales

Sixty powers of ten separate the present era from the Planck time; correspondingly, the particle horizon has grown since t_{Planck} by rather more than $\sim 10^{30}$, in *comoving* coordinates. The microwave background isotropy tells us about the fluctuations only on scales $\gtrsim 10^{-3}\ \ell_H$; one might then ask whether there are any complementary lines of evidence about ε on lengthscales spanning the other 27 orders of magnitude which are of interest. I have discussed and reviewed this topic elsewhere (Rees 1981) so will be brief here.

An upper limit on ε comes from constraints on the number of primordial black holes that could have been produced. An overdense region coming within the horizon could evolve into a black hole if pressure were unable to stabilise it before it collapses. A region entering the horizon with $(\delta\rho/\rho) \simeq 1$ could collapse even if $p = \frac{1}{3}\rho c^2$. Even if the mean amplitude of fluctuations corresponds to a value of ε substantially less than unity, the high-amplitude tail of the distribution may be vulnerable to collapse. Moreover, if primordial black holes now contribute Ω_{bh} to the density parameter the fraction of mass-energy collapsing into primordial holes at early times must not merely be $\lesssim 1$, but for $t \lesssim 10^{12}$ sec (when the expansion is radiation-dominated) must satisfy the stronger requirement of being $\lesssim (\rho_{rad}/\rho_{matter})^{-1}\Omega_{bh} \simeq (t/10^{12}\ \text{sec})^{\frac{1}{2}}\Omega_{bh}$. This is because the radiation mass-energy per comoving volume decreases $\propto R^{-1}$ in the radiation-dominated phase, whereas that due to black holes stays constant. If the fluctuations have a gaussian distribution of amplitudes when entering the horizon (an assumption which is not mandatory, as we do not yet know what originally causes them), then these considerations set a limit on the mean ε in the general range 0.1 - 1, which tends to be slightly more stringent (because the relevant ρ_{rad}/ρ_{matter} factor is larger) on smaller scales. Scales << 1 $M_\odot$ have been discussed by Carr (1976). On scales where black hole evaporation occurs, the limit on Ω_{bh} is $\lesssim 10^{-8}$, set by the γ-ray background, rather than $\lesssim 1$. However, if we are dealing with the exponential tail of a gaussian distribution of amplitudes, this reduces the permitted ε only by a logarithmic factor.

Adiabatic fluctuations on scales << $10^6\ M_\odot$ which do not have sufficient amplitude to collapse, but instead enter an oscillatory phase, would be damped by viscosity (or shock waves) at such an early time that the excess energy could be fully thermalised. On scales $10^6\ M_\odot \lesssim M \lesssim 10^{13} M_\odot$

the damping occurs sufficiently late that the energy input cannot be properly thermalised, and would show up as a distortion in the microwave background (Sunyaev & Zel'dovich 1970). These constraints, and the lower limit on the mass scale where they apply, are somewhat sensitive to Ω_b.

The fact that ε is constrained to be $\lesssim 0.1$ on the scale of mini-black holes (comoving scale $\sim 10^{-24}\ \ell_H$) means that, if the primordial curvature fluctuations obey a power law, and contribute $\varepsilon \simeq 10^{-4}$ on galactic scales, then this power law cannot fall off towards larger scales more steeply than $\propto \ell^{-0.15}$ (i.e. $\propto M^{-0.05}$).

5 QUANTITATIVE LOWER LIMITS TO PRIMORDIAL INHOMOGENEITY IMPLIED BY THE EXISTENCE OF GALAXIES AND CLUSTERS

5.1 Behaviour of fluctuations

Despite the uncertainty in z_* and the nature of the last scattering surface, the background isotropy on larger angles means that ε is constrained to be $\lesssim 10^{-4}$ on the scales to which (6) applies. We also have some (albeit rather crude) upper limits on much smaller scales. Let us now consider the *lower* limits to ε on the scales where gravitationally bound systems have condensed out. These depend on the particular "scenario" for galaxy formation

In a one-component pressure-free universe, perturbations would grow according to the standard law

$$\frac{\delta\rho}{\rho} \propto R$$

This applies at all epochs if $\Omega = 1$, and for $z > \Omega_o^{-1}$ in a low density universe. If pressure is important, this growth law applies only on scales above the Jeans length $\ell_J = \ell_H(c_s/c)$, c_s being the thermal velocity. For scales $\ell < \ell_J$, the perturbations oscillate if the material is collisional (e.g. plasma with short mean-free path) and are phase-mixed away if it is collisionless (e.g. neutrinos).

A further important point is that in a multicomponent universe there is essentially no growth in the density contrast of any component if it is only a small fraction of the total density, the rest of which is expanding homogeneously. (For this reason, primordial black holes or heavy "inos" would not start to cluster at early epochs where the mass-energy density was in radiation.)

5.2 Galaxy formation from adiabatic curvature fluctuations

How does galaxy formation proceed if the initial fluctuations are purely adiabatic? There are three distinctive variants which are worth summarising separately (see, for instance, Silk (1982) for a fuller review).

(a) If there are no leptons of significant mass (i.e. above a few ev), photon viscosity would have attenuated all fluctuations on mass-scales below the "Silk (1968) mass" of

$$M_s \simeq 3 \times 10^{13} (\Omega h^2)^{-5/4} M_\odot \qquad (7)$$

The first fluctuations to "turn around" would be just above this scale, and galaxies would form dissipatively from gas compressed into sheets and filaments. In this mode, there would seem no possibility of early re-heating. The value of ε would need to be $\gtrsim 10^{-3}$ on the scale ℓ_s corresponding to the Silk mass because growth starts only after decoupling of matter and radiation at t_{rec}.

(b) If the Universe is now dynamically dominated by primordial neutrinos, a new mass-scale plays a rôle: this is essentially the mass-scale within the particle horizon at the epoch when $kT \simeq m_\nu c^2$, where m_ν is the mass of the dominant neutrino species (cf (2)). Neutrinos will be collisionless at (and after) this epoch and so cannot sustain oscillatory modes; phase-mixing and Landau damping consequently homogenise scales which can be traversed by a free-streaming neutrino in the expansion timescale. When the thermal energies fall below $m_\nu c^2$ the neutrino proper velocities decrease as R^{-1}, (i.e. $\propto R^{-2}$ in terms of comoving coordinates) and the distance they can move (in comoving coordinates) in the expansion time-scale diminishes. Thus the damping is most severe when $kT \simeq m_\nu c^2$, and attenuates mass-scales up to $\sim (m^3_{Planck}/m_\nu^2)$. More detailed calculations (Bond, Efstathiou & Silk 1980, Bisnovatyi-Kogan & Novikov 1981, Doroshkevich *et al.* 1980, Wasserman 1981) show that the relevant mass-scale is

$$M_\nu \simeq 5 \times 10^{14} (\Omega_\nu h^2)^{-2} M_\odot \qquad (8)$$

where Ω_ν is the contribution of the heaviest neutrino to Ω (related to m_ν by (2)). Again, as in (a), the first scales to turn around will be entire clusters or superclusters rather than individual galaxies. However, growth according to the law $(\delta\rho/\rho) \propto R$ starts not at t_{rec}, but at the earlier stage when the Universe becomes dominated by neutrino rest mass. This means that $\varepsilon \simeq 10^{-4}$ is sufficient to form bound systems by the present time.

(c) <u>If the Universe is now dynamically dominated by primordial non-baryonic particles ("...inos") of rest mass $\gtrsim$ 1 Kev</u>, if there were other kinds of primordial leptons (e.g. right handed neutrinos), or gravitinos or photinos which were somewhat less numerous than neutrinos but had masses m such that they contributed much of the "unseen" mass, then the damping mass due to phase-mixing, etc. is much less than the value (8) for neutrinos: the horizon scale is smaller when kT drops to the rest mass energy of these somewhat heavier particles. However, the growth of fluctuations in this component would be inhibited while T drops through the range $mc^2 \gtrsim kT \gtrsim kT_{eq}$, where T_{eq} (corresponding to time t_{eq}) is the temperature when these particles start to be dynamically dominant over the radiation. The likely outcome is that there will be a preferred mass scale of order the horizon mass at t_{eq} (see Bond & Szalay (1982), or Peebles (1982b) for detailed discussion), corresponding to a small group. However, because the primordial fluctuations in the "inos" on subgalactic scales are not damped out, there will be substructure within these condensations, rendering the processes of galaxy formation like those in the "hierarchical" picture. Since the growth of fluctuations can start at t_{eq}, rather than being delayed until t_{rec}, an initial fluctuation amplitude $\varepsilon \simeq 10^{-4}$ is sufficient to form bound systems by the present time. One possible difference between this variant and (b) above is that the heavier "inos" would find it easier to cluster on smaller scales; the neutrinos may therefore be preferable candidates for yielding $\Omega \simeq 1$, since this is only consistent with the virial theorem considerations if the missing mass does not participate fully in the clustering of galaxies.

The most appealing class of initial fluctuations, in the context of currently fashionable theories, would be *scale-independent* (or almost so), with $\varepsilon \simeq \varepsilon_o$ on all scales (cf. Harrison 1970, Zel'dovich 1972), and purely adiabatic. The background isotropy limits certainly constrain $\varepsilon(\ell)$ to be $\lesssim 10^{-4}$ for $0.3 \lesssim \ell/\ell_H \lesssim 1$ (cf equation (6)). If this is an upper limit to a scale-independent ε_o, we can then ask whether this value is large enough to allow galaxy formation to have occurred. We see from the above that, of the three variants of the adiabatic scenario, *only*

Figure 3. This diagram depicts the astrophysical limits on the amplitude ε of adiabatic metric fluctuations, on various scales ℓ. On large scales, the microwave background isotropy offers stringent upper limits (see (4) and (6)). On very small scales, ε must merely not be so large that too much of the Universe collapses as the relevant scales enter the horizon; the absence of distortions in the microwave background sets a slightly better limit for mass-scales $10^4 - 10^{13}\ M_\odot$. The requirement that bound systems have "turned around" by the present epoch gives *lower* limits (also plotted). A spectrum with $\varepsilon \simeq 10^{-4}$ on all scales is acceptable *if* the Universe is dominated by non-baryonic matter which can start clustering before t_{rec}. If ε has a power-law dependence on ℓ and is $\sim 10^{-4}$ on the scales relevant to galaxy formation, then it cannot fall off more steeply than $\propto \ell^{-0.15}$ without causing excessive production of primordial mini-holes. Also plotted (dotted lines) are the amplitudes ε_g of primordial gravitational waves that can, within the next few years, be probed by: (a) doppler tracking of spacecraft, (b) timing of "quiet" pulsars, and (c) timing of the orbit of the binary pulsar. This diagram is drawn assuming a "flat" background universe with $\Omega = 1$. If $\Omega < 1$ the limits on large scales are modified. Indeed the appropriate definition of ε is ambiguous on scales exceeding the Robertson-Walker curvature radius (cf. equation (5)).

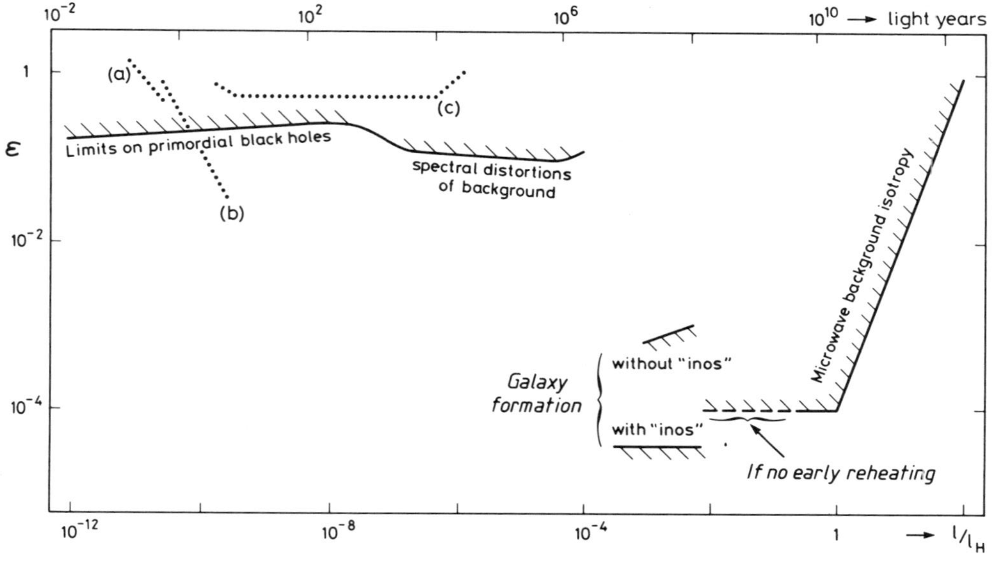

those dominated by non-baryonic matter (i.e. (b) and (c)) pass this test. In fact, variant (a) almost runs up against a contradiction independently of the form of the primordial fluctuation spectrum $\varepsilon(\ell)$, because the upper limit to $\Delta T/T$ on the actual angular scale corresponding to $M \simeq M_s$ is only marginally compatible with the value of $(\delta\rho/\rho)_{rec}$ which is required in order that bound systems should form (Press & Vishniac 1980, Hoffman, Salpeter & Wasserman 1983). Figure 3 depicts the various limits on ε.

5.3 Isothermal (entropy) perturbations

Fluctuations in the baryon-to-photon ratio are not damped out by photon viscosity in the same way as purely adiabatic perturbations. There is therefore no reason why sub-galactic scales, if present initially, should not survive until recombination. Moreover, a large isothermal $(\delta\rho/\rho)_{baryon}$ generates a curvature fluctuation which is less by $(\rho_{baryon}/\rho_{total})$, a small ratio at early times, than for adiabatic perturbations of the same amplitude.

The recent trend has been to disfavour isothermal perturbations because of optimism that n_b/n_γ may be determined by baryosynthesis. In the context of such ideas, the only ways of generating entropy perturbations (short of invoking variations in the CP-violation parameter from place to place) would seem to be the following:

(i) Variations in the expansion rate (inhomogeneous shear), or in the reheating temperature after the GUT phase transition, might cause the quantitative details of baryosynthesis to vary from place to place. The problem here is that the fluctuations must be purely decaying modes, in order to have significant amplitudes at $\sim 10^{-35}$ sec without leading to chaos at later times (Bond, Kolb & Silk 1982).

(ii) There may be irregular dissipation of entropy due to non-linear damping effects, or to phase transitions (e.g. at the "electroweak" symmetry-breaking transition).

Neither of the above seems a natural way of generating entropy perturbations on astrophysically interesting scales — indeed the idea of invoking entropy perturbations now seems perhaps as heretical as baryon non-conservation was ten years ago. To form a bound system by the present time, one needs $(\delta\rho/\rho) \gtrsim 10^{-3}$ at t_{rec} (and therefore, since there is no growth before t_{rec}, this same amplitude is required *ab initio*). However, because the baryons are only a small fraction of the mass-energy density at early times, this is compatible with curvature fluctuations $<< 10^{-4}$.

Since entropy perturbations do not undergo an acoustic phase before t_{rec}, there is no reason why the amplitudes at t_{rec} should have a gaussian distribution. In principle, therefore, one could imagine that the perturbations were restricted to a small fraction of the volume. It is then interesting to pose a question which is the inverse of that discussed in § 4: instead of asking how chaotic the Universe might have been, we can instead ask what are the *minimal* departures from complete homogeneity that could have generated the present structures that we see.

5.4 How smooth could the early Universe have been?

Completely homogeneous initial conditions lead, after 10^{10} yrs, to a quiescent universe devoid of galaxies. If the initial fluctuations were purely adiabatic, then $\varepsilon \gtrsim 10^{-4}$ is required in order for bound galactic-mass systems to have condensed out — indeed, $\varepsilon \gtrsim 10^{-3}$ is needed unless the Universe is dominated dynamically by particles (neutrinos, gravitinos, etc.) which are decoupled from the photons and baryons and can undergo gravitational instability before recombination. But if we envisage isothermal perturbations, then the initial *metric* fluctuations could have been smaller. Isothermal fluctuations survive undamped on all relevant scales (right down to $\sim 1\ M_\odot$ (Hogan 1978)). Moreover, there is then no reason why $(\delta\rho/\rho)$ should have a gaussian distribution of amplitudes on a given mass-scale.

In principle, to initiate galaxy formation it may be sufficient for a few small "seeds" to condense from primordial fluctuations. These seeds could then contribute an energy release which generates larger-scale fluctuations. To take an extreme example, one $\sim 100\ M_\odot$ star could trigger a chain reaction culminating in the formation of an entire cluster of galaxies, so one needs only *one* small density enhancement in every volume of $10^{15}\ M_\odot$. Such might occur, if phase transition led to an inhomogeneous deposition of entropy in such a fashion that a few "islands" were left on a lower adiabat (Doroshkevich, Zel'dovich & Novikov 1967, Ostriker & Cowie 1981, Ikeuchi 1981, Carr & Rees 1983).

5.5 Primordial gravitational waves

Associated with the curvature fluctuations which initiate growing-mode perturbations we would expect gravitational waves to be generated. These waves would lead to a stochastic background. If this were detectable it would provide direct evidence on how chaotic the Universe was on sub-galactic scales — scales where we now only have the rather crude limit $\varepsilon \lesssim 0.2$ from constraints on primordial black hole formation.

The wave amplitude when a given scale enters the horizon can be defined as ε_g, which will be related to the curvature fluctuation ε. As the Universe expands, the amplitude of a given mode varies as R^{-1} and the energy density goes as R^{-4}, like electromagnetic radiation. The energy density of these waves now would be $\Omega_{rad}\ \varepsilon_g^2$ (i.e. $\sim 10^{-4}\ \varepsilon_g^2$). Suggestions have been made that there may be a primordial background with ε_g independent of scale (Starobinsky 1979, Vilenkin 1981).

There are not yet any measurements sensitive enough to detect this maximal level of background, at any wavelength. However, precise timing of pulsars, including the binary pulsar and the newly-discovered fast pulsar, whose timing may be quiet with microsecond precision (Backer, Kulkarni & Taylor 1983) offers the prospect of attaining interesting levels for wave periods 1 - 10^5 yrs within a few years (Bertotti, Carr & Rees 1983); doppler tracking of spacecraft may extend the limits to somewhat shorter periods (see Figure 3).

5.6 Does the evidence favour any particular scheme for galaxy formation?

At first sight, one might regard the evidence for "superclusters" with apparent filamentary structure as unambiguous support for the so-called "pancake" model, where the dominant length scale of irregularity surviving after t_{rec} is $\gtrsim 10$ Mpc (Zel'dovich, Einasto & Shandarin (1982) give a persuasive "manifesto" for this line of thought, with full references to earlier work). Recent N-body simulations (e.g. Shandarin 1982, Frenk, Davis & White 1982, Dekel 1982) show that the filamentary structure is a natural outcome of gravitational clustering if the initial distribution is smoothed at high wave-numbers (i.e. homogenised on small scales). However, there is no need to invoke dissipation to account for the persistence of linear structure: the galaxies could have formed first and clustered later; they need not have condensed from cool gas compressed into sheets or filaments *after* the large-scale structure developed. A further uncertainty in interpreting the observed large-scale clustering is that we do not know whether the distribution of galaxies traces the overall distribution of gravitating matter. The apparent "void" regions may not be deficient in "unseen" mass; they may contain gas which has, for some reason, been prevented from condensing into galaxies, and thus not be deficient even in baryonic matter.

We can hope in the next few years for more clues to the nature and distribution of "unseen" mass. Studies of remote regions (by infrared and X-ray techniques, as well as optical) may tell us at what epoch the galaxies formed, and how they clustered. Further improvements in the sensitivity of microwave background observations may reveal a wealth of detail on primeval inhomogeneities, and perhaps evidence for energy input from pregalactic objects. Only after such developments will we be able to narrow down the range of models whereby the Universe's present structure and contents could have developed from amorphous beginnings.

6 SUMMARY AND CONCLUSIONS

(1) A quantitative statement of the Universe's scale and flatness is that, if it has been Friedmannian since $t \simeq t_{Planck}$, the initial Robertson-Walker curvature radius must exceed the initial radius of the particle horizon by at least $\sim 10^{30}$ ($\sim \alpha_G^{-3/4}$).

(2) The mean density of all forms of gravitating matter corresponds to $\Omega \gtrsim 0.1$. The full critical density $\Omega = 1$ is not required by the observations, and indeed is permissible only if the "unseen" mass is less "clumped" than the galaxies, even on scales as large as ~ 10 Mpc - in other words, we would need to average over a region of larger dimensions than this in order to ensure that it contained the true overall ratio of luminous and "unseen" matter.

(3) Calculations of primordial nucleosynthesis of D and ^{4}He favour a baryon density $\Omega_b h^2$ in the range 0.01 - 0.025 (permitting Ω_b of 0.1, but no more, even if the Hubble constant has its lowest likely value of 50 km sec^{-1} Mpc^{-1}). Non-baryonic forms for the "unseen" mass are therefore attractive in the context of the standard hot big bang model. The favoured value of Ωh^2 corresponds to photon-to-baryon ratio of $\mathcal{S}$ $\sim 2 \times 10^9$.

(4) The microwave background isotropy unambiguously constrains the primordial metric fluctuations to be $\varepsilon \lesssim 10^{-4}$ on scales $0.3\,\ell_H \lesssim \ell \lesssim \ell_H$. If the "cosmic photosphere" is at a redshift $z_* \simeq 1000$, rather than at a smaller redshift, this limit applies down to $\sim 10^{-2}\,\ell_H$. The $\sim 10^{-4}$ limit

on the quadrupole component implies $\varepsilon \lesssim 10^{-4} (\ell/\ell_H)^2$ on scales ℓ exceeding the present Hubble radius.

(5) If the Universe is baryon-dominated, rather than dominated by "inos" of non-zero mass, the formation of galaxies from adiabatic fluctuations at $M \gtrsim M_s$ (eqn (7)) is only marginally consistent with the limits to $\Delta T/T$ on angular scales of a few arc minutes. (This statement is *independent* of any assumption about the ℓ-dependence of ε.)

(6) If the Universe contains mean fluctuations with ε essentially constant ($\sim \varepsilon_o$) on all scales, then $\varepsilon_o \lesssim 10^{-4}$. If the initial fluctuations are purely adiabatic, then ε_o must be close to 10^{-4}; galaxy formation schemes involving "inos" are then favoured, because growth can then begin at t_{eq} rather than t_{rec}.

ACKNOWLEDGEMENTS

I am grateful to many colleagues, especially Dick Bond, Bernard Carr, George Efstathiou, Craig Hogan & Nick Kaiser, for discussions about topics mentioned in this article, and acknowledge support under NSF grant PHY79-19884 at Princeton.

REFERENCES

Audouze, J. (1982). in "Astrophysical Cosmology" ed. H.A. Brück, G.V. Coyne & M.S. Longair, p. 395 (Vatican Publications).
Backer, D.C., Kulkarni, S.R. & Taylor, J.H. (1983). Nature, 301, 314.
Bertotti, B., Carr, B.J. & Rees, M.J. (1983). Mon.Not.R.astr.Soc. (in press).
Bisnovatyi-Kogan, G. & Novikov, I.D. (1981). Sov.Astron.Lett., 24, 516.
Bond, J.R., Efstathiou, G. & Silk, J.I. (1980). Phys.Rev.Lett., 45, 1980.
Bond, J.R., Kolb, E. & Silk, J.I. (1982). Astrophys.J., 255, 341.
Bond, J.R. & Szalay, A. (1982). Astrophys.J. (in press).
Carr, B.J. (1976). Astrophys.J., 206, 8.
Carr, B.J. (1978). Comments on Astrophys., 7, 161.
Carr, B.J., Bond, J.R. & Arnett, W.D. (1983). Astrophys.J. (in press).
Carr, B.J. & Rees, M.J. (1979). Nature, 278, 605.
Carr, B.J. & Rees, M.J. (1983). Mon.Not.R.astr.Soc. (submitted).
Cowsik, R. & McLelland, J. (1972). Phys.Rev.Lett., 29, 669.
David, Y. & Reeves, H. (1980). in "Physical Cosmology", eds. R. Balian, J. Audouze & D.N. Schramm (North-Holland) p. 443.
Davis, M., Huchra, J., Latham, D.W. & Tonry, J. (1982). Astrophys.J., 253, 423.
Dekel, A. (1982). Astrophys.J. (in press).
Dicke, R.H. (1982). Nature, 192, 440.
Dolgov, A.D. & Zel'dovich, Y.B. (1981). Rev.Mod.Phys., 43, 1.
Doroshkevich, A.G., Zel'dovich, Y.B. & Novikov, I.D. (1967). Sov.Astron. 41, 233.
Doroshkevich, A.G., Zel'dovich, Y.B., Sunyaev, R.A. & Khlopov, M.Y. (1980). Sov.Astron.Lett., 6, 457.

Dyer, C. (1976). Mon.Not.R.astr.Soc., 175, 429.
Einasto, J., Joeveer, M. & Saar, E. (1980). Mon.Not.R.astr.Soc., 193, 353.
Einasto, J., Klypin, A. & Shandarin, S.F. (1982). Paper presented at IAU Symposium 104.
Frenk, C., Davis, M. & White, S.D.M. (1982). Astrophys.J. (in press).
Grishchuk, L. & Zel'dovich, Y.B. (1978). Sov.Astron., 22, 125.
Gunn, J.E., Lee, B.W., Lerche, I., Schramm, D.N. & Steigman, G. (1978). Astrophys.J., 223, 1015.
Hanes, D.A. (1982). Mon.Not.R.astr.Soc., 201, 145.
Harrison, E.R. (1970). Phys.Rev., D1, 2726.
Hoffmann, G.L. & Salpeter, E.E. (1982). Astrophys.J., 263, 485.
Hoffmann, G.L., Salpeter, E.E. & Wassermann, I. (1983). Astrophys.J., (in press).
Hogan, C.J. (1978). Mon.Not.R.astr.Soc., 185, 889.
Hogan, C.J., Kaiser, N. & Rees, M.J. (1982). Phil.Trans.R.Soc., A307, 97.
Ikeuchi, S. (1981). Proc.Astr.Soc. Japan, 33, 211.
Kaiser, N. (1982). Mon.Not.R.astr.Soc., 198, 1033.
Kolb, E. & Wolfram, S. (1980). Nucl.Phys., B172, 224.
Marx, G. & Szalay, A.S. (1972). Proc. "Neutrino 72" conference (Techno-inform, Budapest) Vol. I, 191.
Oort, J.H. (1982). Ann.Rev.Astr.Astrophys. (in press).
Ostriker, J.P. & Cowie, L.L. (1981). Astrophys.J.Lett., 243, L127.
Pagel, B.E.J. (1982). Phil.Trans.R.Soc., A307, 19.
Partridge, R.B. (1972). in "Origin and Evolution of Galaxies" ed. B.J.T. Jones & J.E. Jones (Reidel, Dordrecht).
Peebles, P.J.E. (1980). The Large-Scale Structure of the Universe. Princeton University Press (Princeton USA).
Peebles, P.J.E. (1981). Astrophys.J., 248, 885.
Peebles, P.J.E. (1982a). Astrophys.J., 258, 415.
Peebles, P.J.E. (1982b). Astrophys.J. (Lett.), 263. L1.
Press, W.H. & Vishniac, E.T. (1980). Astrophys.J., 236, 323.
Rees, M.J. (1981). in "Quantum Gravity II" eds. C.J. Isham, R. Penrose, & D.W. Sciama (Oxford University Press) p. 273.
Rees, M.J. & Ostriker, J.P. (1977). Mon.Not.R.astr.Soc., 179, 541.
Rees, M.J. & Sciama, D.W. (1968). Nature, 217, 511.
Sachs, R.K. & Wolfe, A.M. (1967). Astrophys.J., 147, 73.
Sandage, A.R. & Tammann, G.A. (1982). in "Astrophysical Cosmology" ed. H.A. Brück, G.V. Coyne & M.S. Longair (Vatican Publications) p.23.
Shandarin, S.F. (1982). Paper presented at IAU Symposium 104.
Schwartzman, V.F. (1969). JETP Letters, 9, 184.
Silk, J.I. (1968). Astrophys.J., 151, 459.
Silk, J.I. (1977). Astrophys.J., 211, 638.
Silk, J.I. (1982). in "Astrophysical Cosmology" ed. H.A. Brück, G.V. Coyne & M.S. Longair (Vatican Publications) p. 427.
Starobinsky, A.A. (1979). Pisma JETP, 30, 719.
Sunyaev, R.A. & Zel'dovich, Y.B. (1970). Astrophys.Sp.Sci., 7, 3.
Tarbet, P.W. & Rowan-Robinson, M. (1982). Nature, 298, 711.
Vilenkin, A. (1981). Phys.Lett., 107B, 47.
Wagoner, R.V., Fowler, W.A. & Hoyle, F. (1967). Astrophys.J., 143, 3.
Wasserman, I., (1981). Astrophys.J., 248, 1.
Weinberg, S. (1982). in "Astrophysical Cosmology", ed. H.A. Brück, G.V. Coyne & M.S. Longair (Vatican Publications) p. 503.
Wilkinson, D.T. (1982). Phil.Trans.R.Soc., A307, 55.
Yang, J., Turner, M.S., Steigman, G., Schramm, D.N. & Olive, K.A. (1983). Astrophys.J. (submitted).
Zel'dovich, Y.B. (1972). Mon.Not.R.astr.Soc., 160, 1P.
Zel'dovich, Y.B., Einasto, J. & Shandarin, S.F. (1982). Nature, 300, 407.

QUANTUM COSMOLOGY AND THE EARLY UNIVERSE

James B. Hartle
Enrico Fermi Institute, University of Chicago
Chicago, IL 60637 U.S.A.

1 INTRODUCTION

In reviews of quantum gravity it is conventional to argue that the issues raised by the theory will become important in cosmology. As this is a volume on cosmology it seems appropriate to make this argument the other way around - that quantum gravity is important for the issues raised by cosmology. This is the case which we shall present in this chapter. We will not be able to make it completely because we lack the tools, the insight and the fundamental theory to go much beyond suggestive model calculations in the direction of rigorous results. In an examination of these calculations in all their variety, however, the argument becomes increasingly persuasive.

The singularity theorems of classical general relativity argue that quantum gravitational effects are important at a sufficiently early stage in the history of the universe. These theorems suggest that any reasonable model of the universe, evolved backwards towards the past using classical dynamics, will pass through regions of arbitrarily strong curvature before encountering an initial singularity (Hawking and Ellis 1968). When the curvature varies significantly on the scale of a Planck length

$$\ell = (16\pi G)^{1/2} \approx 10^{-32} \text{ cm}, \tag{1.1}$$

quantum gravitational effects are certainly important. The interesting question, however, is not whether quantum gravitational effects are significant for the structure and evolution of the universe at some very early time. Rather, the interesting question is whether these effects are important for the universe we see today. Indeed, the central issue in the physics of the early universe is to determine to what extent the large scale structure of the universe that we see today is a consequence of the laws of microscopic physics operating at early times.

In this volume a good deal will be found about how processes occuring at temperatures characteristic of the unification of the strong, electromagnetic and weak interactions can explain such basic properties of our universe as the approximate homogeneity, isotropy and spatial flatness, the baryon to photon ratio, and perhaps even the spectrum of density fluctuations. We learn, for example, how inflation occuring at the GUT epoch can turn a broad class of natural initial conditions into the present unnaturally smooth and flat universe; how the present distribution of matter and radiation could arise from particle production and decay given an initial state of thermal equilibrium; and how a reasonable initial spectrum of fluctuations can be amplified to the one we observe today.

For every one of these processes occuring at the GUT epoch there are processes occuring at the Planck time which may be equally significant for the large scale structural features of the universe. There may be inflation at the GUT epoch on the GUT time scale but there may also be inflation at the Planck epoch on the Planck time scale. There is production of baryons in the GUT epoch but there is also the gravitational production of both matter and radiation at the Planck time. If there is the growth of fluctuations at the GUT epoch these will have as their source the spacetime fluctuations at the Planck epoch. These quantum gravitational processes may be as effective in explaining the large scale structural features of the universe as the ones which follow later. There is certainly no evidence that they act weakly. Further, the quantum gravitational processes act naturally and universally. There is nothing to fine tune. There is thus every reason to investigate quantum gravitational effects for their impact on the large scale structural features of the present universe.

Whatever the action of physical processes, there can be no explanation of the large scale structural features of our universe without some kind of initial condition. Every process acting at the GUT epoch requires some restriction on the initial state. Inflation can act to smooth and flatten a universe but not one which is initially arbitrarily irregular. Particle interactions can explain the baryon number but require an assumption of initial matter in the thermodynamic equilibrium. The natural place to pose initial conditions or to examine their possible range is at the big bang itself. Thus whatever the relative merits of the various physical processes for determining the evolution of the early universe, quantum gravity will be important in determining the initial conditions.

To argue the importance for cosmology of physical processes at the Planck epoch is one thing, to calculate their effect is another. The fundamental obstacle is the lack of a complete and managable quantum theory of gravity. The difficulty is not simply the nonrenormalizability of the perturbation theory, but more fundamentally with the theory's interpretative structure. What is the Hilbert space of states or should we even have one? What variable plays the role of time so essential to most formulations of quantum mechanical theories? (For a review of these issues see Kuchar 1981). These difficulties of the fundamental theory show up everywhere when one tries to apply it and are always in the background of any model calculation. How are we to proceed in the absence of a complete theory? One answer would be to stop working until quantum gravity is understood. The issues in cosmology, however, seem too important for this. A reasonable answer is to build simple models whose properties we can understand, follow the simplest extrapolation of existing physics, paramatrize our theoretical uncertainties and treat these parameters phenomonologically. One is led thereby not to definite predictions but rather to an idea of the range of possibilities for the quantum behavior of the very early universe.

A significant number of quantum cosmological models have been constructed. Unfortunately, this enterprise has not been pursued in a systematic or complete way. A variety of techniques, approximations and starting points have been used and the connection between them is not always clear. In this chapter we shall review some of these models with the aims of displaying this diversity and of indicating what these models suggest for how physical processes at the Planck epoch can influence the universe we see today.

2 TECHNIQUES FOR BUILDING QUANTUM COSMOLOGICAL MODELS

To discuss the various ways of constructing quantum cosmological models we begin with the basic quantum mechanical law for gravity coupled to matter fields. For simplicity we shall consider just a single free scalar matter field. The path integral prescription for the amplitude to go between an initial state $|a\rangle$ and a final state $|b\rangle$ is, schematically,

$$\langle a|b\rangle = \int \delta g \delta\phi \exp(iS[g,\phi]) . \tag{2.1}$$

Here, S is the total action, the sum of the gravitational action

$$S_E = \ell^{-2} \int d^4x \ (-g)^{\frac{1}{2}} R + \binom{\text{surface}}{\text{terms}} , \tag{2.2}$$

and the action for the matter fields coupled to geometry. For example, for a free conformally invariant scalar field the matter action is

$$S_m = -\tfrac{1}{2}\int d^4x \ (-g)^{\frac{1}{2}} [(\nabla_\alpha \phi)(\nabla^\alpha \phi) + \tfrac{1}{6} R\phi^2] + \binom{\text{surface}}{\text{terms}} \tag{2.3}$$

The geometries summed over in eq. (2.1) and the form of the surface terms included in the actions depend on the states $|a\rangle$ and $|b\rangle$. Suppose, for example, one were interested in the amplitude $\langle h''_{ij}, \phi'' | h'_{ij}, \phi' \rangle$ to go from a definite three geometry and value of the matter field on some initial spacelike surface to different definite values on some final surface. One would then sum over all spacetime geometries and field configurations which match those given on the initial and final surfaces. If states of definite momenta were of interest then certain derivatives should match. If the states were particle vacua then one would integrate over field configurations which vanish asymptotically at slightly complex times, etc.

Equation (2.1) is a schematic representation of the fundamental quantum mechanical law not least because the machinery for dealing with the gauge freedom of the gravitational field - ghost fields, gauge fixing action and the measure - has not been explicitly indicated. More fundamentally, however, it is a schematic representation because on the left hand side of eq. (2.1) we have not spelled out specifically the Hilbert space of states and their associated probability interpretation. This is the central interpretative difficulty in quantum gravity and our considerations will have meaning only insofar as they do not depend on having a complete solution to this problem.

2.1 The Cosmological Wave Functional

The basic quantum cosmological problem is to construct the wave functional

$$\Psi[h_{ij},\phi] = \langle h_{ij},\phi|\Psi\rangle, \tag{2.4}$$

which describes the state of the universe on a particular spacelike surface. This wave functional can be expressed as a functional integral

$$\Psi[h_{ij},\phi] = \int\delta g\delta\phi\ e^{iS[g,\phi]}, \tag{2.5}$$

where the sum is over field configurations and four geometries which match the prescribed values on the spacelike surface of interest and which in the past of that surface satisfy an appropriate condition which defines the cosmological state. The form of this condition is the central problem of quantum cosmology. As illustrations of ideas for this principle one might cite, for example, the idea that in field theory in fixed cosmological background spacetimes the analog of the initial particle vacuum state could be defined by summing over only those particle paths which lie between the initial singularity and the surface of interest. In certain simple models this prescription seems to give reasonable results for cosmological pair creation (Chitre and Hartle 1977, Wade and Sachs 1979, Mensky and Karmanov 1980, Charach and Parker 1981). As a second illustration there is Stephen Hawking's idea that states for closed cosmologies are to be constructed by carrying out functional integrals over compact Euclidean four geometries (Hawking 1981, Hartle and Hawking 1983). This is an attractive idea because it would mean that on the Euclidean section there are no boundary conditions at all for cosmological problems!

In the absence of a compelling or completely worked out principle for specifying the wave functional we can only examine the possible candidates and discuss their consequences. What restrictions must the wave functional satisfy? It must first of all obey the constraints of the Hamiltonian theory written in configuration space operator form (see, e.g. DeWitt 1967 and Kuchar 1971 for review). One can derive these constraints from the functional integral starting point as follows: The classical gravitational equations

$$\delta S/\delta g_{\mu\nu} = 0, \tag{2.6}$$

should be satisfied in the quantum theory as operator identities

$$<h_{ij},\phi|\frac{\delta S}{\delta g_{\mu\nu}}|\Psi> = \int \delta g \delta\phi \frac{\delta S}{\delta g_{\mu\nu}} e^{iS[g,\phi]} = 0. \tag{2.7}$$

On the right hand side this follows directly if the measure is translationally invariant and after suitable regularization if it is not. Evaluated on a given hypersurface with normal n^{μ}, four of the classical equations

$$n_{\mu} [\delta S/\delta g_{\mu\nu}] = 0, \tag{2.8}$$

represent constraints in the sense that they do not involve second derivitives of the metric or matter field in the time direction normal to the surface. These classical equations can thus be completely expressed in terms of the metric on the three surface $h_{ij}(x)$, its conjugate momentum $\pi_{ij}(x)$, the field on the surface $\phi(x)$ and its conjugate momentum $\pi_{\phi}(x)$. They are:

$$H = G_{ijkl}\pi^{ij}\pi^{kl} - h^{\frac{1}{2}}\,{}^{3}R + 16\pi h^{\frac{1}{2}}T^{nn} = 0 \tag{2.9}$$

$$H^{i} = -2\pi^{ij}{}_{|j} + 16\pi h^{\frac{1}{2}}T^{in} = 0 \tag{2.10}$$

where ${}^{3}R$ is the scalar curvature of the three metric h_{ij}, h is its determinant and a slash denotes a covariant derivitive constructed from it. The supermetric G_{ijkl} is

$$G_{ijkl} = \frac{1}{2}h^{-\frac{1}{2}}(h_{ik}h_{jl} + h_{il}h_{jk} - h_{ij}h_{kl}) \tag{2.11}$$

The quantities T^{nn} and T^{in} are, respectively, the energy and momentum densities of the matter fields expressed in terms of the field variables and conjugate momenta.

The operator identities corresponding to these classical equations of motion can be written out in the field representation by replacing the canonical momenta by functional derivatives with respect to the field variables as follows

$$<h_{ij},\phi|\pi^{kl}|\Psi> = -i \frac{\delta}{\delta h_{kl}} <h_{ij},\phi|\Psi>, \tag{2.12a}$$

$$<h_{ij},\phi|\pi_{\phi}|\Psi> = -i \frac{\delta}{\delta\phi} <h_{ij},\phi|\Psi>. \tag{2.12b}$$

We thus obtain four functional differential equations for the wave functional .

The three equations corresponding to eq. (2.10) express the independence of the wave functional on the choice of gauge used to express the three geometry. For example, in the case of pure gravity these equations are

$$\left(\frac{\delta\Psi}{\delta h_{ij}}\right)\Big|_j = 0. \tag{2.13}$$

One can make an infinitesimal change of the coordinates in the three surface, calculate the change in the metric components and the consequent change in the wave functional. Equation (2.13) is just the condition that this change vanishes.

The operator equation corresponding to the constraint of eq. (2.9) reads

$$\left(G_{ijkl} \frac{\delta}{\delta h_{ij}} \frac{\delta}{\delta h_{kl}} + h^{\frac{1}{2}}\,{}^3R - 16\pi h^{\frac{1}{2}} T^{nn}\right) \Psi[h_{mn},\phi] = 0, \tag{2.14}$$

where T^{nn} will involve both ϕ and functional derivitives with respect to it. It is this equation which contains the dynamical information important for cosmology. It is called the Wheeler-DeWitt equation (DeWitt 1967, Wheeler 1968) and plays a role analogous to that of the Schrödinger or Klein-Gordan equation in flat spacetime quantum theories. One would like to solve this equation to obtain possible wave functionals for cosmology. One can do this in simple models in which only a few of the infinite number of matter and spacetime degrees of freedom are quantized. This is called the minisuperspace approach. Many models of this kind have been considered by many authors. For example, the geometry might be restricted to be homogeneous and isotropic (e.g. DeWitt 1967) or to be homogeneous and anisotropic (e.g. Misner 1970, 1972, 1973), or one might consider homogeneous, isotropic geometries with a single homogeneous field degree of freedom (e.g. Isham and Nelson 1974, Blyth and

and Isham 1975, Hartle and Hawking 1983). In all of these cases, the wave functional is then a function on the space of restricted parameters describing the remaining degrees of freedom (the minisuperspace) and the Wheeler-DeWitt equation becomes a partial differential equation in these parameters which one can hope to solve. We shall illustrate this approach to constructing quantum cosmological models below, but for greater detail and further references the reader is referred to the lucid reviews by MacCallum (1975) and by Ryan (1972).

2.2 Matrix Elements

A knowledge of the cosmological wave functional and a suitable interpretation of it would in principle allow us to answer all quantum cosmological questions of interest. While these questions are often crude: "Is there a singularity?", "Does the universe evolve towards isotropy?", etc. - the models for which one can solve for the wave functional are often cruder still in the sense that they suppress degrees of freedom which may bear importantly on these issues. Progress can be made in a different way by keeping more degrees of freedom but attempting to approximate simpler quantities than the full wave functional - typically matrix elements and expectation values of the geometry. Instead of attempting simple calculations of the wave functional which has all information we attempt more complete calculations of matrix elements which have less information. It is important to emphasize that these two approaches to constructing quantum cosmological models do not represent different physics but only different approximations to the anticipated underlying theory.

A further advantage to dealing directly with matrix elements such as the metric expectation value is that these are the type of quantities (e.g. tensor fields) in terms of which the classical theory is formulated. They can be shown to obey dynamical equations which are the classical equations plus quantum corrections. A good deal of classical technique can then be carried over in their analysis so that the interpretative problem is easier. In a certain sense in this approach we are never very far from the classical description or calculational methods. In this sense all these types of methods can be called semiclassical.

One of the simplest approximations in which to illustrate

the derivation of a dynamical equation for metric matrix elements is the leading order 1/N approximation (see, e.g. Tomboulis 1977, 1980 and Hartle and Horowitz 1981 whose treatment we follow here). In this approximation one imagines that in addition to gravity one has N identical matter fields which for simplicity we take to be N conformally invariant massless scalar fields ϕ_j, j = 1, ..., N. The matter action is then

$$S_m^N = -\tfrac{1}{2}\int d^4x(-g)^{\frac{1}{2}}[\nabla_\alpha\phi_j\nabla^\alpha\phi^j + \tfrac{1}{6}R\phi_j\phi^j] \tag{2.15}$$

and the basic functional integral defining matrix elements is

$$\langle a|b\rangle = \int\delta g \int\prod_j\delta\phi_j \exp\{i(S_E[g] + S_m^N[g,\phi_j])\}. \tag{2.16}$$

This integral can be done in two steps. First, for a given four geometry and a single scalar field, define a functional Y[g] by

$$\exp(iY[g]) = \int\delta\phi \exp(iS_m[g,\phi]), \tag{2.17}$$

where S_m is the action for a single conformally invariant scalar field. We note that both S_E and Y depend on the states $|a\rangle$ and $|b\rangle$ although we have not indicated this dependence explicitly. Since the matter fields are independent, eq. (2.16) becomes

$$\langle a|b\rangle = \int\delta g \exp\{i(S_E[g] + NY[g])\}, \tag{2.18}$$

in which only the integral over the geometry remains. In order for the large N limit of this theory to exist we must rescale the gravitational coupling constant ℓ^2, setting $\ell^2 = N\lambda^2$, and consider the limit of N large, λ fixed. Under this rescaling eq. (2.18) can be written in a way which shows the dependence on N explicitly

$$\langle a|b\rangle = \int\delta g \exp\{iN(S_E[g] + Y[g])\}, \tag{2.19}$$

S_E now being the rescaled gravitational action. For large N the dominant contribution to this integral will come from the extremum of the effective action

$$\Gamma[g] = S_E[g] + Y[g]. \tag{2.20}$$

This effective action is the same functional as introduced by other means in field theory (see, e.g. DeWitt 1981). It depends on the states $|a\rangle$ and $|b\rangle$.

The value of the metric at the extremum of the effective action is the matrix element $\langle a|g_{\alpha\beta}|b\rangle$ in the leading order 1/N approximation. One can see this by expressing the matrix element as the functional integral

$$\langle a|g_{\alpha\beta}(x)|b\rangle = \int \delta g \delta\phi \; g_{\alpha\beta}(x) \; \exp\{i(S_E[g] + S_m[g,\phi])\}, \tag{2.21a}$$

$$= \int \delta g \; g_{\alpha\beta}(x) \; \exp\{i(S_E[g] + Y[g])\} \tag{2.21b}$$

and then evaluating it by translating the integration variable so that it is centered at the extremum of the effective action. The leading order 1/N approximation for the matrix element is equivalent to the stationary phase approximation for the integral. In this approximation $\langle a|g_{\alpha\beta}(x)|b\rangle$ is the extremum of the exponent in eq. (2.21b), that is, the extremum the effective action.

We shall call the matrix element $\langle a|g_{\alpha\beta}(x)|b\rangle$ the effective geometry. In general it is a complex quantity except for especially simple situations when the initial and final states can be made the same. The more easily interpretable expectation values (say in a given initial state) could be computed if one knew all the initial to final matrix elements.

The two step process for evaluating matrix elements in the leading order 1/N approximation gives rise to a simple interpretation of the functional integral in eq. (2.17) over the matter degrees of freedom. This can be thought of as the functional integral defining a matter quantum field theory in a given background spacetime g. To emphasize this we write

$$\langle a|b\rangle_g = \int \delta\phi \; \exp(iS_m[g,\phi]) \tag{2.22}$$

where the subscript g indicates a matrix element between states of the matter field in the background geometry g. For example, if the geometry has initial and final flat regions then we can compute the particle vacuum persistence amplitude by integrating over field configurations which vanish at values of the time slightly rotated into the complex plane. The imaginary part of Y is then directly related to the total probability to produce pairs of particles

$$P = 1 - \exp(-2\mathrm{Im}Y). \tag{2.23}$$

In this way we can calculate particle production amplitudes.

For the case of free particles under discussion, the integral in eq. (2.17) is quadratic and can be done immediately to give

$$Y = -\frac{i}{2}\mathrm{Tr}\{\log G(x,x')\}_{reg}, \tag{2.24}$$

where G is the matter field Green's function and the subscript reg indicates that the divergent trace must be suitably regulated.

Certain matrix elements of matter field operators may be calculated directly from Y. An important example is the matrix elements of the stress energy tensor whose definition we can take to be

$$T^{\alpha\beta} = \frac{2}{(-g)^{\frac{1}{2}}}\frac{\delta S_m}{\delta g_{\alpha\beta}}. \tag{2.25}$$

Recalling the analog of eq. (2.21) we see

$$\frac{\langle a|T^{\alpha\beta}|b\rangle_g}{\langle a|b\rangle_g} = \frac{2}{(-g)^{\frac{1}{2}}}\frac{\delta Y}{\delta g_{\alpha\beta}}. \tag{2.26}$$

If, for example, the geometry g has initial and final flat regions and the integral is taken over field configuration ϕ which vanish asymptotically at times slightly rotated into the complex plane then we obtain the matrix elements of $T_{\alpha\beta}$ between the initial and final particle vacua.

This interpretation of Y leads to a simple interpretation of the condition for the extremum of the effective action. This becomes

$$G^{\alpha\beta}(g) = \frac{\ell^2}{2}\frac{\langle a|T^{\alpha\beta}|b\rangle_g}{\langle a|b\rangle_g} \tag{2.27}$$

This is a dynamical equation for the matrix element $<a|g_{\alpha\beta}|b>$. It has the form of Einstein's equations plus quantum corrections from the matter fields. The asymmetry between the matter fields and the gravitational field arises because we are working in the leading order 1/N approximation. There is only one gravitational field but N matter fields.

In the semiclassical approach one attempts to solve this equation to construct model cosmologies. In general this can only be done self-consistently. One guesses a form for $g_{\alpha\beta}$, defines appropriate particle states, calculates the matrix elements of the stress-energy tensor and sees whether these act as the source of the assumed $g_{\alpha\beta}$. In certain simple situations we can calculate the matrix elements of $T^{\alpha\beta}$ for a wide class of geometries. Conformally invariant fields in conformally flat spacetimes is the most important example. Then the dynamical equation can be solved directly. Probably the most important thing to stress is that the problem of determining initial particle states and initial behavior of the cosmological geometries are coupled. One cannot solve one without the other. A given cosmological geometry will allow certain natural initial matter states and these will determine how much particle production takes place. In turn the energies of these particles will influence the initial behavior of the geometry.

In the following section we shall illustrate the construction of model quantum cosmologies through the solution of this equation.

3 SOME ILLUSTRATIVE QUANTUM COSMOLOGICAL MODELS

Can quantum gravitational effects cause the universe to bounce rather than collapse to a singularity? Can they eliminate particle horizons? Can they cause inflationary expansion? Can they dissipate anisotropy and inhomogeneity? Can they produce the matter in the universe? These are the kind of dynamical questions that one would like to answer in quantum cosmology. They are the kind of questions which have been discussed in the context of the many quantum cosmological models constructed using the techniques described in the previous section. In this section we shall discuss the status of some of these models and in particular how they bear on the questions raised above. We will be able to discuss only a small fraction of the quantum cosmological models which have been constructed. Even this discussion will not be very systematic because as yet we neither have a set of models which are general enough to supply answers to all the questions raised above nor has every model been analyzed with all the techniques which are available. We shall attempt, however, to discuss some of the more illustrative examples. For reviews of the larger situation the reader is referred to reviews of Hu (1980), Parker (1977), Lukash, Novikov, Starobinsky and Zel'dovich (1976), and MacCallum (1975).

By far the largest class of quantum cosmological models which have been considered are the homogeneous universes. We shall restrict our attention to models in this class. The geometry of these models is described by a line element of the form,

$$ds^2 = -dt^2 + a^2(e^{2\beta})_{ij}\omega^i\omega^j, \quad (3.0)$$

where the scale factor a and the 3 x 3 traceless anisotropy matrix β are functions of the time alone and the ω^i are three time independent one forms appropriate to the group of spatial homogeneity. In the mini-superspace approach, restricting attention to geometries of this form means solving the Wheeler-DeWitt equation for wave functionals whose only dependence on the metric is through the parameters a and β_{ij}. In the case of the semiclassical methods it means solving the dynamical equations with quantum corrections [eq. (2.27)] for four-geometries which have the form of eq. (3.0).

We shall now review some of these homogeneous models as they apply to the interesting questions of quantum cosmology.

3.1 Existence of a Singularity

The simplest models to examine the question of whether quantum effects can influence the existence of a singularity are the homogeneous and isotropic ones. For these, the line element can be written in Robertson-Walker form using proper time as

$$ds^2 = -dt^2 + a^2(t)\,d\sigma_k^{\,2}, \quad (3.1a)$$

or using the conformal time as

$$ds^2 = a^2(\eta)[-d\eta^2 + d\sigma_k^{\,2}], \quad (3.1b)$$

where $d\sigma_k^2$, $k = -1, 0, 1$ are the line elements corresponding to open, flat and closed spatial sections, respectively. Classically there are no pure gravitational solutions which are homogeneous and isotropic so it is appropriate to introduce some matter degrees of freedom into the model in order that there be a well defined classical limit. We shall consider universes which contain both classical radiation and a quantized conformally invariant scalar field. There is no deep reason for these choices. They are common in model building. The conformally invariant scalar field is a simplified representative for the quantized matter degrees of freedom all of which one expects to be exactly or approximately conformally invariant at early stages. It gives a contribution to the stress-energy, which can be constructed from the appropriate action [eq. (2.3)] in the standard way. The classical radiation is a simplified representative for the real radiation at relatively late stages ($t > 10^{-34}$ sec.). Its contribution to the stress energy tensor (in an orthonormal frame) is

$$T^{rad}_{\alpha\beta} = \left(\frac{\tilde{\rho}_r}{a^4}\right) \operatorname{diag}\left(1, \frac{1}{3}, \frac{1}{3}, \frac{1}{3}\right), \tag{3.2}$$

where $\tilde{\rho}_r$ is a constant.

Let us now investigate the issue of the singularity in the minisuperspace model corresponding to these assumptions. [We follow Hartle and Hawking 1983, however, see also Blyth and Isham 1975 and DeWitt 1967 for earlier treatments of closely connected models.] Since the geometry is homogeneous the scalar field is homogeneous as well and $\phi = \phi(t)$. The wave function then depends on only two variables a and ϕ or equivalently or a and the conformally scaled scalar field $\tilde{\phi} = a\phi$,

$$\Psi = \Psi[a, \tilde{\phi}]. \tag{3.3}$$

In the quantum theory one might say that the singularity is avoided if the wave functional behaves in a sufficiently regular way at $a = 0$ and in particular if it vanishes there. We can now ask whether the Wheeler-DeWitt equation for this model admits solutions of this type. With the restricted degrees of freedom and a simplifying rescaling of the variables involved, the Wheeler-DeWitt equation [eq. (2.14)] reduces to the following:

$$[a\frac{\partial}{\partial a}(a\frac{\partial}{\partial a}) - ka^4 + a^2\tilde{\rho}_r + a^2(-\frac{\partial^2}{\partial\tilde{\phi}^2} + \tilde{\phi}^2]\Psi = 0 \quad . \tag{3.4}$$

The first term arises from the momentum part of the equation (a particular operator ordering has been chosen) and the second is from the spatial curvature. The last three terms are the contributions of the classical radiation and the conformally invariant scalar field. The equation can then be separated by expanding it in harmonic oscillator eigen-functions for $\tilde{\phi}$. We write

$$\Psi = \Sigma_n c_n(a)\psi_n(\tilde{\phi}) \quad , \tag{3.5}$$

where

$$\frac{1}{2}(-\frac{\partial^2}{\partial\tilde{\phi}^2} + \tilde{\phi}^2)\ \psi_n(\tilde{\phi}) = (n + \frac{1}{2})\ \psi_n(\tilde{\phi}) \quad . \tag{3.6}$$

Each $c_n(a)$ then obeys its own ordinary differential equation. For small a (near the singularity) we find the possible behaviors

$$c_n(a) \sim c_1 + c_2 \log a \quad , \tag{3.7}$$

where c_1 and c_2 are arbitrary constants. No choice of these constants makes the wave functional vanish at the singularity although the choice $c_2 = 0$ makes it non-singular there. These behaviors at $a = 0$ are, however, strongly dependent on the operator ordering. If we had written the first term in a different fashion we could have made one solution or even both vanish at $a = 0$ or arranged a situation where both were singular. The question of avoidance of the singularity is thus left ambiguous if sufficiently regular behavior of the wave functional near singular geometries is taken as the deciding criterion. One can exhibit solutions for which this criterion is satisfied, and one can also exhibit solutions for which it is not. This situation seems to be generally the case for the minisuperspace models which have been considered. (Compare the discussions in DeWitt 1967, Blyth and Isham 1975, Liang 1972 and Misner 1973. For another approach in which the quantization of restricted degrees of freedom can sometimes lead to cosmological models without singularity see Padmanabhan 1981, and Padmanabhan and Narlikar 1981.)

In another sense, however, the analysis of a model like this one shows that the problem of singularities is less severe in quantum cosmology than in the classical theory. Provided there is a principle which specifies some boundary condition near a = 0, eq. (3.4) will predict the propagation of wave packets into the singularity and back out again. The evolution of the universe through a singularity can thus be followed quantum mechanically when it cannot be so followed classically. This is a situation familiar from other areas of quantum physics, for example, the s-wave scattering of electrons from the Coulomb potential.

Let us now consider the issue of the singularity in the same model but using the semiclassical approach. Here we want to write down and solve the dynamical equation (2.27). As has been appreciated by many authors (e.g. Fulling and Parker 1973, Davies 1977), the regulated stress-energy tensor of the quantized matter fields is not restricted by the positive energy theorems of the classical theory, and the potential therefore exists for solutions to the dynamical equation which are singularity free. A number of instances of this have been displayed (e.g. Fulling and Parker 1973, Fischetti, Hartle and Hu 1979, Berger 1982). The case of conformally invariant fields in the conformally flat Robertson-Walker backgrounds is one of the few where we can obtain an explicit expression for the matrix elements of the stress-energy tensor in terms of the geometry and an explicit equation to solve. It is the case we shall examine here.

Consider first the particle states. Since the field is conformally invariant it has classical solutions of the form

$$\phi(x) = \frac{1}{a(\eta)} e^{\pm i\omega_k \eta} f_k(\vec{x}), \tag{3.8}$$

where the functions f_k labeled by an index k are the eigenfunctions of the Laplacian on the spatial slice and ω_k^2 are its eigenvalues. Eq.(3.8) exhibits a natural definition of positive and negative frequency solutions of the conformally invariant wave operator for these spacetimes. A natural choice for a particle vacuum is therefore to take the state annihilated by the positive frequency part of the field defined in this sense. Since positive and negative frequency solutions never mix there is no creation of particles associated with the scalar field. This is in accord with the general result of Parker (1969) that there is no creation of conformally invariant particles in conformally flat spacetimes. Let us use this vacuum for both the states $|a\rangle$ and $|b\rangle$ in the dynamical equation, eq. (2.27) and solve this equation for the vacuum expectation value of the scale factor. The expectation value $<0|T_{\mu\nu}|0>$ has been calculated by many authors. (See, e.g. Brown and Cassidy 1977, Dowker and Critchley 1977, Christiansen 1978, and Birrell and Davies 1982 for a review.) It follows entirely from the geometry's symmetry, the conservation law and the nonvanishing trace of $<0|T_{\mu\nu}|0>$. Although the trace of the classical stress-energy tensor of a conformally invariant field vanishes identically, the quantum mechanical expectation value $<0|T_{\mu\nu}(x)|0>$ is formally divergent and must be regulated. The popular methods of regularization break the conformal invariance and lead to a non-vanishing trace - the trace anomaly. Its general form is (Duff 1977):

$$<0|T(x)|0> = \alpha\nabla^2 R + \beta(R_{\alpha\beta}R^{\alpha\beta} - \tfrac{1}{3}R^2) + \gamma C_{\alpha\beta\gamma\sigma}C^{\alpha\beta\gamma\delta}, \qquad (3.9)$$

where the constants α, β, and γ are related by $3\alpha - \beta - 2\gamma = 0$. The values of α, β, and γ depend on the number of matter fields in the early universe. In one popular scheme, the dimensional one,

$$\alpha = \frac{1}{2880\pi^2}[N_S + 3N_N + 12N_V], \qquad (3.10a)$$

$$\beta = \frac{1}{2880\pi^2}[N_S + \frac{11}{2}N_N + 62N_V] \qquad (3.10b)$$

where N_S are the number of elementary scalars, N_N the number of two component neutrinos and N_V the number of vector particles. Other regularization schemes might give different answers for these numbers and in particular one can find schemes (unpopular ones) for which they vanish altogether [e.g. Brown (unpublished); Fradkin and Vilkovisky (1978)]. In view of this, and because we do not know the exact nature of the particle theory at very high energies it seems best to regard these constants as phenomenological parameters and investigate their entire range.

Using the Robertson-Walker line element eq. (3.1), the stress-energy tensor deduced by conservation and symmetry from the anomalous trace [eq. (3.9)], and including the contribution of the classical radiation, the dynamical equation

$$G_{00} = \frac{\ell^2}{2}\left(T_{00}^{rad} + \langle 0|T_{00}|0\rangle\right), \tag{3.11}$$

reads as follows in the case k = 0:

$$\frac{6}{\ell^2}(a')^2 = \tilde{\rho}_r + 12\alpha\left[\frac{1}{2}\frac{a'a'''}{a^2} - \frac{a''a'^2}{a^3} - \frac{1}{4}\left(\frac{a''}{a}\right)^2\right] + 3\beta\left(\frac{a'}{a}\right)^4 \tag{3.12}$$

Here, a prime denotes a derivative with respect to η. Equation (3.12) is a third order differential equation for a. Even if we demand that the solution approach the classical solution for a radiation dominated universe at late times

$$a \sim (\tilde{\rho}_r \ell^2/6)^{1/2}\eta, \qquad \eta \to \infty \tag{3.13}$$

this condition will not uniquely determine the behavior at early times. The most complete analysis of the solutions which do become asymptotically classical has been given by Anderson (1983). To illustrate his results let us consider the case $\alpha > 0$, $\beta > 0$ and $\beta/3\alpha > 1$. This is the range of values associated through eq. (3.10) with the particle content of typical grand unified field theories, such as the SU(5) and SO(10) models. For this range and k = 0 Anderson finds a one parameter family of asymptotically classical models shown in Figure 1. All of his solutions deviate significantly from the classical solution at early times. Some solutions are singular. In general, however, the singularity is weaker than the classical singularity at least in the sense that, while classically the time-time component of the Ricci curvature in an orthonormal frame diverges as

$$R_{00} \sim a^{-4}, \tag{3.14}$$

in these quantum models it behaves as

$$R_{00} \sim a^{-2}, \tag{3.15}$$

a less singular behavior. There is another class of solutions which "bounce" at some minimum radius and never become singular. Most of these run away towards the past exponentially expanding to larger and

Figure 1. A one parameter family of possible scale factors for the effective geometries of homogeneous, isotropic, spatially flat model cosmologies driven by classical radiation and the quantum effects of conformally invariant matter fields. These curves are those solutions of eq. (3.12) with the values of $\alpha = (2880\pi^2)^{-1}$, $\beta/3\alpha = 2$, which approach the classical behavior of a radiation dominated universe at late times. They are plotted here in terms of the dimensionless scale invariant variables $b = a/(\ell\tilde{\rho}_r^{1/4})$ and $\chi = 6^{-1/2}\tilde{\rho}_r^{1/4}$ so that the classical behavior is $b = \chi$. At early times all solutions deviate significantly from the extrapolated classical solution (dotted line). The deviation is insignificant after about a Planck time from the extrapolated classical singularity. There are three possible types of early time behavior consistent with classical behavior at late times: (i) Singular solutions for which b vanishes as $\chi \to -\infty$. (ii) A nonsingular time symmetric bounce for which $b = |\chi|$ as $\chi \to -\infty$. (iii) Nonsingular solutions which bounce and expand exponentially towards the past on the Planck time scale. None of the solutions contain cosmological particle horizons. The figure is adapted from Anderson (1983).

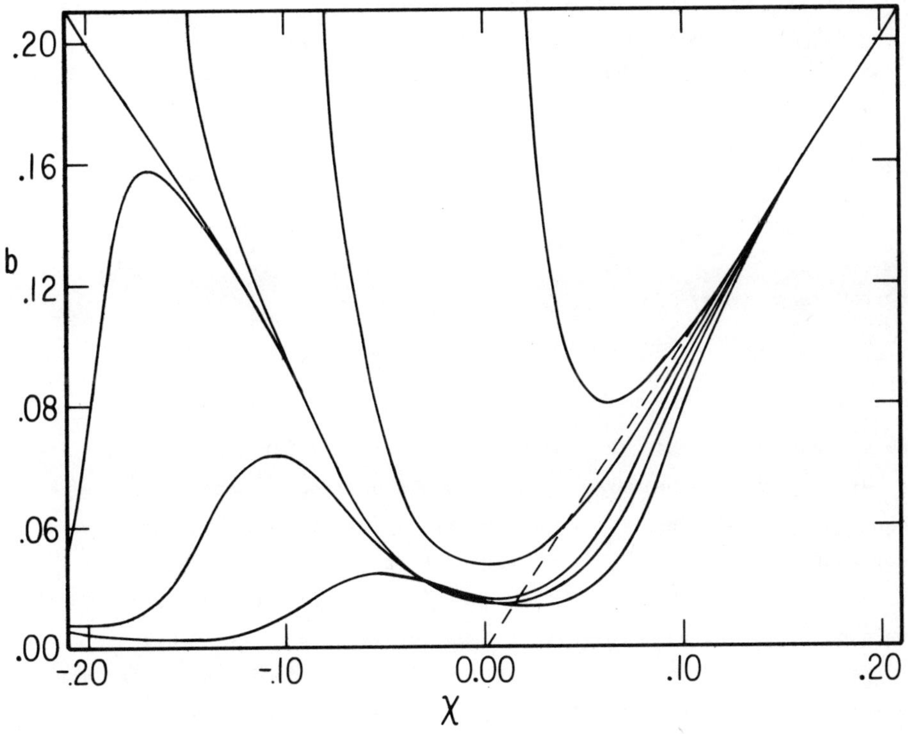

larger volumes on the Planck time scale. This seems an unphysical behavior. There is one time symmetric solution which begins classically in the distant past, collapses to a minimum radius and then reexpands to the present classical epoch. Thus, requiring only classical behavior at late times leaves the universe a wide variety of possible but interesting behaviors at early times. This in some ways recalls the ambiguity in the behavior of the wave functional in the minisuperspace version of this model.

From the present perspective a perhaps even more interesting model than the one we have been discussing, although closely related to it, is the one put forward by Starobinsky (Starobinsky 1980; Gurovich and Starobinsky 1979). This model assumes that there is no radiation in the earliest stages but that the dynamics of the universe in a pure particle vacuum is driven solely by the vacuum stress of the scalar field. It further assumes that the constant α in eq. (3.9) is negative as some regularization schemes do suggest. Then the dynamical equation has initial solutions which are deSitter space

$$a(t) = \frac{1}{H}\cosh Ht \qquad k = +1, \tag{3.16a}$$

$$a(t) = a_o e^{Ht} \qquad k = 0, \tag{3.16b}$$

$$a(t) = \frac{1}{H}\sinh Ht \qquad k = -1, \tag{3.16c}$$

where H is given by

$$H^{-2} = \text{const.} \times (-\alpha\ell^2). \tag{3.18}$$

For the case $k = 0$ this follows easily from eq. (3.12) with $\tilde{\rho}_r = 0$. In early epochs the universe thus follows a deSitter expansion on the Planck time scale. There is no singularity. Of course, this model cannot represent the behavior of the universe at late times. Starobinsky's suggestion is that the decay of instabilities in the quantum expansion could populate the universe with the matter we have today. This is another story; for the present considerations this model is yet another example of how quantum effects can remove the cosmological singularity implied by classical relativity.

On the issue of the existence of a big bang singularity one concludes that there are a variety of models - some singular, some not - which could represent the initial quantum dynamics, and we are as yet unable to implement a physical principle to distinguish between them. Freedom from singularities could, in and of itself, be a compelling requirement for physically reasonable models. One should not overlook, however, the possibility that the quantum theory could provide us with a theory of dynamics which includes singularities and that in doing so perhaps provide some restriction on the possible initial conditions. From this point of view the singular models are of equal interest with the nonsingular ones.

3.2 Horizons and Inflation

The same models that we have used to illustrate that quantum effects can alter the existence and strength of the initial singularity can also be used to illustrate that they can alter the causal structure of the early universe as well. In particular, they will affect the existence of cosmological particle horizons.

Standard classical models must face up to the problem of horizons - the existence of spacetime regions visible from today but causally disjoint from each other because their past light cones intersect the singularity before they intersect each other. (See Figure 2.) The existence of horizons is a fundamental obstacle to any dynamical explanation of the homogeneity of the universe. No physical process can enforce the same physical properties in one spacetime region as in another which is causally disjoint from it. The existence of horizons, however, is not a law of geometry; it is a consequence of a particular classical dynamics. Altering this dynamics to remove horizons is an essential prerequisite to any dynamical explanation of the homogeneity.

In homogeneous, isotropic universes we have already seen that the quantum effects of conformally invariant matter fields can weaken or remove the singularity. Even if the singularity is not removed, these effects can still remove horizons. This is illustrated by the models calculated by Anderson (Figure 1). The singular ones all have their singularity at conformal time $\eta = -\infty$ (although the proper time to the singularity is always finite). As a glance at Figure 2 shows,

Figure 2. The problem of cosmological particle horizons illustrated by the k = 0 homogeneous and isotropic models. These have the line element

$$ds^2 = a^2(\eta)[-d\eta^2 + dx^2 + dy^2 + dz^2]$$

so that in an η-x diagram light rays move along 45° lines. Events A and B are both visible from the present P, but they themselves do not look back on any common past. No physical mechanism which affected physical properties at A could therefore have also affected those at B. In the standard classical cosmological model this is the situation for any two events A and B on the last scattering surface of the cosmic background radiation which are today separated by more than a few degrees on the celestial sphere. The remarkable isotropy of this radiation cannot, therefore, find any dynamical explanation in a cosmology with this geometry.

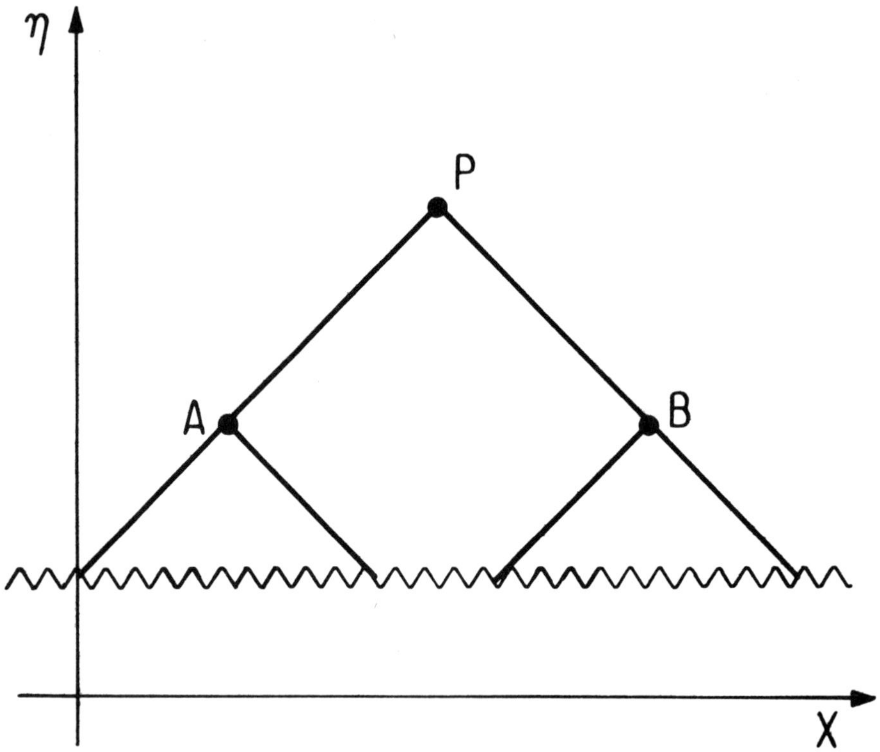

in this case the past light cone of every point intersects the past light cone of every other point and there are thus no horizons. There are then no horizons in any of his models if $\beta/3\alpha > 1$. If there is anyone who believes that horizons are essential to an understanding of the early universe, it should also be noted if $\beta/3\alpha < 1$ it is possible to obtain models which have them [Fischetti, Hartle and Hu (1979)].

There is a great deal in this volume about the uses of inflation - a rapid expansion of the universe on a time scale short compared to its age. Inflation is a potential explanation of homogeneity, isotropy, spatial flatness, the low density of monopoles and the spectrum of fluctuations which produced galaxies. Current grand unified theories of the particle interactions predict such an inflationary phase ending when the temperature is about 10^{14} GeV driven by the energy of the vacuum before a phase transition. It is important to emphasize, however, that, as Starobinsky's model shows, the universe could have had a much earlier inflationary phase at the Planck time driven by quantum gravitational effects. Of course, this inflation is too early to help with the monopole problem - there are no monopoles, only vacuum. It is unclear whether it is relevant to the problem of homogeneity and isotropy since these are central assumptions to producing the model. It can, however, offer an explanation of the spatial flatness problem and the origin of fluctuations in the same way that inflations due to phase transitions can. Some of the cosmological issues for which the inflationary picture offers such a handy resolution may have been resolved by a similar quantum gravitational effect occuring much earlier.

3.3 Dissipation of Anisotropy

As was first suggested by Zel'dovich (1970) the creation of elementary particle pairs in the time dependent geometry of the big bang can drive an initially homogeneous and anisotropic universe towards a homogeneous and isotropic one. The physical argument is simple: As we have already seen, there is no production of conformally invariant particles in homogeneous universes which are isotropic. These geometries are conformally flat and the amplitudes for pair creation of conformally invariant fields vanish because they vanish in flat space. If, on the other hand, the homogeneous universe possesses some anisotropy, then there will be pair creation. This dissipative process, like friction, will tend to drive the universe towards a state where the dissipation ceases, that is a state of isotropy. In this way pair production can dissipate anisotropy. Semiclassical model calculations by Zel'dovich and Starobinsky (1971) by Parker and Hu (1978) and by Hartle and Hu (1980) have confirmed this simple physical picture. One looks for solutions of the semiclassical equation [e.g. eq. (2.27)] for homogeneous cosmological geometries with anisotropy. For example, one can assume the Bianchi I form

$$ds^2 = - dt^2 + a^2(t)\ (e^{2\beta})_{ij}\ dx^i dx^j, \tag{3.19}$$

where the anisotropy is described by the traceless 3 x 3 matrix $\beta_{ij}(t)$. One could then take, for example, the analog of eq. (2.27) with the expectation value of $T_{\mu\nu}$ in the initial particle vacuum on the right hand side and try to solve for the expectation value of the metric (Zel'dovich and Starobinsky 1971, Parker and Hu 1978). Alternatively, one could take eq. (2.27) with the initial particle vacuum for the initial state, the final particle vacuum for the final state, and try and solve for the corresponding matrix element of the geometry (Hartle and Hu 1980). In either case the calculation of matrix elements of $T_{\mu\nu}$ which enters into the dynamical equation can no longer be carried out exactly as a functional of the metric parameters a and β. Rather, the matrix elements of $T_{\mu\nu}$ can either be expressed approximately (Parker and Hu 1978) or developed perturbatively in the anisotropy (Hartle and Hu 1980). To give a concrete example, let us consider the latter case.

As demonstrated in Section 2 the dynamical equations governing the initial-final matrix elements of the geometry in the leading order 1/N approximation can be derived from the effective action

$$\Gamma = S_E - \frac{i}{2}\ \{\mathrm{Tr}\ \log G(x,x')\}_{reg}. \tag{3.20}$$

One can calculate this action functional for geometries of the form (3.19) by expanding G in powers of β. Recall first, that G is the Green's function for a conformally invariant scalar field moving in a background geometry of the form (3.19). In the limit of exact isotropy G is the Green's function appropriate to the vacuum for conformally invariant fields that we have already defined. It is conformally related to the flat space Feynman Green's function. For small anisotropy we can construct G perturbatively, regarding β as a small change in the external potential in which the scalar field moves. One can then obtain the action as a series in β (Figure 3). Stopping after the quadratic order as the lowest approximation we can extremize the effective action to obtain effective geometry. When the action is evaluated at this extremum one finds from eq. (2.23) the total probability P to produce a particle pair in the tube T of given spatial cross section extending over the whole history of the universe. This is:

Figure 3. The expansion of the effective action for a conformally invariant scalar field in a homogeneous, spatially flat geometry in powers of the anisotropy. Γ_0 is the action in the isotropic limit which gives rise to eq. (3.12). In the subsequent terms the propagators are those in the isotropic limit and are conformally related to the flat space propagators. The vertices represent a given order in the interaction of the scalar field with the anisotropy found by expanding the classical action [eq. (2.3)] in powers of β.

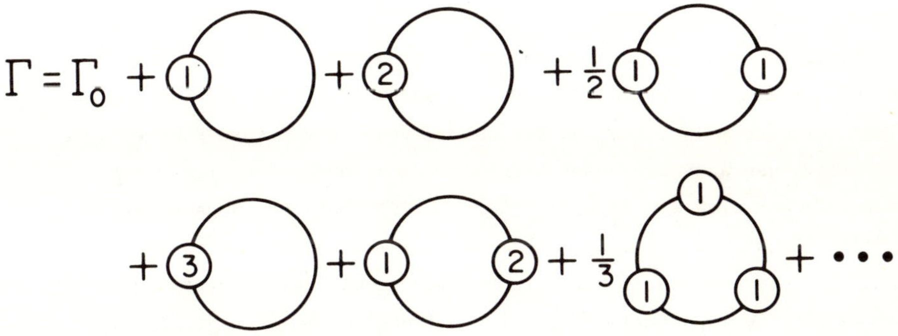

$$P = \frac{1}{1920\pi}\int_T d^4x\,(-g)^{1/2}\bar{C}_{\alpha\beta\gamma\delta}C^{\alpha\beta\gamma\delta}. \tag{3.21}$$

(There is a complex conjugate because the effective geometry is complex.) If the effective geometry had the classical behavior near the singularity this integral would diverge. In fact the softening of the singularity produced by the effects of the trace anomaly in the isotropic limit is enough to make this integral finite.

The anisotropy in the effective geometry in such models decays rapidly away on a fraction of the Planck time as Figure 4 shows. This simple calculation thus demonstrates that pair creation can dissipate anisotropy but being perturbative is limited to small anisotropy. The more elaborate calculations of Hu and Parker (1978) and Zel'dovich and Starobinsky (1971) show that even large anisotropies are dissipated in about a Planck time.

3.4 Production of Matter Fields

Particle production in general proceeds efficiently in the early universe. This is shown clearly by the typical expressions for the total production probabilities [e.g. eq. (3.21)] which diverge if the classical behavior is assumed for the geometry. It is a natural question to ask whether this process could give rise to all the matter in the universe. That is, whether the universe could have started in a particle vacuum and have quantum gravitational processes create the matter content with its observed thermal distribution. Such creation would be essential in a model like Starobinsky's which starts from a vacuum. The mechanism he suggests assumes growing oscillatory instabilities in the exponential expansion. These can be thought of as gravitational excitations of the spacetime characterized by the Planck mass. When coupled to the matter fields these excitations can decay into matter and thus the universe can be populated.

Outside of this one case we do not have complete, self consistent calculations of model universes which start from a particle vacuum and create their own matter. We do, however, have some suggestive calculations. Leonard Parker (1976) considered models in which one smoothes out the initial singularity of the Robertson-Walker spacetimes by joining them onto initial flat domains. He showed in the test field

Figure 4. The decay of a small anisotropy in the effective geometry of a homogeneous cosmology driven by the leading order 1/N quantum effects of conformally invariant scalar fields. For the values $\alpha = \beta = (2880\pi^2)^{-1}$ there is a homogeneous, isotropic solution of the semiclassical equation of motion, [eq. (3.12)] which is singular but horizon free. This figure shows the evolution of a small homogeneous anisotropic perturbation in that effective geometry. The figure shows a quantity W^2 proportional to the absolute square of the Weyl tensor of the effective geometry plotted in arbitrary units as a function of proper time from the singularity in units of the Planck time, ℓ. Several curves are shown corresponding to different values of the scale introduced when regulating the graphs in Figure 3. The results, however, are not very sensitive to the value of this parameter. In all cases the anisotropy decays rapidly in a fraction of a Planck time. The figure was taken from Hartle (1980).

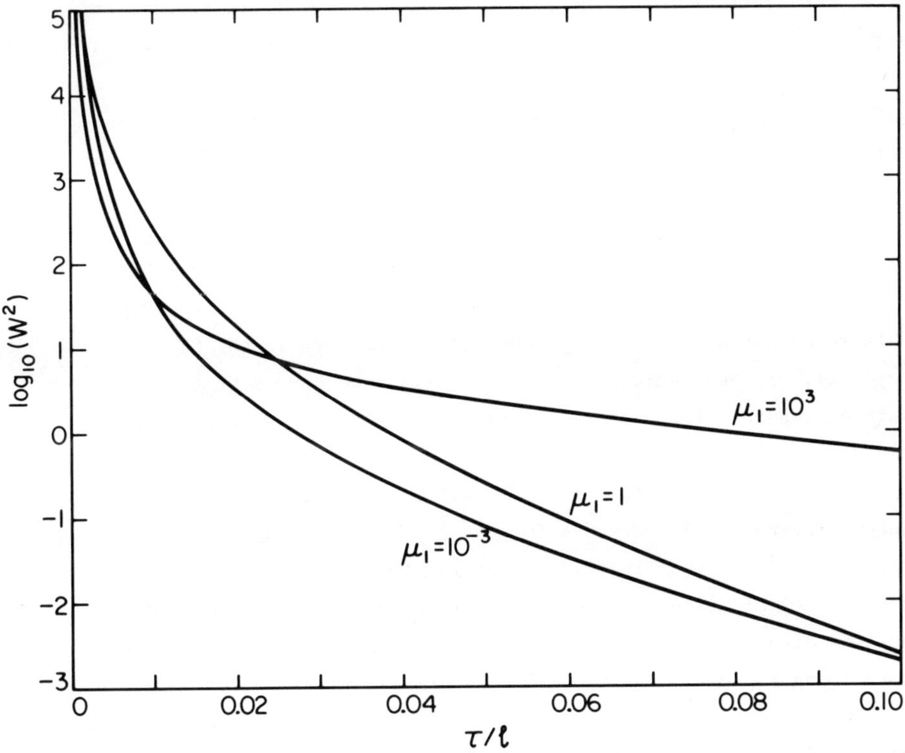

approximation that there would be enough conformally non-invariant particles produced to support the universe in its later stages and further that over a wide range of smoothings these would be produced with an approximate thermal spectrum. Hu and Parker (1977) showed that the production of gravitons in similar models was sufficient to rapidly convert an expansion characterized by a matter equation of state over to one that was appropriately radiation dominated. There is thus considerable encouragement for the point of view that the universe may have created its own matter but we as yet lack a complete set of self-consistent models.

4. CONCLUSIONS

Despite the absence of a complete and managable quantum theory of gravity our survey shows that considerable progress has been made in constructing models displaying the possible implications such a theory might have for the dynamics of the very early universe. We have been able to make this progress because the principles used in constructing the models seem reasonable candidates for consequences of a complete theory and because the models generally do not go far beyond the semiclassical regime in incorporating quantum effects. Even here the absence of a clearly understood quantum gravitational theory is keenly felt. All of the models lack either essential interpretative features or essential constraints which would make them unique. For the minisuperspace models we do not have a clear probability interpretation of the wave functional. In the semiclassical calculations we have a plethora of possible behaviors for even the simplest homogeneous and isotropic geometries.

Even in the face to these deficiencies the importance of quantum gravitational processes for the structure and dynamics of the early universe emerges clearly. Taken together the models display a wide range of significant early behavior. We have models which are free from singularities, singular but free from horizons, anisotropic but evolving rapidly to isotropy, and initially empty but creating their own matter. What we now need is progress on two fronts: We need progress with the fundamental theory, particularly to classify the possible cosmological states and clarify the issue of boundary conditions. We also need further model building to incorporate more general degrees of freedom and initial conditions. Whatever the outcome of this future work the evidence to date suggests that in the enterprise of connecting initial conditions with the universe we see today, quantum gravity will play a central role.

ACKNOWLEDGEMENT

The preparation of this review was supported in part by the National Science Foundation under grants PHY 80-26043 and PHY 81-07384.

REFERENCES

Anderson, P. (1983). (To be published).

Berger, B. (1982). Phys. Lett. 108B, 394.

Blyth, W. F. and Isham, C. (1975). Phys. Rev. D, 11, 768.

Brown, L. S. and Cassidy, J. P. (1977). Phys. Rev. D, 16, 1712.

Brown, M. (unpublished).

Birrell, N. D. and Davies, P. C. W. (1982). Quantum Fields in Curved Space, Cambridge: Cambridge University Press.

Charach, Ch. and Parker, L. (1981). Phys. Rev. D, 24, 3023.

Chitre, D. M. and Hartle, J. B. (1977). Phys. Rev. D, 16, 251.

Christiansen, S. M. (1978). Phys. Rev. D, 17, 946.

Davies, P. C. W. (1977). Phys. Lett. 64B, 402.

DeWitt, B. (1967). Phys. Rev., 160, 1113.

DeWitt, B. (1981). A Gauge Invariant Effective Action, In Quantum Gravity 2, ed. C. J. Isham, R. Penrose and D. W. Sciama, Oxford: Clarendon Press.

Dowker, J. S. and Critchley, R. (1977). Phys. Rev. D, 16, 3390.

Duff, M. J. (1977). Nucl. Phys., B125, 334.

Fischetti, M., Hartle, J. B. and Hu, B.-L. (1979). Phys. Rev. D, 20, 1757.

Fradkin, E. S. and Vilkovisky, G. A. (1978). Phys. Lett., 73B, 209.

Fulling, S. and Parker, L. (1973). Phys. Rev. D, 7, 2357.

Gurovich, V. T. and Starobinsky, A. A. (1979). Zh. Eksp. Teor. Fiz., 77, 1699 [Sov. Phys. JETP 50, 844].

Hartle, J. B. (1980). Phys. Rev. D 22, 2091.

Hartle, J. B. and Hawking, S. W. (1983). (To be published.)

Hartle, J. B. and Horowitz, G. T. (1981). Phys. Rev. D 24, 257.

Hartle, J. B. and Hu, B.-L. (1980). Phys. Rev. D 21, 2756.

Hawking, S. W. (1981). Boundary Conditions of the Universe. Talk Given at the Vatican Study Week on Cosmology and Fundamental Physics, unpublished.

Hawking, S. W. and Ellis, G. F. R. (1968). Ap. J. 152, 25.

Hu, B.-L. (1980). Quantum Field Theories in Relativistic Cosmology. In Recent Developments in Relativity, ed. R. Ruffini, Amsterdam: North Holland.

Hu, B.-L. and Parker, L. (1977). Phys. Lett. 63A, 217.

Isham, C. and Nelson, J. E. (1974). Phys. Rev. D 10, 3226.

Kuchar, K. (1973). Canonical Quantization of Gravity. In Relativity, Astrophysics and Cosmology, ed. W. Israel, Dordrecht-Holland: R. Reidel.

Kuchar, K. (1981). Canonical Methods of Quantization. In Quantum Gravity 2, ed. C. J. Isham, R. Penrose and D. W. Sciama, Oxford: Clarendon Press.

Liang, E. P. T. (1972). Phys. Rev. D 5, 2458.

MacCallum. M. A. H. (1975). Quantum Cosmological Models. In Quantum Gravity, ed. C. J. Isham, R. Penrose and D. W. Sciama, Oxford: Clarendon Press.

Mensky, M. B. and Karmanov, O. Yu (1980). Gen Rel. and Grav. 12, 267.

Misner, C. W. (1970). Classical and Quantum Dynamics of a Closed Universe. In Relativity, New York: Plenum Press.

Misner, C. W. (1972). Minisuperspace, In Magic without Magic, ed. J. Klauder, San Francisco: W. H. Freeman.

Misner, C. W. (1973). Phys. Rev. D 10, 3271.

Padmanabhan, T. (1981). Gen. Rel. and Grav. 13, 451.

Padmanabhan, T. and Narlikar, J. V. (1981). Phys. Lett. 84A, 361.

Parker, L. (1969). Phys. Rev. 183, 1057.

Parker, L. (1976). Nature 261, 20.

Parker, L. (1977). Production of Elementary Particles by Strong Gravitational Fields. In Asymptotic Structure of Spacetime, ed. F. P.

Esposito and L Witten, New York: Plenum Press.

Parker, L. and Hu, B.-L. (1978). Phys. Rev. D 17, 933.

Ryan, M. (1972). Hamiltonian Cosmology, Berlin: Springer-Verlag.

Starobinsky, A. A. (1980). Phys. Lett. 91B, 99.

Tomboulis, E. (1977). Phys. Lett. 70B, 361.

Tomboulis, E. (1980). Phys. Lett. 97B, 77.

Wade, A. E. and Sachs, R. K. (1979). (Unpublished.)

Wheeler, J. A. (1968). Superspace and the Nature of Quantum Geometrodynamics. In Battelle Rencontres, ed. C. DeWitt and J. A. Wheeler, New York: W. A. Benjamin.

Zel'dovich, Ya. B. (1970). Pis'ma Zh. Eksp. Teor. Fiz. 12, 443. [JETP Lett. 12, 307.]

Zel'dovich, Ya. B. and Starobinsky, A. A. (1971). Zh. Eksp. Teor. Fiz. 61, 2161. [Sov. Phys. JETP 34, 1159.]

GUTS AND SUPERSYMMETRIC GUTS IN THE VERY EARLY UNIVERSE

John Ellis

Stanford Linear Accelerator Center
Stanford University, Stanford, California 94305

0 INTRODUCTION

This talk is intended as background material for many of the other talks treating the possible applications of GUTs to the very early universe. I start with a review of the present theoretical and phenomenological status of GUTs before going on to raise some new issues for their prospective cosmological applications which arise in supersymmetric (susy) GUTs. The first section is an update on conventional GUTs (Ellis 1981), which is followed by a reminder of some of the motivations for going supersymmetric (Dimopoulos & Georgi 1981; Sakai 1982). There then follows a simple primer on susy and a discussion of the structure and phenomenology of simple susy GUTs. Finally we come to the cosmological issues, including problems arising from the degeneracy of susy minima, baryosynthesis and supersymmetric inflation, the possibility that gravity is an essential complication in constructing susy GUTs and discussing their cosmology, and the related question of what mass range is allowed for the gravitino. Several parts of this write-up contain new material which has emerged either during the Workshop or subsequently. They are included here for completeness and the convenience of the prospective reader. Wherever possible, these anachronisms will be flagged so as to keep straight the historical record.

1 STANDARD GUTS

Presumably you are familiar with the motivations and guiding principles of conventional GUTs (Georgi & Glashow 1974; Georgi *et al.* 1974). They will not be discussed here, but just an update given on their phenomenological status (for more details see Ellis 1981). GUTs make several low energy predictions, some of which work and some of which are less successful. First the good news: standard GUTs predict (Marciano & Sirlin 1981; Llewellyn Smith *et al.* 1981)

$$\sin^2\theta_W^{eff}(\langle Q^2\rangle = 20\ \text{GeV}^2) = 0.215 \pm 0.002 \tag{1}$$

for $\Lambda_{\overline{MS}}$ (4 operational flavors) = 0.1 to 0.2 GeV. This is to be compared with the experimental value for the effective value of $\sin^2\theta_W$ (including radiative corrections) at an average momentum transfer $\langle Q^2\rangle = 20\ \text{GeV}^2$ in deep inelastic νN scattering (Marciano & Sirlin 1980; Llewellyn Smith & Wheater 1981):

$$\sin^2\theta_W^{eff}(\langle Q^2\rangle = 20\ \text{GeV}^2) = 0.216 \pm 0.012 \quad . \tag{2}$$

Another successful prediction (Chanowitz et al. 1977; Buras et al. 1978; Nanopoulos & Ross 1982) is for the b quark mass, deduced from the τ lepton mass:

$$m_b \approx 5\ \text{GeV} \tag{3}$$

if one assumes the existence of 6 quark flavors corresponding to 3 light neutrinos. Now for the less good news: in the same way that (3) was deduced from m_τ one can also deduce from m_μ that (Buras et al 1978):

$$m_s \approx \tfrac{1}{2}\ \text{GeV} \quad (?) \tag{4}$$

whereas many theorists believe that the true short distance value of $m_s \approx 150$ MeV (Weinberg 1977). I am not convinced by their arguments and prefer to wait and see what value of m_s will emerge from lattice QCD calculations. Finally the bad news: on the same basis as the predictions (3) and (4) one can also predict (Buras et al. 1978)

$$m_d/m_s = m_e/m_\mu \tag{5}$$

while experimentally we know that $m_e/m_\mu \approx 1/200$ and conventional QCD phenomenology suggests that $m_d/m_s \approx 1/20$. Despite the quantitative difficulties (4,5) the quark and lepton masses are qualitatively in the correct relationship, and perhaps some small effect can come in at the level of MeV to cure the ratio (5) for m_d/m_e (Ellis & Gaillard 1979).

Thus, undiscouraged, we do not yet lose faith in the most exciting prediction of conventional GUTs, namely for baryon decay. From Figure 1 we find a decay amplitude $\propto g^2/m_X^2$, where g is the GUT coupling constant and m_X the superheavy boson mass, from which we deduce

$$\Gamma(B \to \bar{\ell} + X) \propto 1/m_X^4 \quad , \quad \tau_B \propto m_X^4 \quad . \tag{6}$$

In simple models such as minimal SU(5) or SO(10) one can estimate

$$m_X \simeq (1\ \text{to}\ 2) \times 10^{15} \times \Lambda_{\overline{MS}} \tag{7}$$

where $\Lambda_{\overline{MS}}$ is the strong interaction scale parameter estimated to be

$$\Lambda_{\overline{MS}} \simeq (0.1 \text{ to } 0.2) \text{ GeV} \quad . \tag{8}$$

Combining Equations (7) and (8) we infer that

$$m_X = (1 \text{ to } 4) \times 10^{14} \text{ GeV} \quad . \tag{9}$$

Various estimates of the constants of proportionality in the dependence (6) on m_X then suggest (Ellis et al. 1980) that

$$\tau_B \simeq (10^{27} \text{ to } 10^{31}) \text{ years} \tag{10}$$

in simple GUTS. There are now two experiments (Krishnaswamy et al. 1981, 1982; Battistoni et al. 1982) reporting positive evidence for baryon decay -- one since the Workshop-corresponding to a nucleon lifetime of a few times 10^{30} years. It will be interesting to see whether these preliminary indications are confirmed, and whether the branching ratios for baryon decay conform to the conventional GUT predictions. If so, it would be a dramatic confirmation of GUT ideas.

Fig. 1. Diagrams contributing to the dimension 6 operators responsible for baryon decay in minimal SU(5).

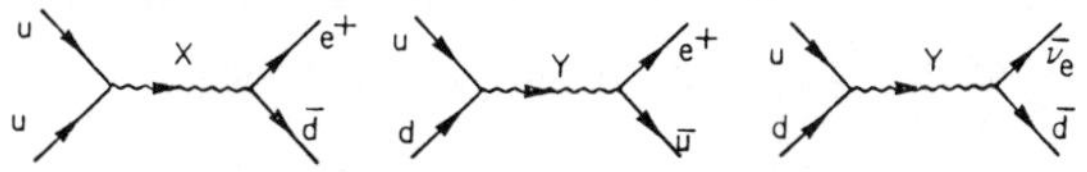

Ironically, even as our experimental colleagues make these fascinating observations, theoretical fashion is deserting conventional GUTs. The reason is the hierarchy problem: why/how is $m_W/m_X << 1$? Conventional GUTs contain two vastly different energy scales m_W and m_X, and m_W in particular is much less than the natural candidate for a basic mass scale, namely the Planck mass $m_P \simeq 10^{19}$ GeV. In conventional GUTs the two mass scales m_W and m_X are associated with two sets of Higgs fields -- e.g. the $\underline{24}$ ϕ and the $\underline{5}$ H of Higgses in minimal SU(5). These are to have a ratio of vacuum expectation values of order 10^{-12}:

$$m_W = \frac{gv}{2}\ , \ m_X = \frac{5gV}{2\sqrt{2}} \ : \ \langle 0|H|0\rangle \equiv v \ , \ \langle 0|\phi|0\rangle \equiv \frac{V}{\sqrt{30}}\begin{pmatrix} 2 & & & & 0 \\ & 2 & & & \\ & & 2 & & \\ & & & -3 & \\ 0 & & & & -3 \end{pmatrix} \tag{11}$$

To achieve this gymnastic feat we need a scalar potential $V(\phi,H)$ in which the effective mass of the SU(2) doublet components of the H multiplet is very small:

$$|m_H^2| \lesssim O(10^{-24})\, v^2 \quad . \tag{12}$$

This feat is difficult to arrange, since there are many contributions to the scalar masses which are a priori much larger than (12), suggesting the necessity of some concealed symmetry if bizarre and unmotivated cancellations are to be avoided. For example, there is the effect of propagating scalar particles through space-time foam (see Figure 2(a)):

$$\delta m_H^2 = O(m_P^2) \tag{13}$$

according to the local experts (Hawking et al. 1979, 1980). Then there are quadratic divergences in perturbation theory (see Figure 2(b)) which yield

$$\delta m_H^2 \propto \int^{\Lambda} d^4k \, \frac{1}{k^2} \propto \Lambda^2 : \Lambda^2 = O\left(m_X^2 , m_P^2\right) ? \tag{14}$$

There are also contributions to the light Higgs masses from the Higgses with large vacuum expectation values (see Figure 2(c)):

$$\delta m_H^2 \propto V^2 \quad . \tag{15}$$

Interactions like the $\phi^2 H^2$ shown in Figure 2(c) are generated by radiative corrections (see Figure 2(d)) even if they were unlawfully excluded from the tree-level Lagrangian (Gildener 1976, Buras et al. 1978). One must adjust the parameters of the Lagrangian taking into account at least 12

Fig. 2. (a) A scalar particle propagating through space-time foam may acquire a mass $O(m_P)$ (Hawking et al. 1979, 1980). (b) A quadratically divergent contribution to the scalar boson (mass)^2. (c) A large contribution to the "light" Higgs (mass)^2 from the condensation of "heavy" Higgs in the vacuum. (d) One-loop contributions to the Higgs self-couplings.

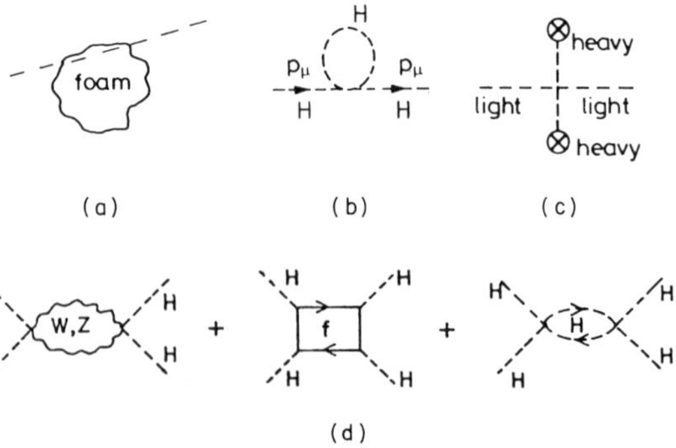

loop diagrams in perturbation theory so as to reduce the right-hand-side of Equation (15) by $O(\alpha^{12}) = O(10^{-24})$ to make it acceptably small.

Theorists have toyed with several possible solutions to this hierarchy problem. The latest and most promising is supersymmetry (susy), so let us now look at the construction of susy GUTs.

2 SUPERSYMMETRY PRIMER

Before constructing susy GUTs, let us first refresh our recollections of susy and of its phenomenology. What is susy? It is a symmetry between fermions and bosons which is generated by charges Q with spin ½ (Gol'fand & Likhtman 1971; Volkov & Akulov 1973; Wess & Zumino 1974a). They obey an algebra of anticommutators, as one would expect for fermion operators, namely

$$\left\{Q^i_\alpha, Q^{\dagger\beta}_j\right\} = (\sigma^\mu)^\beta_\alpha P_\mu \delta^i_j \tag{16}$$

where the $i = 1,\dots,N$ are the different types of extended supersymmetry, the α and β are spinorial indices, and P_μ is the momentum operator. Each operator Q^i_α changes spin by half a unit. Since gauge theories only have spins (or helicities) between ± 1, they can accommodate at most $N = 4$ (global) supersymmetries. If one makes the supersymmetry transformations local, then Equation (16) means that one necessarily must include general coordinate transformations. Therefore one must include gravity and accommodate spins between ± 2, in which case $N \leq 8$ supersymmetries are allowed (see Section 6) (Van Nieuwenhuizen 1981). In all these cases one has equal numbers of boson and fermion states $|B\rangle$ and $|F\rangle$, since

$$Q|B\rangle = |F\rangle \quad \text{and} \quad Q|F\rangle = |B\rangle \tag{17}$$

respectively. We will restrict ourselves to simple $N = 1$ susy in the discussions that follow. In this case there are two classes of supermultiplets (representations of the global supersymmetry algebra) which are relevant. They are the

$$\text{gauge supermultiplet} : \begin{pmatrix} 1 \\ \tfrac{1}{2} \end{pmatrix} \tag{18}$$

which must lie in an adjoint representation of the gauge group, and the

$$\text{chiral supermultiplet} : \begin{pmatrix} \tfrac{1}{2} \\ 0,0 \end{pmatrix} \tag{19}$$

which may lie in any representation of the gauge group. If $N \geq 2$ we are forced to have fermions lying in real representations of the gauge group, which conflicts with the phenomenological (and GUT) necessity of a complex fermion representation.

Why susy? This is a question with many answers, one of which is simply because susy is beautiful. But this is a rather subjective argument -- mother aardvarks also believe that their offspring are beautiful, a point of view that can be debated. Susy is the only known type of symmetry which has not yet been exploited in fundamental physics, so perhaps we should look for an application. It reduces the number of divergences in quantum field theory: for example the $N = 4$ gauge theory appears to be completely finite (Mandelstam 1982), while the $N = 8$ supergravity theory is probably finite up to 6 loops (Grisaru & Siegel 1982). Susy provides a home for scalar fields, as can be seen from the chiral supermultiplet (19) where they are linked together with spin ½ fermions. Susy has the general property of linking together matter and radiation -- witness the gauge supermultiplet (18) relating spin 1 gauge bosons to spin ½ fermions, and the $N = 1$ supermultiplet containing the graviton and the gravitino discussed in Section 6. Extended supergravity theories are the only available candidates for unifying particle physics and gravity, though it must be confessed that present models still leave something to be desired (Ellis et al. 1982b).

All the previous motivations are somewhat philosophical. The practical reason for much current interest in susy is the prospect of "solving" the hierarchy problem in GUTs (Dimopoulos & Georgi 1981; Sakai 1982). "Solving" is in quotation marks because susy does not (yet) cast any convincing light on the question why $m_W/m_P \simeq O(10^{-17})$ or $m_W/m_X \simeq O(10^{-13})$, but it can alleviate the technical difficulties of maintaining this hierarchy once it has been imposed. This is because of the reduction in the number of divergences mentioned earlier. There are no quadratic divergences in susy theories, which removes one of the worst corrections (14) to scalar (Higgs) boson masses. Also, the only logarithmic divergences are wave function and gauge coupling renormalizations. There are no intrinsic renormalizations of the Yukawa couplings or the Higgs self-couplings (Wess & Zumino 1974b; Iliopoulos & Zumino 1974; Ferrara et al. 1974). This means that if one sets the contribution (15) to m_H^2 equal to zero at the tree level, it will not be generated by radiative corrections.

Perhaps we will eventually need to worry about quantum gravity (13) as well, but at least some progress has been made.

How susy? Susy cannot be exact in the real world, as it would imply equal masses for bosons and fermions $m_B = m_F$, just as conventional isospin invariance implies $m_p = m_n$. One must therefore ask how badly susy can be broken, i.e., how heavy can unseen supersymmetric partners be? Presumably the scale of susy breaking must be less than the scale $O(m_P \approx 10^{19}$ GeV) associated with gravitation. A more stringent constraint is that of "solving" the hierarchy problem. Let us suppose God in her infinite wisdom fixes $m_{Higgs} = 0$ (why? this is the reason that the hierarchy problem is not really solved). In an exactly susy world m_{Higgs} would stay zero, but if susy is broken there are corrections due to imperfect cancellations between the diagrams shown in Figure 3:

$$\delta m_H^2 \simeq O(1/16\pi^2)(g^2 \text{ or } \lambda^2)(m_B^2 - m_F^2) \tag{20}$$

where $g(\lambda)$ is a gauge (Yukawa) coupling. In order to maintain the Weinberg-Salam Higgs boson sufficiently light: $|\delta m_H^2| \lesssim O(1)\ \text{TeV}^2$ we must require

$$|m_B^2 - m_F^2| \lesssim \frac{O(16\pi^2)}{(g^2 \text{ or } \lambda^2)} \times O(1)\ \text{TeV}^2 \quad . \tag{21}$$

Thus we see that while the susy breaking may be relatively small, it can be considerably larger than $O(1)\ \text{TeV}^2$ in a supermultiplet which is weakly coupled to the Weinberg-Salam Higgs multiplet ($\lambda << 1$) (Ellis et al. 1982c). There is no necessity for the susy breaking mass splittings $|m_B^2 - m_F^2|$ to be universal for different supermultiplets.

In all theories of broken susy we expect to find the light susy partner particles listed in Table I. Alongside each sparticle entry is indicated the most stringent present lower limit on its mass. All of

Fig. 3. One-loop contributions to the $(\text{mass})^2$ of scalar particles.

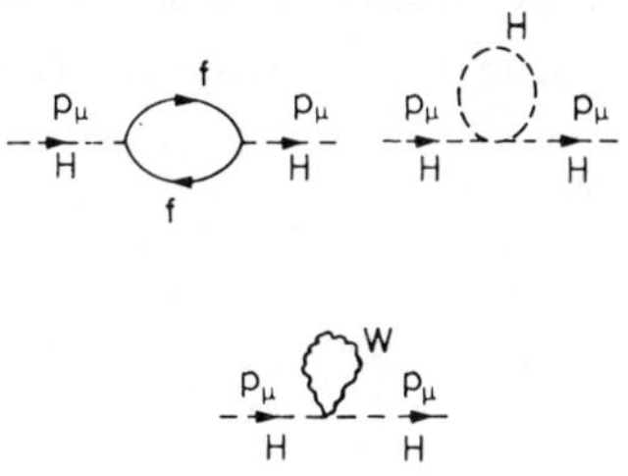

Table I. Supersymmetric Particles

	Particle	Partner	Spin	Lower Limit on Mass
Gauge Supermultiplets	gluon g	gluino $\tilde{g}$	½	3 GeV
	$W^{\pm}$	wino $\tilde{W}^{\pm}$	½	15 GeV
	γ,Z	photino $\tilde{\gamma}$, zino $\tilde{Z}$ or wino $\tilde{W}^{o}$, bino $\tilde{B}$	½	0 GeV
Chiral Supermultiplets	quarks q	squarks $\tilde{q}$	0	15 GeV
	leptons ℓ	sleptons $\tilde{\ell}$	0	15 GeV
	Higgses H	shiggses $\tilde{H}^{\pm}$	½	15 GeV
		$\tilde{H}^{o}$	½	0 GeV

the lower limits come from unsuccessful searches for new particles at PETRA, with the exception of the gluino mass limit which comes from hadron-hadron collisions and perhaps heavy quarkonium decays (Farrar & Fayet 1978 a,b; Campbell et al. 1982).

There are several possible philosophies for breaking susy. One is to do it explicitly, introducing arbitrary "soft" mass terms for unseen particles which are sufficient to push them above the lower bounds in Table I. This strategy is aesthetically unattractive. It introduces many new parameters whose magnitude is not explained, and does not respect naturally the constraints imposed by flavor-changing neutral interactions (Ellis & Nanopoulos 1982a) and by CP violation (Ellis et al. 1982a). It is more appealing to break susy spontaneously.

Either of two strategies may be followed. One approach has spontaneous susy breaking arising in the gauge sector, and is called D-breaking (Fayet & Iliopoulos 1975). In the other approach soft susy breaking arises in the Yukawa interactions of the chiral superfields (F-breaking) (O'Raifeartaigh 1975; Fayet 1975). These names derive from the conventional notation for the scalar field potential (Fayet & Ferrara 1977)

$$V = \sum_{\alpha} |D_{\alpha}|^2 + \sum_{i} |F_{\phi_i}|^2 \tag{22}$$

where

$$D_{\alpha} \equiv g_{\alpha} \sum_{i} \bar{\phi}_i T_{\alpha} \phi_i + \xi_{\alpha} \tag{23}$$

with the g_α the different gauge couplings and ξ_α an arbitrary constant appearing if α is a U(1) factor, and

$$F_{\phi_i} \equiv \partial P/\partial\phi_i \tag{24}$$

where

$$P = \sum_{i,j,k} \lambda_{ijk}\phi_i\phi_j\phi_k + \sum_{i,j} m_{ij}\phi_i\phi_j + \sum_i a_i\phi_i + b \tag{25}$$

is a cubic polynomial in the chiral superfields ϕ_i called the superpotential. D-breaking models require a new U(1) factor in the gauge group with an associated gauge boson U, and attempt to get realistic mass spectra at the tree level (Fayet 1981). F-breaking models rely on radiative corrections to feed the susy breaking through to all the known supermultiplets of Table I, and are technically more complicated to analyze.

These two classes of models have rather different phenomenologies, and it is advisable to keep both in mind. In the D-breaking models one can derive some upper limits of order 40 GeV on the masses of squarks and sleptons (Fayet 1981), but only under simplifying assumptions which need not be valid. Conventional neutral current phenomenology forbids $m_U/m_{Z^o} = O(1)$, but $m_U << m_{Z^o}$ and $m_U >> m_{Z^o}$ are both tenable hypotheses (Fayet 1981; Barbieri et al. 1982 a,b). It is a noteworthy feature of F-breaking models that the sparticles of Table I are naturally heavier than their familiar partners. Gauginos are heavier than gauge bosons because they are not protected by gauge invariance, while squarks and sleptons are heavier than quarks and leptons because they are not protected by chiral symmetry. In either class of models the primordial scale of m_S susy breaking could in principle be as large as $O(m_X$ or $m_P)$, though we will see later that this may give problems when we make susy local and go to a supergravity theory. Figure 4(a),(b) illustrate two possible scenarios for the spectroscopy of "light" susy particles, based on D- and F-breaking models respectively (Fayet 1981; Ellis et al. 1982 c,d).

Fig. 4. Possible spectroscopies (a) in susy broken à la Fayet & Iliopoulos (1975), and (b) in susy broken à la Fayet (1975) and O'Raifeartaigh (1975).

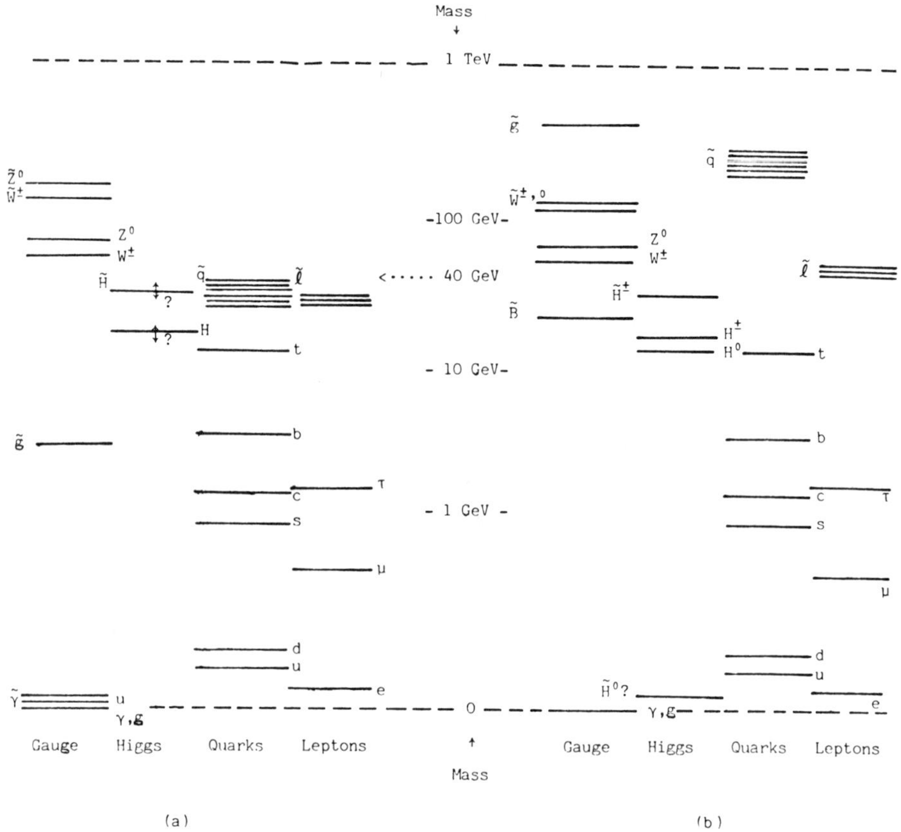

3 SUPERSYMMETRIC GUTS

Now that we have developed some notions about the spectroscopy of spontaneously broken susy, we will now proceed to construct simple susy GUTs and discuss their phenomenology. The fermions and Higgs of conventional GUTs are now assigned to chiral supermultiplets. For example, the conventional $\underline{\bar{5}}\ \psi^{\alpha}$ and $\underline{10}\ \chi_{\alpha\beta}$ of fermions in a normal SU(5) generation now become chiral supermultiplets with additional spin-zero components (see Equation (19)). Likewise the adjoint $\underline{24}$ of Higgses ϕ^{α}_{β} also acquires spin ½ partners. It turns out that one requires an additional doubling of the "light" Higgses, so that one has both $\underline{5}$ and $\underline{\bar{5}}$ supermultiplets H_{α} and $\bar{H}^{\alpha}$.

These are both necessary if one is to give masses to all the quarks and leptons via terms in the superpotential (Dimopoulos & Georgi 1981; Sakai 1982):

$$P \ni m_{2/3}\ \varepsilon^{\alpha\beta\gamma\delta\varepsilon}\ \chi_{\alpha\beta}\ \chi_{\gamma\delta}\ H_{\varepsilon} + m_{-1/3}\ \psi^{\alpha}\ \chi_{\alpha\beta}\ \bar{H}^{\beta} \tag{26}$$

and we thereby also cancel the chiral anomalies which would otherwise be generated by the fermions in a single Higgs chiral supermultiplet. It is easy enough in principle to arrange the superpotential couplings so that the effective masses of the SU(2) doublet Higgses in the supermultiplets H, $\bar{H}$ vanish:

$$P \ni \lambda_1 \left[\frac{1}{3}\,\mathrm{Tr}(\phi^3) + \frac{m}{2}\,\mathrm{Tr}(\phi^2) \right] + \lambda_2\ \bar{H}(\phi + 3m')H \tag{27}$$

All one has to do is choose $m = m'$. This condition is unaesthetic and its origin is unclear. However it is technically natural because of the no-renormalization theorem (Wess & Zumino 1974b; Iliopoulos & Zumino 1974; Ferrara et al. 1974). For more details of the construction of susy GUTs, see Ellis et al. (1982d) and papers cited therein.

We now turn to the phenomenology of the minimal susy GUTs. A first remark is that the rate of approach of the gauge couplings is modified (Dimopoulos et al. 1981; Ibáñez & Ross 1981; Einhorn & Jones 1982) thereby increasing the expected value of the grand unification mass scale: in leading loop order

$$\frac{1}{\alpha_3(Q)} - \frac{1}{\alpha_2(Q)} = \left(\frac{11 + (N_H/2)}{12\pi}\right) \ln\left(\frac{Q^2}{m_X^2}\right) \xrightarrow[\text{susy}]{} \left(\frac{9 + (3N_H/2)}{12\pi}\right) \ln\left(\frac{Q^2}{m_X^2}\right) \tag{28}$$

The decrease from 11 to 9 is due to gaugino loops partially counteracting gauge boson loops. The increase in sensitivity to light Higgs multiplets is due to the apparition of light shiggs fermions. Since N_H is even in a susy GUT, the Higgs effects are at least six times greater than in minimal conventional GUTs. With $N_H = 2$ one finds when 2-loop effects are included (Einhorn & Jones 1982; Ellis et al. 1982j)

$$m_X = 6(\pm 3?) \times 10^{16} \times \Lambda_{\overline{MS}} \tag{29}$$

so that $\Lambda_{\overline{MS}} = 150$ MeV corresponds to m_X about 1×10^{16} GeV. The corresponding value of $\sin^2\theta_W$ is

$$\sin^2\theta_W = 0.236 \pm 0.002 \tag{30}$$

for $\Lambda_{\overline{MS}}$ = 100 to 200 MeV. This is less comfortably close to the best experimental value (2) than was the conventional GUT prediction (1), but judicious adjustment of the susy particle thresholds may dissolve this conflict. If one goes to $N_H = 4$, Equation (28) shows that m_X is decreased, but then $\sin^2\theta_W$ exceeds 0.255 (Einhorn & Jones 1982; Ellis et al. 1982j). Remarkably enough, the value of m_b/m_τ calculated in susy GUTs is numerically essentially identical with that in conventional GUTs:

$$\left.\frac{m_b}{m_\tau}\right|_{\substack{\text{conv}\\ \text{1 loop}}} : \left[\left.\frac{m_b}{m_\tau}\right|_{\substack{\text{susy}\\ \text{1 loop}}} = \left[\frac{\alpha_3(2m_b)}{\alpha_3(m_\tau)}\right]^{12/23}\left[\frac{\alpha_3(m_\tau)}{\alpha_3(m_W)}\right]^{4/7}\left[\frac{\alpha_3(m_W)}{\alpha_{SUM}}\right]^{8/9}\right]$$

$$= 1 : 1.0 \quad . \qquad (31)$$

Looking at the increased value (29) of m_X, one might expect the nucleon lifetime to be longer in susy GUTs than in conventional GUTs, if $\tau_B \propto m_X^4$ as before.

However, Weinberg (1982c) and Sakai & Yanagida (1982) have pointed out that there is a new class of baryon decay diagrams in susy theories, shown in Figure 5, which may give a much shorter lifetime. They involve a dimension 5 $\Delta B = \Delta L$ operator (Figure 5(a)) coupling quarks, squarks, leptons and sleptons arising from Higgs supermultiplet exchange, which has a coefficient $O(1/m_X)$:

$$\frac{g_Y^2}{m_{H_X}}\left(\tilde{q}\,\tilde{q}\,q\,\ell \text{ or } q\,q\,\tilde{q}\,\ell\right) \quad . \qquad (32)$$

The operator (32) does not yet give baryon decay, but must be dressed as in Figure 5(b) by the exchange of a gaugino with a supersymmetry breaking mass which we take to be $m_{\tilde{W}} = O(m_W)$, to give a conventional effective $qqq\ell$ operator. The coefficient of this operator will be

$$O\left(\frac{\alpha}{4\pi}\right)\frac{g_Y^2}{m_{H_X} m_{\tilde{W}}} \qquad (33)$$

Fig. 5. (a) Diagrams contributing to the dimension 5 operators responsible for baryon decay in minimal susy GUTs, and (b) the dimension 5 operator dressed by an external gaugino loop (Weinberg 1982c; Sakai & Yanagida 1982).

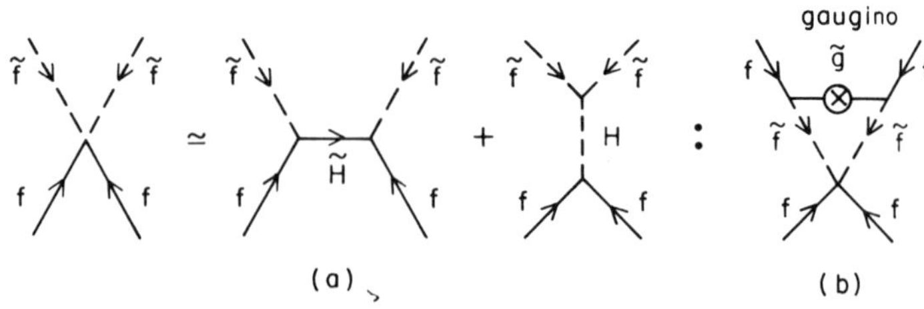

and we therefore have $\tau_B \propto m_X^2 m_{\tilde{W}}^2$ rather than m_X^4. One might worry that the baryon lifetime may now be too short, but this appears not to be the case (Dimopoulos et al. 1982; Ellis et al. 1982j). If one compares the coefficients of the effective operators for baryon decay in susy and conventional GUTs, one sees that the susy interaction is boosted by a factor $O(m_X/m_{\tilde{W}}) = O(10^{12})$ but that this is compensated by Yukawa couplings g_Y^2 (32) of $O(10^{-8})$, a loop factor (33) of $O(10^{-2})$, a short distance suppression factor of $O(10^{-1})$, and the change in m_X^{-1} (29) of $O(10^{-1})$. Thus the conventional and susy baryon decay rates are comparable, and a susy baryon has $\tau_B \gtrsim O(10^{30})$ years if

$$m_{\tilde{W}}\, m_X > O(10^{18})\ \text{GeV}^2 \tag{34}$$

or

$$m_{\tilde{W}} \gtrsim O(100)\ \text{GeV} = O(m_W) \tag{35}$$

if $m_X = O(10^{16})$ GeV as suggested by Equation (29).

A possible snag, however, is that susy baryon decay modes are very different from conventional baryon decay modes (Dimopoulos et al. 1982; Ellis et al. 1982j):

$$B \to \bar{\nu}K >> \bar{\nu}\pi >> \mu^+\pi >> e^+K >> e^+\pi \tag{36}$$

and do not match very well the characteristics of the baryon decay candidates reported (Krishnaswamy et al. 1981, 1982) from the Kolar Gold Fields which look more like $B \to e^+\pi, \rho, \omega$. Many of the proposed new baryon decay experiments are optimized for these decay modes, but would be less efficient at detecting $B \to \bar{\nu}K$. An idea (Rozanov 1982) for a detector to look for $B \to \bar{\nu}K$ decays is shown in Figure 6. One looks for $n \to \bar{\nu}K_L^o$ decays in the rock surrounding a cave, with counters around the walls to veto incoming particles, and a detector inside to pick up any penetrating $K_L^o \to \pi e\nu$ or $\pi\pi$ or $\pi\pi\pi$ decays, reconstruct them and look for a monochromatic K_L^o signal coming from baryon decay. It seems likely that one could search for susy baryon decays (36) with a sensitivity to lifetime of order 10^{32} years.

Fig. 6. The idea of Rozanov (1982) for detecting $n \to \bar{\nu}K_L^o$ decays.

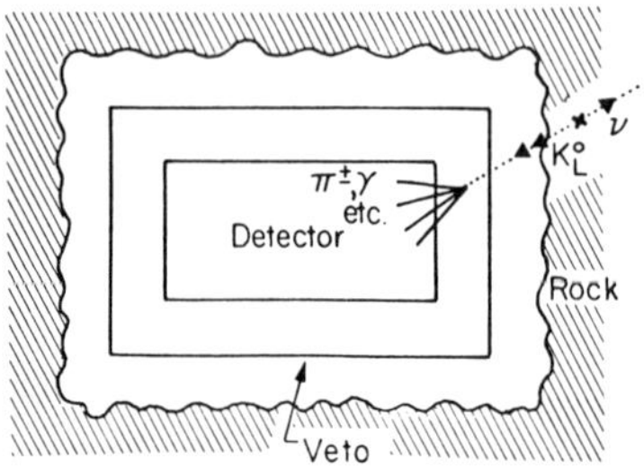

4 SUSY GUTS AND COSMOLOGY

Life in the very early Universe is certainly more complicated and potentially interesting in susy GUTs. We have already seen (22) the form of the effective scalar potential in a susy theory at zero temperature. The potential (22) is clearly positive semi-definite, and is zero if and only if susy is unbroken. Realistic models often have several susy minima which are degenerate with zero vacuum energy, at least in the absence of radiative corrections. For example, the minimal SU(5) model (26,27) discussed earlier has degenerate susy minima which are

$$\left.\begin{array}{lll} \text{SU(5) invariant} & : & \langle 0|\phi|0\rangle = 0 \\ \text{SU(4)} \times \text{U(1) invariant} & : & \langle 0|\phi|0\rangle = \frac{V}{2\sqrt{2}}\begin{pmatrix} 1 & & & & 0 \\ & 1 & & & \\ & & 1 & & \\ & & & 1 & \\ 0 & & & & -4 \end{pmatrix} \\ \text{SU(3)} \times \text{SU(2)} \times \text{U(1) invariant} & : & \langle 0|\phi|0\rangle = \frac{V}{\sqrt{30}}\begin{pmatrix} 2 & & & & 0 \\ & 2 & & & \\ & & 2 & & \\ & & & -3 & \\ 0 & & & & -3 \end{pmatrix} \\ \text{Dragon (1982) minimum} & : & \langle 0|\phi|0\rangle = \frac{V}{\sqrt{30}}\begin{pmatrix} 2 & & & & 0 \\ & 2 & & & \\ & & 2 & A & \\ & & & -3 & \\ 0 & & & & -3 \end{pmatrix} \end{array}\right\} \quad (37)$$

and more complicated possibilities exist in other susy GUTs (Frampton & Kepkart 1982; Buccella et al. 1982 a,b). The qualitative situation is shown in Figure 7: at high temperatures $T = O(m_X)$ the Universe may not

Fig. 7. Degenerate susy potential minima in minimal susy SU(5) (Dimopoulos & Georgi, 1981; Sakai 1982).

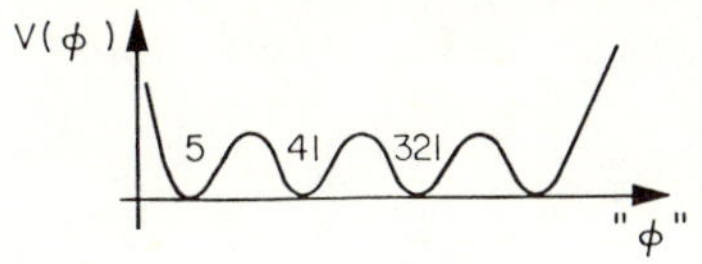

know which of these minima to choose, and different causally separated regions of the very early Universe may choose differently.

There are several possible ways of breaking the degeneracy between the different susy minima. Clearly susy breaking effects do, but they give

$$\delta V \sim m_S^2\, m_X^2 \quad \text{or} \quad m_S^4 \tag{38}$$

where m_S is a susy breaking parameter which is possibly much less than m_X. Even if the SU(3) × SU(2) × U(1) minimum is favored by these effects (38), the rate at which parts of the Universe which had cooled into one of the other minima (37) then tunnel into the energetically preferred one may be unacceptably slow (Ellis et al. 1982f; Srednicki 1982b):

$$\text{Tunnelling probability} \propto e^{-B} : B = O\left(\frac{m_X}{m_S}\right)^6 \lambda^4 \tag{39}$$

where λ is a generic chiral superfield coupling. Either one wants $m_S \gtrsim m_X$ (which may run into other difficulties) or else one must take $\lambda << 1$ if one is to get an adequately short lifetime for the false minima (Nanopoulos et al. 1982).

In point of fact, finite temperature fluctuations may get one out of the false minima more quickly than quantum tunnelling. Finite temperature effects give energy shifts proportional to the number of light particles. Unfortunately, in minimal SU(5) this means that the SU(5) invariant minimum (37) is the lowest, and the SU(4) × U(1) and SU(3) × SU(2) × U(1) minima are an equal amount higher. However, it is possible to construct variants of the simplest susy GUT (26,27) in which SU(5) invariant minimum actually breaks supersymmetry, and hence is higher than the SU(4) × U(1) and SU(3) × SU(2) × U(1) minima (Ellis et al. 1982d), while finite temperature effects favor the desired SU(3) × SU(2) × U(1) minimum as shown in Figure 8.

Fig. 8. Possible form of the potential in non-minimal susy SU(5) (Ellis et al. 1982d).

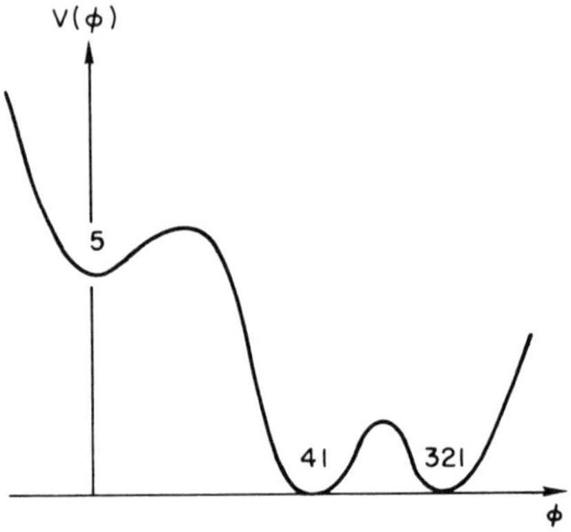

There is another cute idea for getting out of an undesirable SU(5) minimum should it exist (Nanopoulos & Tamvakis 1982a; Srednicki 1982a). The point is that in the SU(5) invariant phase the gauge coupling tends to become strong at an energy or temperature $O(10^{10})$ GeV. This may lead to the degeneracy between the various minima being broken at this level. For example, when the SU(5) coupling is strong, presumably it confines, and the number of light degrees of freedom may be reduced, so that it becomes energetically disfavored by comparison with the SU(4) × U(1) and SU(3) × SU(2) × U(1) minima. If the generic Yukawa coupling λ is sufficiently small, the Universe may then find its way to the desired SU(3) × SU(2) × U(1) minimum as indicated in Figure 9. It may be that the transition passes through an intermediate SU(4) × U(1) phase with the previous history recycled at a strong coupling temperature $O(10^6)$ GeV.

Fig. 9. Possible non-perturbative SU(5) effect on the minima in a susy GUT (Nanopoulos & Tamvakis 1982a; Srednicki 1982a).

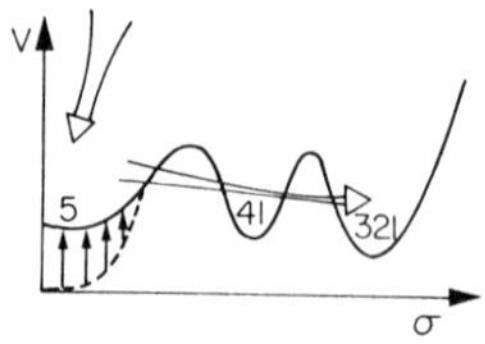

Since SU(5) is broken at temperatures $<< m_X$ in this scenario, one must rethink Big Bang baryosynthesis. An attractive option is to have relatively light "heavy" color triplet Higgses with masses $O(10^{10}$ to $10^{11})$ GeV. This is the lowest mass compatible with $d = 6$ Higgs exchange operators letting baryons live longer than 10^{30} years. If dominant, they give (Nanopoulos & Tamvakis 1982b)

$$B \to \mu^+ K \quad \text{or} \quad \bar{\nu} K \quad . \tag{40}$$

If there are such light color triplet Higgses, one must find some extra symmetry to forbid them from making a $d = 5$ operator contribution to baryon decay (cf. Figure 5(a)).

Either of the scenarios in Figure 8 or Figure 9 can probably generate sufficient baryon number in the very early Universe.

5 SUPERSYMMETRIC INFLATION

The other currently exciting application of GUTs in the very early Universe is to generate cosmological inflation (Guth 1981; Linde 1982a; Albrecht & Steinhardt 1982). This can be driven by a vacuum potential energy density which is much larger than T^4. *A priori* it might seem to be a disadvantage that in susy theories the vacuum energies at (broken) susy minima are small (38) which might make it more difficult to drive inflation. Furthermore, the smallness of the vacuum energy may make it difficult to reheat to a high enough temperature to achieve baryosynthesis. However, both these difficulties can be resolved if one recalls that susy is automatically broken at finite temperatures, and hence there is no conclusive reason why the Universe should be in a supersymmetric configuration while it is inflating. If susy is broken the vacuum energy can be much larger than suggested by (38). Indeed, susy may even be beneficial in achieving sufficient inflation without generating unacceptably large fluctuations (Ellis *et al*. 1982 g,h,i). This is because susy allows one to adjust or fine-tune parameters without having to worry about them being disturbed by radiative corrections.

To see this, let us first (Ellis *et al*. 1982g) look at the conventional Coleman-Weinberg potential

$$V(\phi) = A\phi^4(\ln \phi^2/\sigma^2 - \tfrac{1}{2}) + D\phi^2 \tag{41}$$

where

$$A = (1/64\pi^2\sigma^4)\left(\sum_B g_B m_B^4 - \sum_F g_F m_F^4\right) \tag{42}$$

with $g_{B(F)}$ the number of boson (fermion) helicity states coupled to the field ϕ. In minimal SU(5) one has

$$A = \frac{5625}{1024\pi^2} g^2 \quad . \tag{43}$$

The parameter D is Equation (41) is an effective $(\text{mass})^2$ parameter which can get contributions from many sources:

$$D = \frac{1}{2}\left(m_0^2 + cT^2 + bR - 3\lambda \langle\phi^2\rangle\right) \tag{44}$$

where m_0 is the flat-space, zero-temperature mass, $c = (75/8)g^2$ and T is probably the Hawking temperature $T_H = H/2\pi$ during an inflationary De Sitter epoch, b is an unknown parameter which is probably O(1) and specifically 1/6 for conformally coupled scalar fields, R is the scalar curvature, $-\lambda/4$ is an effective scalar self-coupling and $\langle\phi^2\rangle$ the quantum expectation value of ϕ^2 (Linde 1982b). To get sufficient inflation we must ensure that D is small:

$$\frac{3H^2}{2D} > 65 \tag{45}$$

where

$$H = \left(\frac{4\pi A\sigma^4}{3m_P^2}\right) \tag{46}$$

is the Hubble parameter during the inflationary epoch. The condition (45) requires a delicate suppression of the D term (44) which seems unnatural in a theory which is not supersymmetric. However, susy allows us to "set and forget" m_0 at any value we like, the cT^2 term can be arbitrarily small if the field driving inflation is a gauge singlet, the bR term is absent in a conventionally Weyl-rescaled model of chiral supermultiplets coupled to simple N = 1 supergravity (Cremmer et al. 1978, 1979, 1982 a,b), and the effective scalar coupling λ can be chosen to respect the Linde (1982b) condition

$$\lambda \leq \frac{\pi^2}{1800} \simeq 5 \times 10^{-3} \tag{47}$$

which is not the case in the conventional GUT model (41) where A (43) is typically O(1/10). Even if we get enough inflation, we must guard against having excessively large fluctuations. In our earlier work (Ellis et al. 1982g) mentioned at this meeting, we took the point of view that one only had a chance of suppressing fluctuations to sufficiently low levels if the Hawking & Moss (1982) action

$$B = \frac{3m_P^4}{8}\left[\frac{1}{V_0} - \frac{1}{V_1}\right] \tag{48}$$

was considerably larger than unity. In this case it might be legitimate to view the flip of an entire De Sitter horizon volume from the origin $\phi = \phi_0$, $V = V_0$ to the local maximum of the potential (41) at $\phi = \phi_1$, $V = V_1$ shown in Figure 10. If B were only of order unity, there would be no strong reason for large regions of space to become hung up at the local maximum, and one would expect the phase transition to be very inhomogeneous and yield unacceptably large fluctuations. The condition that B (48) for the potential (41) be large is:

$$\frac{3m_P^4 D^2}{16A^3\sigma^8} >> 1 \quad . \qquad (49)$$

The conditions of sufficient inflation (45) and of smoothness (49) can only be reconciled if A is very small (Ellis et al. 1982g):

$$A < 3(\pi/130)^2 = 0(10^{-3}) \qquad (50)$$

which is certainly not the case for minimal SU(5) for which A = 0(1/10) from Equation (43). Indeed, two recent calculations (Hawking 1982; Guth & Pi 1982) discussed at this meeting suggest that the magnitude of fluctuations in a new inflationary Universe based on minimal SU(5) are indeed much too large.

We have now seen several reasons why susy may help the inflationary Universe: it gives us more freedom to suppress the effective mass parameter D (44) in a natural way, and it enables us to tune down A (42)

Fig. 10. Sketch of the effective potential for ϕ, including the local minimum V_0 at the origin, the barrier provided by the local maximum V_1, and the global minimum $V = 0$ at $\phi = \mu$.

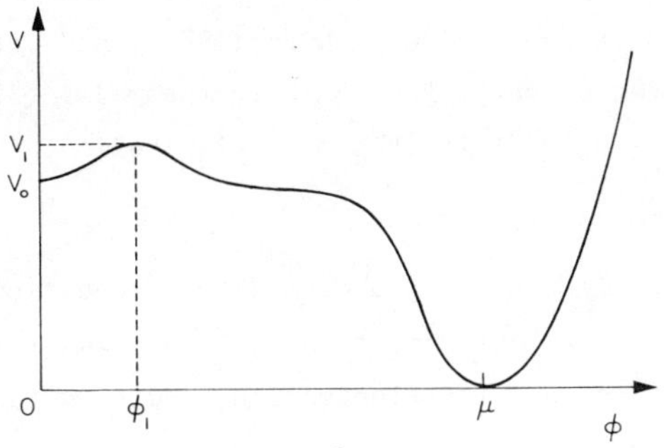

as desired (50) for solving the homogeneity problem. Therefore, since the Workshop we have constructed (Ellis et al. 1982h) a toy supersymmetric model which one can use as a test-bed for supersymmetric inflation. It is defined by a superpotential

$$P(\phi,X,Y) = aX\,\phi(\phi-\mu) + GY(\phi^2-\mu^2) \tag{51}$$

whose associated potential (22,24) has a zero-temperature supersymmetric minimum at $\phi=\mu$, and a non-zero value at $\phi=0$ which can be used to drive inflation. The conditions (45,49) can be satisfied if

$$a,b \sim O\left(10^{-3}\mu/m_P\right) \quad , \tag{52a}$$

$$\frac{a^2}{b^2} - 2 = O\left(10^{-1}\mu^2/m_P^2\right) \quad . \tag{52b}$$

It is certainly possible to impose these conditions on the supersymmetric model (51), and they are technically natural, but they look like rather inelegant fine-tuning, at least in the case of (52b) when $\mu \ll m_P$. For this reason, it looks more appealing to postulate that inflation took place close to the Planck epoch, which might entail taking $\mu \gg m_X$. In this case we would want the Higgs field driving inflation to be a SU(5) gauge singlet, which is indeed how we chose ϕ in the superpotential (51) and incidentally enables us to suppress the finite temperature contribution to D in Equation (44). The hypothesis of inflation driven by an order parameter much larger than m_X we term "primordial inflation" (Ellis et al. 1982h).

There is another reason for favoring this hypothesis which is related to the fluctuation problem. We have recently used (Ellis et al. 1982i) the formalism of Olson (1976) developed by Guth & Pi (1982) to analyze the spectrum of perturbations expected in the supersymmetric toy model (51). We find that they are approximately scale invariant in a way not very dissimilar from that found by Guth & Pi (1982), and with a magnitude which can be adjusted by varying a and b in the model (51). We get the desired spectrum $\delta\rho/\rho = O(10^{-4})$ if (Ellis et al. 1982i)

$$a,b = O\left(10^{-6}\mu/m_P\right) \quad . \tag{53}$$

This indicates that the previous Hawking & Moss (1982) condition (49) which becomes (52a) for our toy model, while necessary to suppress inhomogeneities, is not in itself sufficient. The fluctuation condition (53) appears to push us in the direction of very small a,b unless $\mu = O(m_P)$. Another motivation for primordial supersymmetric inflation?

One snag of primordial inflation is that it reopens the grand unified monpole problem. Inflation no longer suppresses the monopole density and one must appeal to another mechanism. Two possibilities are the supercosmology (Nanopoulos & Tamvakis 1982 a,b; Srednicki 1982 a,b) scenario outlined in Section 5, and the possibility (Bais & Rudaz 1980) that monopole production may be suppressed by thermodynamic factors $\exp[-(m_M/T_G)]$ during a second order GUT phase transition (Ellis et al. 1982h).

6 IS GRAVITY AN ESSENTIAL COMPLICATION?

So far we have neglected gravity in our discussion of GUTs and susy GUTs, and generally one thinks that this is likely to be a reasonable first approximation. However we have already seen a couple of ways in which gravitational effects may show up in inflationary cosmology. One is the R term in the expression (44) for the effective Higgs $(\text{mass})^2$ D. The other is the T^2 term in (44) which may be expected to become the square of the Hawking temperature $T_H = H/2\pi$ during the De Sitter inflationary epoch. The coefficient b of the R term in (44) is arbitrary until we know how scalar fields couple to gravity, while treating finite curvature effects solely through this and a T_H^2 term may not be completely satisfactory. Another reason why we might worry about gravitational effects is the impetus (52a,53) that we received towards inflation with $\mu = O(m_P)$. Finally, there is the more abstract theoretical point that if one wants to make a theory with local supersymmetry then one must incorporate gravity and construct a supergravity theory, which is indeed the only consistent framework for combining susy and gravity.

As mentioned in Section 2, there are supergravities with $N = 1,2,3,\ldots,8$ extended supersymmetries (Van Nieuwenhuizen 1981). Even if we believe that the ultimate physical theory has $N > 1$, for the reasons of chirality discussed in Section 2 the only possible effective low-energy theory is one based on $N = 1$. In this theory the supermultiplets (18,19) are supplemented by the

$$\text{graviton-gravitino supermultiplet} : \begin{pmatrix} 2 \\ 3/2 \end{pmatrix} . \qquad (54)$$

In addition to the new gravitino particle and its couplings, there are additional low-energy non-renormalizable effective interactions among familiar particles (18,19) which are scaled by inverse powers of the

Planck mass (Cremmer et al. 1978, 1979, 1982 a,b). An important example is the effective scalar potential. It is convenient to write the modified potential in a Weyl-rescaled form where there is no $R\phi^2$ term: $b = 0$ in Equation (44). The potential for gauge-singlet fields A (no D term in Equation (22)) is then

$$V = \exp\left(\tfrac{1}{2}\,|A|^2\right)\left[2|F_A + \tfrac{1}{2}A^*P|^2 - 3|P|^2\right] \tag{55}$$

if we make the simplifying assumption of canonical derivative terms $\tfrac{1}{2}|\partial_\mu A|^2$ in the Lagrangian. (More general forms are possible, but they do not alter qualitatively the subsequent discussion.) Note that we have used "natural" units $\kappa^2 = 8\pi/3m_P^2 \equiv 1$ in writing the expression (55). When we compare and contrast with the globally supersymmetric potential (22) we see the important difference that a negative term is now present. This enables one to cancel the positive vacuum energy otherwise expected in states with susy broken, and even enables us to have states with negative vacuum energy. Weinberg (1982b) has emphasized that the P-dependent terms in (55) can break the degeneracy between different globally supersymmetric minima, but has shown that even if there are minima with negative vacuum energy, a state with zero vacuum energy is stable against decay into them.

Now that we have introduced the gravitino (54), we must ask and answer a few questions about it, notably what is its mass and how many should there be in the Universe? The gravitino acquires a mass by eating the spin ½ goldstino particle associated with the spontaneous breakdown of global supersymmetry, in much the same way that a gauge boson becomes massive by eating a spin zero Goldstone boson (Van Nieuwenhuizen 1981). The gravitino mass is related to the scale of supersymmetry breaking in such a way that

$$m_{\tilde{G}} = \sqrt{m_S} = \exp\left(\tfrac{1}{4}\,|A|^2\right)|P| \tag{56}$$

if the cosmological constant is set to zero by cancelling the positive and negative terms in (55). We see from Equation (55) that there is a contribution

$$\tfrac{1}{2}\,|A|^2|P|^2\exp\left(\tfrac{1}{2}\,|A|^2\right) \tag{57}$$

to the spin-zero field mass matrix which implies a lower bound (Ellis & Nanopoulos 1982b)

$$m_0 \gtrsim m_{\tilde{G}} \tag{58}$$

on the average of spin-zero particle masses in a chiral supermultiplet.

We know that at least some spin-zero particles must be quite light, notably squarks, sleptons and Higgs bosons. It is always possible (Gaillard *et al*. 1982) to set up a spontaneously broken global symmetry in such a way that one of the two spin-zero particles in a given chiral supermultiplet is massless, but we do not know of any realistic examples where both of them are massless at the tree level. Furthermore, in many cases the component which is massless at the tree level is only a pseudo-Goldstone boson because the corresponding global symmetry is broken by gauge or other non-gravitational interactions. In this case the (almost) massless spin-zero component can acquire a $(\text{mass})^2$ which is plausibly $O(\alpha)$ times m_G^2 (Gaillard *et al*. 1982). Since both spin-zero components of all quark and lepton supermultiplets must have masses $\lesssim O(10^2)$ GeV in order to protect the gauge hierarchy, while the "light" Higgs bosons must also have masses $\lesssim O(10^2)$ GeV in order to break the weak gauge symmetry at the desired scale, the potential existence of a few pseudo-Goldstone bosons does not help. The average masses m_0 of squarks, sleptons and Higgses must all be $\lesssim O(10^2)$ GeV and therefore the constraint (58) on the average spin-zero boson mass tells us that (Ellis & Nanopoulos 1982b)

$$m_{\tilde{G}} \lesssim O(10^2)\ \text{GeV} \tag{59}$$

which in turn (56) means that

$$m_S \lesssim O(10^{10})\ \text{GeV} \quad . \tag{60}$$

This range of m_S means that a susy GUT is still essentially supersymmetric on a mass scale $O(10^{16})$ GeV. However, this does not prevent one from achieving inflation, as was exemplified by our toy model (51).

There is a further cosmological complication with the mass of the gravitino which was pointed out by Weinberg (1982a). It is that gravitinos have relatively long lifetimes:

$$\tau_{\tilde{G}} \gtrsim O\left(m_P^2/m_{\tilde{G}}^3\right) \quad . \tag{61}$$

If present in the very early Universe, they may survive embarrassingly long, and either dominate the Universe with their excess mass density, and/or drown it in entropy when they eventually decay (the "gravitino problem"). Weinberg (1982a) suggested that either gravitinos should decay before nucleosynthesis in which case

$$m_{\tilde{G}} \gtrsim O(10^4)\ \text{GeV} \quad , \quad m_S \gtrsim O(10^{11})\ \text{GeV} \tag{62}$$

or else they should be sufficiently light to avoid contributing too much density to the present-day Universe (Pagels & Primack 1982):

$$m_{\tilde{G}} \lesssim O(1)\ \text{keV} \quad , \quad m_S \lesssim O(10^6 \text{ to } 10^7)\ \text{GeV} \quad . \tag{63}$$

The lower limit (62) is in prima facie conflict with the upper limit (59): does this mean we have to choose the lower range (63) of gravitino masses? Not necessarily, because we have shown during this Workshop (Ellis et al. 1982a) that even a relatively modest amount of inflation can suppress the primordial gravitino number density sufficiently low to make the conditions (62,63) unnecessary, while gravitino production after the Planck epoch is not sufficient to recreate the gravitino problem. Our results are exhibited graphically in Figure 11, which shows that a suppression factor $Y_G \lesssim O(10^{-15})$ of the gravitino number density relative to that naively expected in an adiabatically expanding Universe would suffice to respect all Big Bang nucleo- and baryosynthesis constraints, while gravitino production after the Planck epoch is not important. The amount of inflation required to solve the gravitino problem is relatively modest, since an increase in the

Fig. 11. Cosmological constraints on the abundance Y_G of gravitinos relative to the normal cosmological abundance. An inflation factor L would give a primordial gravitino abundance $Y_G = O(1/L^3)$. Also shown in the figure are some typical gravitino production mechanisms which respect the cosmological constraints.

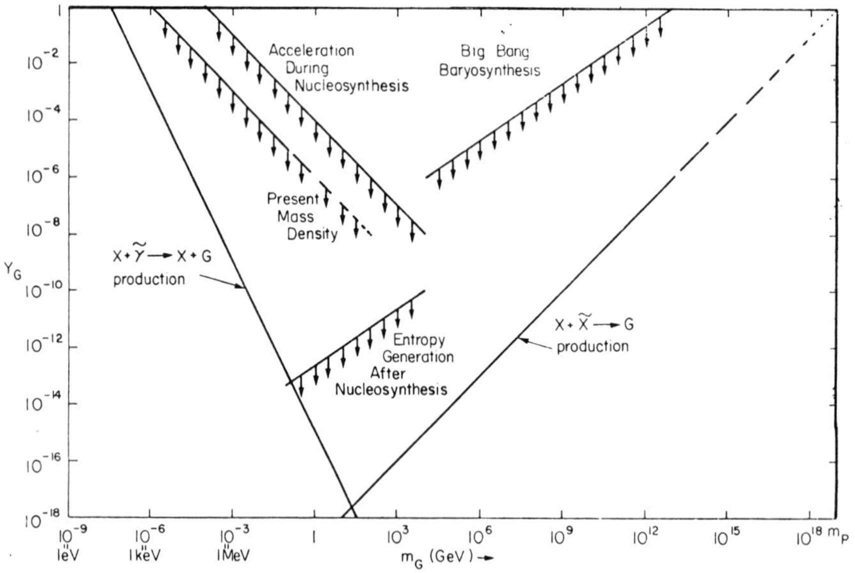

cosmic scale factor by $O(10^5)$ suffices to suppress Y_G by $O(10^{-15})$. We therefore think that all gravitino masses in the particle physics range (59) may be reconciled with the cosmological constraints.

It is clear that the next step should be to rework supersymmetric inflation using the supergravity effective potential (55). Also one should investigate more closely the cosmological production of gravitinos around and shortly after the Planck epoch, to see how close to m_P one can push the order parameter of the Higgs field driving inflation (cf. the parameter μ of Equations (51),(52b) and (53)). So far we do not have a complete answer to the question raised in the heading to this section.

7 CONCLUSIONS

One may briefly summarize as follows the status of GUTs and supersymmetric GUTs in the very early Universe as follows. Conventional GUTs do not (yet?) conflict seriously with experiment. Furthermore, they perform baryosynthesis in an elegant way. However, they are embarrassed by the relative paucity of grand unified monopoles which may be explained away by cosmological inflation. Unfortunately, it is difficult to achieve sufficient inflation in a natural way and with sufficiently small fluctuations $\delta\rho/\rho$. Furthermore, conventional GUTs have technical difficulties with maintaining the gauge hierarchy.

These technical problems are resolved in supersymmetric GUTs, whose conventional phenomenology is (almost) equally satisfactory. Unfortunately, susy GUTs are often plagued by a plethora of degenerate vacua which complicate discussion of the GUT phase transition and baryosynthesis. However, susy may alleviate the fine-tuning and fluctuation problems associated with inflation in conventional GUTs. When combined with gravity, susy gives birth to a massive gravitino which presents some cosmological problems that are fortunately soluble.

Life with susy is difficult but potentially rewarding.

ACKNOWLEDGEMENTS

It is a pleasure to thank D.V. Nanopoulos, K.A. Olive, K. Tamvakis and my other collaborators for the enjoyable times we have spent thinking about the topics discussed here. It is also a pleasure to thank G. Gibbons, S. Hawking and S. Siklos for organizing such a stimulating workshop, and to thank the participants for interesting discussions.

REFERENCES

Albrecht, A. & Steinhardt, P.J. (1982). Phys. Rev. Lett., 48, 1220.
Bais, F.A. & Rudaz, S. (1980). Nucl. Phys., B170 [F51], 507.
Barbieri, R., Ferrara, S. & Nanopoulos, D.V. (1982a). Zeit für Phys., C13, 267.
Barbieri, R., Ferrara, S. & Nanopoulos, D.V. (1982b). CERN preprint TH-3309.
Battistoni, G. et al. (1982). Report at the Paris International Conference on High Energy Physics.
Buccella, F., Derendinger, J.-P., Ferrara, S. & Savoy, C.A. (1982a). CERN preprint TH-3212.
Buccella, F., Derendinger, J.-P., Ferrara, S. & Savoy, C.A. (1982b), CERN preprint TH-3274.
Buras, A.J., Ellis, J., Gaillard, M.K. & Nanopoulos, D.V. (1978). Nucl. Phys., B135, 66.
Campbell, B.A., Ellis, J. & Rudaz, S. (1982). Nucl. Phys., B198, 1.
Chanowitz, M.S., Ellis, J. & Gaillard, M.K. (1977). Nucl. Phys., B128, 506.
Cremmer, E., Ferrara, S., Girardello, L. & Van Proeyen, A. (1982a). CERN preprint TH-3312.
Cremmer, E., Ferrara, S., Girardello, L. & Van Proeyen, A. (1982b). CERN preprint TH-3348.
Cremmer, E., Julia, B., Scherk, J., Ferrara, S., Girardello, L. & Van Nieuwenhuizen, P. (1978). Phys. Lett., 79B, 231.
Cremmer, E., Julia, B., Scherk, J., Ferrara, S., Girardello, L. & Van Nieuwenhuizen, P. (1979). Nucl. Phys., B147, 105.
Dimopoulos, S. & Georgi, H (1981). Nucl. Phys., B193, 150.
Dimopoulos, S., Raby, S. & Wilczek, F.A. (1981). Phys. Rev., D24, 1681.
Dimopoulos, S., Raby, S. & Wilczek, F.A. (1982). Phys. Lett., 112B, 133.
Dragon, N.V. (1982). Phys. Lett., 113B, 288.
Einhorn, M.B. & Jones, D.R.T. (1982). Nucl. Phys., B196, 475.
Ellis, J. (1981). Lectures presented at the Les Houches Summer School, LAPP preprint TH-48/CERN TH-3174.
Ellis, J., Ferrara, S. & Nanopoulos, D.V. (1982a). Phys. Lett., 114B, 231.
Ellis, J. & Gaillard, M.K. (1979). Phys. Lett., 88B, 315.
Ellis, J., Gaillard, M.K., Nanopoulos, D.V. & Rudaz, S. (1980). Nucl. Phys., B176, 61.
Ellis, J., Gaillard, M.K. & Zumino, B. (1982b). Acta Physica Polonica, B13, 253.
Ellis, J., Ibáñez, L.E. & Ross, G.G. (1982c). Phys. Lett., 113B, 283.
Ellis, J., Ibáñez, L.E. & Ross, G.G. (1982d). CERN preprint TH-3382.
Ellis, J., Linde, A.D. & Nanopoulos, D.V. (1982e). CERN preprint TH-3356.
Ellis, J., Llewellyn Smith, C.H. & Ross, G.G. (1982f). Phys. Lett., 114B, 227.
Ellis, J. & Nanopoulos, D.V. (1982a). Phys. Lett., 110B, 44.
Ellis, J. & Nanopoulos, D.V. (1982b). Phys. Lett., 116B, 133.
Ellis, J., Nanopoulos, D.V., Olive, K.A. & Tamvakis, K. (1982g). Phys. Lett., 118B,335.
Ellis, J., Nanopoulos, D.V., Olive, K.A. & Tamvakis, K. (1982h). CERN preprint TH-3404.
Ellis, J., Nanopoulos, D.V., Olive, K.A. & Tamvakis, K. (1982i). CERN preprint TH-3437.
Ellis, J., Nanopoulos, D.V. & Rudaz, S. (1982j). Nucl. Phys., B202, 43.
Farrar, G. & Fayet, P. (1978a). Phys. Lett., 76B, 575.
Farrar, G. & Fayet, P. (1978b). Phys. Lett., 79B, 442.
Fayet, P. (1975). Phys. Lett., 58B, 67.

Fayet, P. (1981). In Unification of the Fundamental Particle Interactions, eds. S. Ferrara, J. Ellis & P. Van Nieuwenhuizen, p. 587. New York: Plenum Press.

Fayet, P. & Ferrara, S. (1977). Phys. Rep., 32C, 249.

Fayet, P. & Iliopoulos, J. (1975). Phys. Lett., 51B, 331.

Ferrara, S., Iliopoulos, J. & Zumino, B. (1974). Nucl. Phys., B77, 413.

Frampton, P.H. & Kephart, T.W. (1982). Phys. Rev. Lett., 48, 1237.

Gaillard, M.K., Hall, L.J., Zumino, B., Del Aguila, F., Polchinski, J. & Ross, G.G. (1982). Private communication and to be published.

Georgi, H. & Glashow, S.L. (1974). Phys. Rev. Lett., 32, 438.

Georgi, H., Quinn, H.R. & Weinberg, S. (1974). Phys. Rev. Lett., 33, 451.

Gildener, E. (1976). Phys. Rev., D14, 1667.

Gol'fand, Y.A. & Likhtman, E.P. (1971). Pis'ma Zh. Eksp. Teor. Fiz., 13, 323.

Grisaru, M.T. & Siegel, W. (1982). Nucl. Phys., B201, 292.

Guth, A.H. (1981). Phys. Rev., D23, 347.

Guth, A.H. & Pi, S.-Y. (1982). Phys. Rev. Lett., 49, 1110.

Hawking, S. (1982). Phys. Lett., 115B, 295.

Hawking, S. & Moss, I.G. (1982). Phys. Lett., 110B, 35.

Hawking, S., Page, D.N. & Pope, C.N. (1979). Phys. Lett., 86B, 175.

Hawking, S., Page, D.N. & Pope, C.N. (1980). Nucl. Phys., B170 [F51], 283.

Ibáñez, L.E. & Ross, G.G. (1981). Phys. Lett., 105B, 439.

Iliopoulos, J. & Zumino, B. (1974). Nucl. Phys. B76, 310.

Krishnaswamy, M.R. et al. (1981). Phys. Lett., 106B, 339.

Krishnaswamy, M.R. et al. (1982). Phys. Lett., 115B, 349.

Linde, A.D. (1982a). Phys. Lett., 108B, 389.

Linde, A.D. (1982b). Lebedev Physical Institute preprint 88.

Llewellyn Smith, C.H., Ross, G.G. & Wheater, J.F. (1981). Nucl. Phys., B177, 263.

Llewellyn Smith, C.H. & Wheater, J.F. (1981). Phys. Lett., 105B, 486.

Mandelstam, S. (1982). U.C. Berkeley preprint UCB-PTH-82/15.

Marciano, W.J. & Sirlin, A. (1980). Phys. Rev., D22, 2695.

Marciano, W.J. & Sirlin, A. (1981). Phys. Rev., Lett., 46, 163.

Nanopoulos, D.V., Olive, K.A. & Tamvakis, K. (1982). Phys. Lett., 115B, 15.

Nanopoulos, D.V. & Ross, D.A. (1982). Phys. Lett., 108B, 351.

Nanopoulos, D.V. & Tamvakis, K. (1982a). Phys. Lett., 110B, 449.

Nanopoulos, D.V. & Tamvakis, K. (1982b). Phys. Lett., 113B, 151.

Olson, D.W. (1976). Phys. Rev., D14, 327.

O'Raifeartaigh, L. (1975). Nucl. Phys., B96, 331.

Pagels, H. & Primack, J. (1982). Phys. Rev. Lett., 48, 223.

Rozanov, A. (1982). Private communication.

Sakai, N. (1982). Zeit. für Phys., C11, 153.

Sakai, N. & Yanagida, T. (1982). Nucl. Phys., B197, 533.

Srednicki, M. (1982a). Nucl. Phys., B202, 327.

Srednicki, M. (1982b). Nucl. Phys., B206, 132.

Van Nieuwenhuizen, P. (1981). Phys. Rep., 68C, 189.

Volkov, D. & Akulov, V.P. (1973). Phys. Lett., 46B, 109.

Weinberg, S. (1977). Trans. N.Y. Acad. Sci. II, 38.

Weinberg, S. (1982a). Phys. Rev. Lett., 48, 1303.

Weinberg, S. (1982b). Phys. Rev. Lett., 48, 1776.

Weinberg, S. (1982c). Phys. Rev., D26, 187.

Wess, J. & Zumino, B. (1974a). Nucl. Phys., B70, 39.

Wess, J. & Zumino, B. (1974b). Phys. Lett., 49B, 52.

MONOPOLES IN THE VERY EARLY UNIVERSE

John Preskill
Lyman Laboratory of Physics, Harvard University
Cambridge, Massachusetts 02138, U.S.A.

1 *INTRODUCTION*

In recent years, it has become increasingly popular in theoretical particle physics to speculate about the structure of matter at extremely short distances, distances of order 10^{-28} cm, or smaller. These speculations are frustrating, because to explore this very-short-distance physics experimentally would require experiments performed at correspondingly high energies, energies of order 10^{15} GeV, or higher. Such experiments will not be performed in any foreseen accelerator. Partly for this reason, some particle physicists have become interested in cosmology. For we believe that the universe was once extremely hot, so hot that collisions with energies of order 10^{15} GeV or above occurred copiously. We can therefore attempt to use the very early universe as our accelerator, and look today for relics of the very energetic past.

We have today a reasonably firm understanding of particle physics up to energies of order 100 GeV. This physics has been explored in accelerator experiments, and the experiments have tended to confirm our theoretical expectations. As a result, we can confidently extrapolate the standard big-bang cosmology (for a review see Peebles 1971, or Weinberg 1972) back to times of order 10^{-10} sec when the temperature of the universe was about 100 GeV. But to discuss the evolution of the universe at earlier times, we must be willing to speculate about particle physics at higher energies.

The currently fashionable grand unified theories are reputed to provide a complete description of particle physics up to energies of order 10^{19} GeV, where quantum gravitational effects, not yet encompassed by these theories, begin to play a significant role. They thus offer us an opportunity to analyze the evolution of the universe after the first 10^{-44} sec, when the temperature was below 10^{19} GeV. According to the most popular version of grand unification, particle physics at energies much greater than 100 GeV closely resembles the physics we already know; only at energies of order 10^{15} GeV do new interactions with qualitatively different properties begin to assert themselves. If this "desert hypothesis" proves correct, it will be discouraging to experimental particle physicists. But it will also enhance the importance of cosmology to particle physicists, for the study of the very early universe will provide one of the few means at our disposal of exploring the new physics at energies of order 10^{15} GeV and beyond.

The universe was sufficiently hot to probe the new short-distance physics predicted by grand unified theories during only the first 10^{-35} sec. Fortunately, there are at least two relics from the first 10^{-35} sec that we have a good chance of observing today. One is the baryon number of the universe, which *has* been observed. The other is the magnetic monopole, which may have been observed. This talk is about the second relic.

Grand unified theories are briefly reviewed in Section 2, and the properties of magnetic monopoles are discussed in Section 3. The role of monopoles in cosmology is the subject of Section 4. The remainder of the talk concerns the prospects for observing monopoles. Theoretical limits on the flux of monopoles in cosmic rays are reviewed in Section 5, and the implications of a recently reported candidate monopole event are considered in Section 6. Remarks about experiments that may be performed in the future are contained in Section 7. Section 8 contains conclusions.

2 *GRAND UNIFICATION*

By a grand unified theory, I mean any particle physics model in which the standard low-energy gauge group $SU(3)_{color} \times [SU(2) \times U(1)]_{electroweak}$ is embedded in a simple gauge group G which is spontaneously broken at a large mass scale. (Actually most of what will be said applies also to the case of unification in a semi-simple gauge group.) The SU(5) model of Georgi and Glashow (1974) is the simplest of many possible models which satisfy this criterion; for a review see Langacker (1981). If one adopts the "desert hypothesis", that is, assumes that no unexpected new interactions or particles appear at intermediate energies, then the mass scale M_{GUT} at which unification occurs can be calculated; the prediction is $M_{GUT} \sim 10^{15}$ GeV (Georgi, Quinn and Weinberg, 1974). But it is important to bear in mind that the desert hypothesis may be wrong while the general idea of grand unification is still correct. Then the unification scale could be different.

If we do accept the desert hypothesis, then the relative strengths of the interactions observed at low energy can also be computed. In particular, the Weinberg angle can be predicted, and the prediction is in striking agreement with the measured value. (See Langacker, 1981.) This agreement is the best reason for believing the desert hypothesis.

Another very attractive feature of grand unification is that quantization of electric charge is naturally explained. Measured electric charges are always found to be integer multiples of the electron charge. In the standard $SU(3) \times SU(2) \times U(1)$ model, this quantization is imposed by hand, but in grand unified theories, it is a necessity; the electric charge operator obeys nontrivial commutation relations with other operators in the theory, and just as the angular momentum algebra can require the eigenvalues of J_z to be integer multiples of $\frac{1}{2}\hbar$, so the commutation relations obeyed by the electric charge operator require its eigenvalues to be integer multiples of a fundamental unit.

Three general properties of grand unified models are particularly important in cosmology. First, the interactions mediated by particles with masses of order M_{GUT} typically fail to conserve baryon number. This property leads on the one hand to the prediction that the proton should decay, with a lifetime which can be estimated, if we accept the desert philosophy, to be 10^{29-33} yrs. Also, during the first 10^{-35} seconds, when the temperature was of order M_{GUT}, baryon number nonconserving processes occurred copiously. Therefore, any pre-existing initial baryon number was wiped out, and the abundance of baryons today can be calculated, in the context of a specific grand unified model, by considering the evolution of the baryon number density in the very early universe. The results of these calculations are quite model-dependent (for a detailed discussion, see Kolb and Wolfram 1980, and Fry *et al.* 1980), but it is not implausible that the observed relative abundance of baryons to photons, about 10^{-9}, was generated in this way.

Another important consequence of grand unification is that the universe is expected to have undergone a phase transition at a critical temperature T_c of order M_{GUT}. For $T > T_c$, the full grand unified gauge symmetry, which is spontaneously broken at low temperature, is restored. As we have heard from previous speakers (Guth 1982; Linde 1982b), this phase transition may have had a very significant effect on the evolution of the universe.

A third general consequence of grand unification turns out to be closely related to quantization of electric charge: The particle spectrum of a grand unified model necessarily contains magnetic monopoles, stable particles which carry magnetic charges. The necessity of magnetic monopoles in grand unified theories was demonstrated by 't Hooft (1974) and Polyakov (1974). But a general connection between electric charge quantization and magnetic monopoles was pointed out by Dirac (1931) many years earlier.

3 *MAGNETIC MONOPOLES*

Dirac envisaged a magnetic monopole as a semi-infinitely long, infinitesimally thin solenoid. One end of the solenoid, viewed in isolation, appears to be a magnetic charge. But it makes sense to identify this object as a magnetic monopole only if no conceivable experiment can detect the solenoid, in the limit in which it is infinitesimally thin.

We might imagine trying to detect the solenoid by doing an electron interference experiment; such an experiment gives a null result only if the phase picked up by the electron wave function, when the electron is transported along a closed path enclosing the solenoid, is trivial. Suppose a monopole with magnetic charge g sits at the origin, so that the magnetic field is

$$\vec{B} = g \frac{\hat{r}}{r^2} , \qquad (1)$$

and that the solenoid lies on the negative z-axis. Then the vector potential can be written in polar coordinates as

$$\vec{A} \cdot d\vec{r} = g(1-\cos\theta)\, d\phi \quad . \qquad (2)$$

The electron interference experiment fails to detect the solenoid if

$$\text{phase} = \exp[-ie \oint \vec{A} \cdot d\vec{r}] = e^{-4\pi i e g} = 1 , \qquad (3)$$

where $-e$ is the electron charge. Hence, we require the magnetic charge g of the monopole to satisfy the quantization condition

$$g = \frac{n}{2e} , \qquad (4)$$

where n is an integer. The minimal allowed charge $g_D = 1/2e$, is called the Dirac magnetic charge.

As Dirac (1931) observed, we can now turn this argument around, as follows: Suppose there exists a particle with electric charge Q and vanishing magnetic charge. Then it is consistent for a magnetic monopole with magnetic charge g_D to exist only if $\exp[4\pi i Q g_D] = 1$ or

$$\frac{Q}{e} = \frac{1}{2e\, g_D} m = m \ , \qquad (5)$$

where m is an integer. Therefore, the existence of a magnetic monopole implies quantization of electric charge.

It may disturb you that, to derive the Dirac quantization, Eq. (4), I have used the electron charge -e. We would like to believe that quarks exist, and the electric charge of a down quark, for example, is $-\frac{1}{3}e$. Will not the same argument as before, applied to a down quark instead of an electron, lead to the conclusion that the minimal allowed magnetic charge is $3g_D$ instead of g_D?

No, not if quarks are confined. For if quarks are permanently confined in hadrons, it makes sense to speak of performing a quark interference experiment only over distances smaller than 10^{-13} cm, the size of a hadron. It is true that when the down quark is transported around Dirac's solenoid, its wave function acquires the nontrivial phase

$$\exp[-i\frac{e}{3}\oint \vec{A}_{em} \cdot d\vec{r}] = e^{-i\,2\pi/3} \neq 1 \qquad (6)$$

due to the coupling of the down quark to the electromagnetic vector potential, if the monopole carries the Dirac magnetic charge g_D. But we must recall that the down quark carries another degree of freedom, color. The solenoid is not detectable if the monopole also has a *color magnetic field*, such that the phase acquired by the down quark wavefunction due to the color vector potential compensates for the phase due to the electromagnetic vector potential, or

$$\exp[\,i\, e_c \oint \vec{A}_{color} \cdot d\vec{r}] = e^{i\,2\pi/3} \qquad (7)$$

where e_c is the color gauge coupling.

The correct conclusion, then, is not that the minimal magnetic charge is $3g_D$, but rather that the monopole carrying magnetic charge g_D must also carry a color magnetic charge. The color magnetic field of the monopole becomes screened by nonperturbative strong interaction effects at distances greater than 10^{-13} cm. (For more details, see Coleman 1982.)

Dirac's argument shows that the existence of magnetic monopoles implies quantization of electric charge. The converse, in a certain sense, is also true. Any grand unified model, in which electric charge is quantized because the electromagnetic U(1) gauge group is embedded in a simple group which is spontaneously broken at the mass scale M_{GUT}, necessarily contains magnetic monopoles. This remarkable result was discovered by 't Hooft (1974) and Polyakov (1974).

The magnetic monopoles constructed by 't Hooft and Polyakov are topologically stable defects in the Higgs field which acts as an order parameter for the spontaneous breakdown of the grand unified gauge symmetry. The statement that these objects are topological means the following: A long distance away from the center of the monopole, the Higgs field asymptotically approaches in each direction a value which can be obtained by performing a gauge transformation on a standard vacuum value; however, it is impossible to perform a gauge transformation such that the Higgs field approaches the same constant value in all directions simultaneously. Under this circumstance, one says that the "sphere at infinity" is wrapped around the manifold of gauge-equivalent vacua in a topologically nontrivial way. As a consequence, the Higgs field configuration cannot fall apart, and is endowed with a conserved quantum number which turns out to be the magnetic charge. (For a much more detailed discussion, see Coleman 1982.)

The properties of a monopole of the 't Hooft-Polyakov type, such as its size and mass, can be calculated in a given grand unified model. By the size r of the monopole, I mean the characteristic distance from the center of the monopole over which the Higgs field approaches its asymptotic value. That part of the monopole inside radius r is called the "core". The size r is determined by a compromise between the energy stored in the core of the monopole, which is of order $4\pi M_{GUT}^2 r$, and the magnetostatic field energy, which is of order $4\pi g^2/r$. The sum of these two terms is minimized for $r \sim M_{GUT}/g$. The magnetic charge of a grand unified monopole is typically the Dirac charge, so its properties may be summarized by

$$\text{Charge:} \quad g = g_D = 1/2e \ ,$$

$$\text{Size:} \quad r \sim e^{-1} M_{GUT}^{-1} \sim M_X^{-1} \sim 10^{-28}\ \text{cm} \ ,$$

$$\text{Mass:} \quad m \sim (4\pi/e^2) M_X \sim 10^{16}\ \text{GeV} \ , \qquad (8)$$

where M_x is the typical mass of one of the heavy gauge bosons in the model. The numerical estimates of r and m in Eq. (8) are based on the standard estimate of M_x, derived from the desert hypothesis. The gauge coupling $e^2/4\pi$ is the running coupling constraint renormalized at the mass scale M_{GUT}, which is somewhat larger than $\alpha \sim 1/137$.

The grand unified monopole is unusually heavy for a stable elementary particle; the mass 10^{16} GeV $\sim 10^{-8}$g is comparable to the mass of a bacterium. We also see from Eq. (8) that the size r of the monopole is larger than its Compton wavelength by a factor of order $4\pi/e^2$. In this sense, the monopole is a nearly classical object; quantum mechanics plays an insignificant role in determining the structure of the monopole over distances of order r, in the limit in which e^2 is small.

Most of the mass of the monopole is concentrated in the tiny core of radius r, but the monopole also has structure on other, much larger, distance scales. In particular, as we have already remarked, the monopole is a hadron; it has a color magnetic field extending out to 10^{-13} cm. At distances larger than 10^{-13} cm, it interacts only electromagnetically.

While the prediction $m \sim 10^{16}$ GeV follows from the desert hypothesis, it is important to realize that the existence of magnetic monopoles is a very general consequence of grand unification, and that the monopole mass could be different. Monopoles occur in any model in which the electromagnetic U(1) is embedded in a spontaneously broken semisimple gauge group. The monopole mass is $m \sim 4\pi g M_{GUT}$, where g is the magnetic charge, and M_{GUT} is the symmetry breaking scale at which the U(1) gauge group first appears.

To put monopoles in the proper perspective, it is helpful to enumerate the stable particle spectrum of a typical grand unified theory. Nothing is stable in these theories unless there is a good reason. The photon is stable; it is presumably exactly massless. The lightest neutrino is stable, because it is the lightest fermion, and its decay is forbidden by angular momentum conservation. The electron is stable, because it is the lightest electrically charged particle, and electric charge is exactly conserved. And the lightest magnetic monopole is stable, because it is the lightest magnetically charged particle, and magnetic charge is exactly conserved. (The graviton is also stable, but I have not mentioned it, because the grand unified theories I have considered do not include gravity.)

4 MONOPOLES AND COSMOLOGY

A large class of particle physics models predict the existence of stable magnetic monopoles. But to say that monopoles exist is not necessarily to say that we have a reasonable chance of observing one. If the monopole mass is really as large as 10^{16} GeV, then there is no hope of producing monopoles in accelerator experiments in the foreseeable future. Let us consider, then, whether monopoles might have been produced in the very early universe (Zel'dovich and Khlopov 1978; Preskill 1979; for another review see Weinberg 1982).

As was mentioned earlier, we expect that the complete gauge symmetry of a grand unified model is restored at temperatures $T > T_c \sim M_{GUT}$; the Higgs scalar field Φ has a vanishing expectation value for $T > T_c$ and a nonzero expectation value for $T < T_c$. But 't Hooft-Polyakov monopoles can exist only when the gauge symmetry is spontaneously broken; no monopoles are present while $T > T_c$. As the universe cools below T_c monopoles can be produced. Because monopoles, unlike the other superheavy particles in grand unified theories, are stable, the density of monopoles per comoving volume established in the phase transition at $T \sim T_c$ can be subsequently reduced only by annihilation of monopole-antimonopole pairs. As the universe expands, monopoles and antimonopoles have a more difficult time finding each other, and an appreciable density of monopoles per comoving volume might persist.

Thus, the problem of estimating the monopole abundance may be separated into two parts. We must estimate the initial density of monopoles established in the course of the phase transition, and we must determine to what extent the monopole density is subsequently reduced by pair annihilation. (I will assume that the densities of monopoles and antimonopoles are equal.)

Let us first consider the production of monopoles during the phase transition. The detailed mechanism by which monopoles are produced depends on the nature of the phase transition; in particular, on whether it is a second-order (or weakly first-order) transition, in which large fluctuations occur, or a strongly first-order transition, in which supercooling occurs. In either case, there is no reason to believe that the monopole density was ever in thermal equilibrium.

In the case of a second-order (or weakly first-order) phase transition, the Higgs field Φ, which acts as an order parameter for the transition, undergoes large random fluctuations for $T \sim T_c$. As the universe expands and cools, the Higgs field is rapidly quenched, and a large density of topological defects becomes frozen in; these defects are the monopoles and antimonopoles. The quenching process can be described in the following way (Kibble 1976): At the time that monopoles are being produced, the Higgs field is uncorrelated over distances larger than some characteristic correlation length ξ. We may thus regard the Higgs field as having a domain structure, with ξ the characteristic size of a domain. At the intersection point of several domains, each with a randomly oriented Higgs field, there is some probability p that the Higgs field configuration is topologically nontrivial; if so, a monopole or antimonopole must form at the intersection point. The probability p depends on the detailed structure of the monopole, but it is not very much less than one. According to the above picture, the density of monopoles initially established in the phase transition is

$$(n)_{\text{initial}} \sim p\,\xi^{-3} \quad . \tag{9}$$

This argument sounds suspicious, because it relies on the notion of a Higgs field domain structure, although it is always possible to make a uniform Higgs field look wiggly by performing a gauge transformation. However, there is no fundamental difficulty. We may always fix the gauge in some appropriate way so that the idea of a Higgs field domain makes sense, and we have reached a conclusion about the density of topological defects, which is a gauge-invariant quantity.

It is not clear how best to go about calculating ξ, but we can easily obtain an upper bound on ξ from a simple causality argument (Einhorn *et al.* 1980; Guth and Tye 1980). The Higgs field must be uncorrelated over distances exceeding the horizon length d_H, the largest distance any signal could have traveled since the initial singularity; thus $\xi < d_H$. In terms of the temperature T, d_H is given by

$$\xi < d_H \sim \frac{Cm_p}{T^2} \quad . \tag{10}$$

Here $m_p \sim 10^{19}$ GeV is the Planck mass, and $C = (0.60)N^{-\frac{1}{2}}$, where N is the effective number of massless spin degrees of freedom in thermal equilibrium at temperature T; in a minimal grand unified theory, $C \sim 1/20$. Combining (9) and (10), we conclude that the initial value of the dimensionless quantity n/T^3 is bounded by

$$(n/T^3)_{\text{initial}} \gtrsim p(T_c/Cm_p)^3 \quad . \tag{11}$$

In a typical grand unified theory with $T_c \sim 10^{15}$ GeV, $Cm_p \sim 10^{18}$ GeV, and $p \sim 1/10$, we obtain $(n/T^3)_{initial} \gtrsim 10^{-10}$.

In the case of a strongly first-order phase transition, supercooling occurs. The phase with unbroken grand unified gauge symmetry becomes unstable when $T < T_c$, but nonetheless persists for a while, until bubbles of the stable broken-symmetry phase eventually begin to nucleate. These bubbles expand, collide, and coalesce, filling the universe with the stable phase (Coleman 1977). Inside each bubble, the Higgs field is quite homogeneous, so that each bubble contains a negligible number of monopoles. But when the expanding bubbles collide, monopoles can be produced.

Although it is not easy to calculate in detail the initial density of monopoles produced by bubble collisions, we can obtain a lower bound on the monopole density by invoking an argument similar to the one applied above to the case of a second-order transition. Now each bubble can be regarded as a Higgs field domain, and the density of monopoles produced by the collisions must exceed the probability factor p times the density of bubbles at the time they collide. Since bubbles cannot expand faster than the speed of light, each bubble must be smaller than the horizon size d_H, and the density of bubbles is greater than d_H^{-3}. Therefore, we conclude again that $n > p\, d_H^{-3}$, and the bound (11) still holds, except that T_c is replaced by a nucleation temperature T_n, at which nucleation becomes probable.

Let us now consider how the initial abundance of monopoles established in the phase transition subsequently evolves. Once the phase transition is complete, the rate of monopole production is negligible compared to the expansion rate of the universe; we may therefore ignore it. We will also neglect correlations between the positions of monopoles and antimonopoles; this approximation is reasonable when the monopole density is sufficiently small. The monopole density n is then governed by a rate equation of the form

$$\begin{aligned} dn/dt &= -(\text{Annihilation Term}) - (\text{Expansion Term}) \\ &= -D(T)n^2 - 3(\dot{R}/R)n\ , \end{aligned} \tag{12}$$

if the monopole and antimonopole densities are assumed to be equal. The factor D(T) is determined by the dominant annihilation process at a given temperature. At high temperature, there is a dense plasma of light electrically charged particles. This plasma aids the annihilation process, because it dissipates the energy of a monopole drifting toward a nearby antimonopole, allowing capture to occur. Once a bound monopole-antimonopole pair forms, it quickly cascades down, emitting many photons and gluons, and finally annihilates into a burst of Higgs particles and X-bosons. To estimate the annihilation rate, then, we must estimate the capture rate.

The drift velocity of a monopole in the plasma is independent of the monopole mass m. Since the coupling of a monopole to a charged particle is eg = O(1), it is easy to guess that $D(T) \sim g^2T/n_{ch}$, where n_{ch} is the density of light electrically charged particles, roughly given by N_cT^3, if N_c is the number of species of light charged particles in thermal equilibrium. The monopole density n will tend to adjust so that the two terms on the right-hand side of Eq. (12) are of the same order of magnitude. Since $\dot{R}/R \sim T^2/Cm_p$, we obtain

$$\frac{n}{T^3} \sim \left(\frac{N_c}{g^2}\right)\left(\frac{T}{Cm_p}\right) . \quad (13)$$

The diffusive capture process described above operates effectively only if the mean free path λ of a monopole in the plasma is shorter than the characteristic distance $\lambda_c \sim g^2/T$ at which capture can occur. But $\lambda \sim (1/N_cT)(m/T)^{\frac{1}{2}}$ becomes comparable to λ_c when $T \sim m/g^4N_c^2$; at lower temperatures this capture process is cut off. When diffusive capture ends, the monopole density is given by (Preskill 1979)

$$\frac{n}{T^3} \sim \frac{1}{N_cg^6}\,\frac{m}{Cm_p} . \quad (14)$$

If the initial value of n/T^3 is smaller than the right-hand side of (14), then the annihilation rate never competes with the expansion rate, and n/T^3 is unchanged.

At lower temperatures, monopoles and antimonopoles must shake off photons or gluons in order to capture each other. This process is too inefficient to keep pace with the Hubble expansion (Preskill 1979; Dicus *et al.* 1982), and n/T^3 remains frozen at the value given in Eq. (14).

Plugging in the standard estimates, $m \sim 10^{16}$ GeV, $Cm_p \sim 10^{18}$ GeV, $N_cg^6 \sim 10^7$, Eq. (14) becomes $n/T^3 \sim 10^{-9}$. Therefore, if the smallest possible initial monopole density per comoving volume consistent with the bound (11), $n/T^3 \sim 10^{-10}$, is established in the phase transition, it is not further reduced by annihilation at all. The only way to reduce n/T^3 further is through nonadiabatic effects which increase the entropy density, but such effects cannot dilute the monopole abundance by many orders of magnitude without at the same time diluting the baryon excess. Neglecting generation of entropy, we conclude that the density of magnetic monopoles today is $n \sim 10^{-10}\,T^3$, which is comparable to the density of baryons. This conclusion is clearly absurd, if the mass of a monopole exceeds the mass of a baryon by a factor of order 10^{16}.

We have uncovered the "monopole problem", an apparently serious conflict between grand unified theories and standard big-bang cosmology. Various attempts have been made to resolve this conflict. These attempts tend to fall into two categories. Some authors have suggested mechanisms which enhance the monopole-antimonopole annihilation rate; others have suggested ways of suppressing the initial production of monopoles in the phase transition.

If the monopole density were really $n \sim 10^{-10}T^3$, the universe would become matter dominated by monopoles at a relatively early time, and the monopoles would be expected to condense into clumps in which the annihilation rate would be enhanced. A detailed analysis of this scenario, however, indicates that this enhancement does not suffice to reduce the monopole abundance to an acceptable level (Goldman *et al.* 1981; Fry 1981; Dicus *et al.* 1982).

Mechanisms have been advanced in which monopoles and antimonopoles form bound pairs connected by flux tubes. Were this to occur, pair annihilation would swiftly reduce the monopole abundance to an acceptable value. One proposal (Linde 1980) is that the color magnetic field of the monopole is confined to a flux tube at high temperature. Another (Lazarides and Shafi 1980) is that, at least in some models, monopoles and antimonopoles become connected by Z^0 flux tubes after the weak interaction symmetries get spontaneously broken at $T \sim 300$ GeV. In both cases, however, the proposed flux tubes are not topologically stable, and their existence has not been demonstrated.

Langacker and Pi (1980)(see also Bais and Langacker 1982) *have* succeeded in constructing a scenario in which monopole-antimonopole pairs become confined by flux tubes. Their proposal is that, over some temperature range, the U(1) gauge group of electromagnetism is spontaneously broken, which causes the U(1) magnetic flux to collapse to a tube. Specifically, they constructed a model which undergoes, as the temperature decreases, the sequence of phase transitions

$$SU(5) \rightarrow SU(3) \times SU(2) \times U(1) \rightarrow SU(3) \rightarrow SU(3) \times U(1) . \qquad (15)$$

While the universe is in the SU(3) phase, any previously produced monopoles are expected to annihilate, and when the transition to the SU(3) × U(1) phase finally occurs, the temperature is so low that an acceptably small monopole abundance is generated. Unfortunately, to achieve the symmetry breaking pattern in Eq. (15), Langacker and Pi must impose on their model a complicated and contrived Higgs structure. Nonetheless, theirs is a viable proposal for reducing the monopole abundance.

Among suggested means of reducing the initial monopole abundance established in the phase transition, one proposal (Bais and Rudaz, 1980) is that the large fluctuations in the Higgs field at temperatures near T_c in a second-order phase transition allow the monopole density to relax to a very small value. But it has not been explained how this mechanism can generate the acausal correlations needed to defeat the bound (11).

Our discussion of monopole production by bubble collisions in the case of a strongly first-order phase transition left open the possibility that the initial monopole density is suppressed if bubble nucleation is delayed long enough. But a detailed analysis (Guth and Weinberg 1982) in the context of the "old" inflationary universe (Guth 1981) indicates that, if nucleation is sufficiently delayed for the monopole abundance to be significantly suppressed, then the phase transition never finishes, and a conventional cosmological scenario cannot be recovered.

Another possibility is that, contrary to the desert hypothesis, the electromagnetic U(1) is embedded in a semisimple gauge group which is spontaneously broken at the mass scale $M_U << 10^{15}$ GeV. For $M_U \leqslant 10^9$GeV, the bound (11) gives an acceptably small monopole density. (Observational limits on the monopole abundance are discussed in Section 5; for light monopoles, $m \leqslant 10^{11}$GeV, $n/T^3 \leqslant 10^{-29}$ is required.) But this suggestion should be regarded with caution, because it requires that we take (11) seriously as an *equality*, not just a bound. That is, we must believe that the relevant correlation length which determines the initial monopole abundance is really the largest possible length d_H, and not a much shorter length. I think that this assumption has a far better chance of being justified in the case of a strongly first-order phase transition than in the case of a second-order (or weakly first-order) phase transition. (Even for light monopoles, annihilation cannot reduce an initially large monopole density to an acceptable level.)

Naturally, there is no monopole problem if there are no monopoles. The monopole problem does not arise if the electromagnetic U(1) is not embedded in a spontaneously broken semisimple gauge group; that is, if the basic premise of grand unification is simply wrong.

Surely the most attractive proposal for suppressing the initial abundance of monopoles established in the grand unified phase transition is the new inflationary universe scenario (Linde 1982a; Albrecht and Steinhardt 1982). As we have heard from some of the previous speakers (Guth 1982; Linde 1982b; Steinhardt 1982), one arranges in this scenario, at the cost of some fine tuning of parameters, that the universe undergoes many e-foldings of inflation *after* the appearance of bubbles or fluctuation regions in which the Higgs field is nonzero. Some monopoles are produced when bubbles or fluctuation regions first form, but they are subsequently "inflated away"; in the course of the exponential inflation, the monopole density is reduced to a negligible value. Eventually, the cosmological constant which drove the exponential inflation becomes rapidly thermalized, and the universe "reheats". During this thermalization process, some monopoles may be produced. Let us try to estimate how many (Turner 1982; Lazarides *et al.* 1982).

For the purpose of an order of magnitude estimate, we suppose that all light degrees of freedom instantaneously assume thermal distributions characterized by a reheating temperature $T_R \ll m$, and ask how the monopole density, chosen to vanish at temperature T_R, subsequently evolves, in a standard radiation-dominated cosmology. Neglecting pair annihilation, the monopole density is determined by the rate equation

$$\frac{dn}{dt} = (\text{Production Term}) - (\text{Expansion Term})$$

$$= D(T)\, n_{eq}^2(T) - 3(\dot{R}/R)n \,, \tag{16}$$

where D(T) is the same function as appears in Eq. (12), and $n_{eq} \sim (mT)^{3/2} e^{-m/T}$ is the equilibrium density of monopoles at temperature T. This form of the monopole production term in the rate equation is obtained by a standard detailed-balance argument; if the monopole density did assume the equilibrium value, then the monopole production and annihilation rates would have to be equal. In response to the objection that the cross section for production of monopole-antimonopole pairs in hard binary collisions is suppressed by a factor of the form e^{-c/e^2}, because monopoles are large compared to their Compton wavelength (Witten 1979a; Drukier and Nussinov 1982), we note that the monopole production rate is dominated by many-particle processes, not binary collisions.

Substituting $\dot{R}/R \sim T^2/m_p$ and $D(T) \sim g^2/T^2$, as in our discussion of monopole-antimonopole annihilation, Eq. (16) may be rewritten

$$\frac{d}{dT^{-1}}\left(\frac{n}{T^3}\right) \sim g^2 m_p \left(\frac{m}{T}\right)^3 e^{-2m/T}, \tag{17}$$

which, when integrated from $T = T_R$ to $T = 0$, yields

$$\frac{n}{T^3} \sim g^2 \frac{m_p}{m}\left(\frac{m}{T_R}\right)^3 e^{-2m/T_R}. \tag{18}$$

Since the ratio m/T_R appears in the exponent, it should be estimated carefully; let us consider an explicit example.

In the new inflationary universe scenario, the reheating temperature T_R can be estimated fairly precisely; the effective potential is required to satisfy the Coleman-Weinberg (1973) condition, and the vacuum energy in the symmetric phase can therefore be expressed in terms of gauge boson masses. Assuming that this vacuum energy is shared among all thermalized radiation species after reheating, the reheating temperature can be computed. In the "minimal" SU(5) model one finds $T_R \sim M_x/6.2$. The monopole mass can also be accurately estimated when the Coleman-Weinberg condition holds; in the minimal SU(5) model it is $m \sim \alpha_{GUT}^{-1} M_x$, where $\alpha_{GUT} \sim (42)^{-1}$ is the gauge coupling, renormalized at the unification scale. For the minimal SU(5) model, then, Eq. (18) becomes

$$\frac{n}{T^3} \sim (6\,\alpha_{GUT}^{-1})^3 \left(\frac{m_p}{M_x}\right) e^{-(12.4)\alpha_{GUT}^{-1}} \sim 10^{-220} . \tag{19}$$

This is discouraging; if the monopole density is really as small as indicated in Eq. (19), we will never see a monopole.

I have presented two theoretical estimates of the monopole abundance. One, Eq. (11), is unacceptably large. The other, Eq. (19) is depressingly small. It is hardly necessary to add that estimates anywhere in between can easily be obtained. For example, I can construct grand unified models in which the gauge coupling at unification is larger than in the minimal SU(5) model, so that m/T_R is smaller, and the abundance given by Eq. (18) is larger. There is no precise theoretical expectation of what the cosmological monopole abundance should be.

The qualitative prediction is nonetheless significant. Grand unified theories require the existence of magnetic monopoles. These monopoles could have been produced in the very early universe. In the new inflationary universe scenario it is possible for the monopole abundance to be both small enough to be acceptable and large enough to be observable. Clearly, we should look for them.

5 *LIMITS ON THE MONOPOLE FLUX*

People have been looking for magnetic monopoles for a long time. But one quickly recognizes that the traditional monopole search experiments do not place significant constraints on superheavy monopoles with mass m of order 10^{16} GeV. These traditional searches have relied on the strong ionization power of relativistic monopoles (Fleischer *et al.* 1969; Price *et al.* 1978) or have sought monopoles trapped in the earth's crust (Kolm *et al.* 1971; Alvarez *et al.* 1971). But a superheavy monopole would be expected to be slowly moving and very penetrating; it need not ionize heavily or stop in the earth (Preskill 1979; Lazarides *et al.* 1981).

How slowly moving? The monopole can be accelerated by either gravitational fields or magnetic fields; which effect is more important depends on the mass of the monopole. From gravitational fields alone, the monopole would acquire a typical galactic infall velocity, of order 10^{-3}c, regardless of its mass. To determine the effect of the magnetic field in our galaxy, recall that the field has a strength B of order 3×10^{-6} gauss and a coherence length L of order 10^{21} cm (see Parker 1979). A monopole with the Dirac charge g_D crossing one coherence length is accelerated to

$$v = \left(\frac{2g_D BL}{m}\right)^{\frac{1}{2}} \sim 3\times10^{-3}c\left(\frac{B}{3\times10^{-6}q}\right)^{\frac{1}{2}}\left(\frac{L}{10^{21}\text{ cm}}\right)^{\frac{1}{2}}\left(\frac{10^{16}\text{ GeV}}{m}\right)^{\frac{1}{2}} . \tag{20}$$

This magnetic acceleration is therefore more important than the gravitational acceleration for $m \lesssim 10^{17}$ GeV. The monopole does not attain a relativistic velocity for $m > 10^{11}$ GeV.

How penetrating? The stopping power of slowly moving magnetic monopoles remains a rather controversial subject, about which a little more will be said later. But the energy loss in rock of a monopole with $v/c \lesssim 10^{-2}$ is surely not much larger than (Ahlen and Kinoshita 1982)

$$\frac{dE}{dx} \sim 100\left(\frac{v}{c}\right)\text{GeV/cm} . \tag{21}$$

Thus the range in rock of a monopole with $m \sim 10^{16}$ GeV is larger than 10^{11} cm;

the monopole passes through the earth without slowing down.

Although superheavy monopoles are not easily stopped or detected, there are nonetheless quite severe limits on the flux of magnetic monopoles in cosmic rays. (For a review, see Fry 1982.) One limit on the number density of monopoles can be obtained because monopoles are very heavy, and has nothing to do with their magnetic charge. Demanding that the mass density due to monopoles not exceed the observational limit $\rho \lesssim 10^{-29}\ \mathrm{g\ cm^{-3}}$, we obtain the bound

$$n \lesssim 10^{-21}\ \mathrm{cm^{-3}} \left(\frac{10^{16}\ \mathrm{GeV}}{m}\right) \tag{22}$$

on the monopole density n. Using the estimate (20) for the monopole velocity, which is valid for $m \lesssim 10^{17}$ GeV, we obtain a limit on the monopole flux,

$$nv \lesssim 10^{-2}\ \mathrm{m^{-2}yr^{-1}} \left(\frac{10^{16}\ \mathrm{GeV}}{m}\right)^{3/2} \qquad (m \lesssim 10^{17}\ \mathrm{GeV}). \tag{23}$$

If the monopole mass exceeds 10^{17} GeV, it is conceivable that monopoles are confined to galactic halos. Thus, the local density could be considerably higher than the bound (22). Assuming that the mass of the halo is less than 10^{12} solar masses, and that the typical monopole velocity is $10^{-3}c$, we obtain the flux limit

$$nv \lesssim 10^{2}\ \mathrm{m^{-2}yr^{-1}} \left(\frac{10^{17}\ \mathrm{GeV}}{m}\right) \qquad (m \gtrsim 10^{17}\ \mathrm{GeV}). \tag{24}$$

A somewhat more restrictive flux limit is obtained by a slightly less direct argument, due to Parker (1970). (See also Turner *et al.* 1982.) The galactic magnetic field accelerates monopoles. Thus, the energy density $U = B^2/8\pi$ stored in the field is dissipated at the rate $dU/dt \sim \langle g n \vec{v}\cdot\vec{B}\rangle$. By demanding that the field energy is not substantially depleted in a time τ, of order 10^8 years, required to regenerate the field, we obtain the limit

$$nv \lesssim \frac{B}{8\pi g \tau} \sim 2\times 10^{-4}\ \mathrm{m^{-2}yr^{-1}} \left(\frac{B}{3\times 10^{-6}\mathrm{gauss}}\right)\left(\frac{10^{8}\mathrm{yr}}{\tau}\right). \tag{25}$$

A nice feature of this flux limit is that it appears to be independent of the mass m of the monopole.

However, it is implicitly assumed in the derivation of (25) that gravitational effects on the trajectory of the monopole are neglibible, and we have already argued that this is not so for $m \gtrsim 10^{17}$ GeV. If a monopole enters a coherent domain of the galactic magnetic field with incident energy $\frac{1}{2}mv^2 > gBL$, then the energy it extracts from the domain is a second-order effect, due to the deflection of the monopole trajectory as it crosses the domain; on the average it is (Turner *et al.*, 1982)

$$\Delta E \sim (gBL)^2/\tfrac{1}{2}mv^2 \ . \tag{26}$$

Therefore, the rate of dissipation of magnetic field energy scales like $1/m$ for $m \gtrsim 10^{17}$ GeV, and our flux limit becomes

$$nv \lesssim 2\times10^{-4}\text{m}^{-2}\text{yr}^{-1}\left(\frac{m}{10^{17}\,\text{GeV}}\right)\left(\frac{B}{3\times10^{-6}\,\text{gauss}}\right)\left(\frac{10^{8}\,\text{yr}}{\tau}\right) \tag{27}$$

$$(m \gtrsim 10^{17}\,\text{GeV}).$$

This crosses the constraint (24) for $m \sim 10^{20}$ GeV; therefore, combining (24) and (27), we can obtain the mass-independent flux limit

$$nv \lesssim 10^{-1}\text{m}^{-2}\text{yr}^{-1} \ . \tag{28}$$

It should be admitted, though, that the above arguments are rather naive, particularly for the case of monopoles which are heavy enough to remain confined to galaxies. We have treated monopoles as test particles riding the galactic magnetic field. If, instead, magnetic charge density fluctuations play a significant part in establishing the field, it is conceivable that the Parker limit (27) can be exceeded (Turner *et al.* 1982; Salpeter *et al.* 1982).

In view of (27) I had been saying for several years that anyone who wished to do a useful experiment to search for superheavy magnetic monopoles should think about how to build a very big detector. It appeared to be necessary to cover a football field in order to have a reasonable chance of seeing one monopole in a year. I was surprised when I heard that a candidate monopole event had been observed in a detector smaller than the palm of my hand.

6 THE CABRERA EXPERIMENT

The experiment was performed by Blas Cabrera (1982a), at Stanford. Briefly, the apparatus consists of a coil of superconducting wire with radius r = 2.5 cm and N = 4 turns. (For a more detailed description see Cabrera 1982b.) The current in the wire is monitored by a SQUID. What happens if a magnetic monopole passes through the coil?

The relevant integrated Maxwell equation, suitably modified for the existence of a magnetic monopole current, is

$$\oint_\Gamma \vec{E} \cdot d\vec{r} = -\frac{1}{c}\frac{d\Phi}{dt} - \frac{4\pi}{c}\frac{dQ_m}{dt} , \tag{29}$$

where Φ is the magnetic flux through a surface S_Γ bounded by the closed path Γ, and dQ_m/dt is the magnetic monopole current flowing through S_Γ. Applying Eq. (29) to a path entirely contained in the superconducting wire, where $\vec{E} = 0$, and integrating over time, we find that the change $\Delta\Phi$ in the magnetic flux linking the coil and the total magnetic charge ΔQ_m which passes through the coil are related by

$$\Delta\Phi = -4\pi N \Delta Q_m . \tag{30}$$

In particular, if a magnetic monopole with magnetic charge g passes through a coil with $N = 4$ turns, the change in the magnetic flux will be

$$\Delta\Phi = -8\Phi_0\left(\frac{g}{g_D}\right) , \tag{31}$$

where $g_D = 1/2e$ is the Dirac magnetic charge, and $\Phi_0 = 4\pi g_D/2$ is the flux quantum of the superconductor; the factor of two arises because the electric charge of a Cooper pair is 2e.

Hence, a monopole passing through Cabrera's coil would cause a sudden change in the flux linking the coil, and a corresponding shift in the DC current level in the wire, which would be detected by the SQUID. One such event has been seen (on 14 February 1982, at 1:53 p.m., P.S.T.) with $|\Delta\Phi| \sim 8\Phi_0$.

An important advantage of Cabrera's detector is that it is sensitive to monopoles of arbitrarily low velocity. It is therefore conceivable that his detector could register a genuine monopole event while a much larger detector with a velocity threshold would see no evidence of monopoles. However, a single event cannot be taken very seriously; confirmation is sorely needed. I will nonetheless suppose in the following discussion that Cabrera has really discovered a monopole, and will consider the implications of this discovery.

If Cabrera's event was caused by a monopole, then he has

measured the magnetic charge of the monopole; it is consistent with the Dirac charge g_D. Ironically, this result appears to be in conflict with the result of another celebrated experiment performed at Stanford, the reported discovery by Fairbank and collaborators (LaRue *et al.* 1977; 1979; 1981) of fractional charge on matter. Dirac's very general reasoning, reproduced in Section 3, indicates that the existence of monopoles with magnetic charge $g_D = 1/2e$ is inconsistent with the existence of *isolated* electrically charged particles carrying electric charge e/3.

Actually, there *are* theoretically acceptable ways of reconciling these two experiments. One amusing possibility is that the objects which carry fractional electric charge are actually "dyons" which also carry magnetic charge. For two dyons with electric and magnetic charges (Q_1, g_1) and (Q_2, g_2), the Dirac consistency condition becomes generalized to (Schwinger 1966; Zwanziger 1968)

$$Q_1 g_2 - Q_2 g_1 = n/2 \ , \tag{32}$$

where n is an integer, and Eq. (32) is satisfied if the electric charge q of an object with magnetic charge g is given by

$$Q/e = m + e\, g\, \theta/\pi \ , \tag{33}$$

where m is an integer; θ is an arbitrary parameter which may take any value from $-\pi$ to π (Witten 1979b). The minimal electric charge carried by a dyon with the Dirac magnetic charge is $Q = e\theta/2\pi$. It is conceivable that Fairbank and Cabrera have discovered the same object; that Fairbank has measured its electric charge, and Cabrera has measured its magnetic charge, although I know of no reason why $\theta/2\pi$ should be close to 1/3. In any event, this theoretical possibility appears to be experimentally excluded. Cabrera (1982b) has measured the magnetic charges of two niobium spheres on which fractional electric charge has been observed; the measured magnetic charges were consistent with zero.

Another possible way to reconcile the two measurements is to postulate the existence of a new long-range force which couples to Cabrera's magnetic monopoles and Fairbank's fractional electric charges, but not to ordinary matter. I will not pursue this possibility here; for a discussion see Preskill 1982; Strominger 1982; Barr *et al* 1982; Pantalone 1982.

A second implication of Cabrera's event, if we assume that it was really caused by a monopole, concerns the monopole flux. Clearly, it is an exaggeration to say that a single event constitutes a measurement of the flux, but it is improbable that an event would be observed in a detector with an area of 20 cm^2 in less than a year of running if the flux is much less than

$$nv \sim 5 \times 10^2 \, m^{-2} \, yr^{-1} \ . \qquad (34)$$

Disturbingly, this flux is more than six orders of magnitude above the Parker limit (25), which applies for $m \lesssim 10^{17}$ GeV, and nearly four orders of magnitude above the mass-independent limit (28). Perhaps it is possible, as suggested earlier, for the limit (28) to be exceeded by several orders of magnitude, if the galactic magnetic field is produced by magnetic charge density fluctuations (Salpeter *et al.*, 1982), but if we take (28) seriously (as I am inclined to), then we are forced to conclude that the monopole flux at Stanford is considerably larger than the ambient flux in our galaxy. Why should Stanford occupy such a privileged position?

A possible answer is that Stanford is close to a star, our sun (Dimopoulos *et al.* 1982a). A flux incident on the solar system which nearly saturates the Parker limit might be able to maintain a monopole "cloud" surrounding the sun with a significantly enhanced monopole density, if monopoles passing through the sun (or Jupiter) are able to lose enough energy to become captured in solar orbit. It is unclear whether this mechanism can provide the necessary flux enhancement of at least four orders of magnitude, but if it can, one is led to the interesting prediction that the monopole flux incident on the earth is anisotropic, and that the typical monopole velocity is of order $10^{-4}c$, the velocity of the earth in its orbit about the sun. The angular distribution and velocity distribution of monopoles should resemble those of meteorites, which are also projectiles in solar orbit.

7 *FUTURE EXPERIMENTS*

Many experiments currently running or planned have been specifically designed to search for superheavy magnetic monopoles. I will not attempt to review them all. For detailed accounts of many of these experiments, see Carrigan and Trower 1982.

Several experiments have recently reported limits on the cosmic ray flux of slowly moving, weakly ionizing particles. (For a review see

Giacomelli 1982.) An understanding of the ionization loss of slowly moving monopoles is necessary in order to extract from these results a limit on the monopole flux. On the basis of the best available estimates of the ionization loss (Ahlen and Kinoshita 1982), it is conservative to conclude that the flux of monopoles with velocity $v \gtrsim 10^{-3}c$ satisfies

$$nv \lesssim 1\,m^{-2}\,yr^{-1} \ . \qquad (35)$$

Therefore, if Cabrera's event was caused by a monopole, it was probably a very slow monopole with $v < 10^{-3}c$. This observation makes the solar cloud scenario seem more attractive.

Is it possible to build a non-induction detector which will reliably respond to a monopole with $v \sim 10^{-4}c$? Unfortunately, the stopping power of a monopole with $v \lesssim 10^{-3}c$ is quite uncertain. One is inclined to say that the response of an atom or molecule to a very slowly moving monopole passing nearby is similar to its response to a magnetic field which adiabatically turns on and off; therefore, it is unlikely to become excited. But this conclusion is not necessarily correct, because the very strong magnetic field of the monopole greatly distorts the energy levels of the atom or molecule; if the ground state and an excited state closely approach each other, the adiabatic approximation may fail badly. To decide whether this occurs, detailed knowledge of the level structure in the inhomogeneous magnetic field of the monopole is needed. In fact, recent calculations (Drell *et al.* 1982) have shown that the energy loss in atomic hydrogen and in helium of a monopole with velocity between $10^{-4}c$ and $10^{-3}c$ is much higher than previously suspected. This energy loss is almost exclusively due to atomic excitation followed by photon emission, with negligible ionization loss. It is not yet clear to what extent this result will generalize to other materials.

The top priority of present and future monopole search experiments is to obtain more events, either in induction devices like Cabrera's or in non-induction detectors. But assuming that enough events are accumulated to confirm the discovery of the monopole to everyone's satisfaction, what more can be learned by doing more intricate experiments? To test the predictions of the solar cloud hypothesis, measurements of the angular distribution and velocity distribution of monopoles incident on the earth are needed. Such measurements could be performed with an array of superconducting coils, each similar to Cabrera's device. To measure velocities up to 10^{-3}c over distances less than one meter, it is necessary to resolve time delays of less than a microsecond.

It is obviously very important to measure the mass of the monopole; to do so, the trajectory of the monopole must be measurably deflected. A monopole with the Dirac magnetic charge, mass m, and velocity v, if exposed to a magnetic field B perpendicular to its path over a path length L, is deflected by

$$\theta \sim 2\times 10^{-6}\,\text{radians}\left(\frac{10^{16}\ \text{GeV}}{m}\right)\left(\frac{10^{-4}c}{v}\right)^2\left(\frac{BL}{100\ \text{kG}-\text{m}}\right) . \tag{36}$$

To place interesting limits on the monopole mass, reasonably good angular resolution is required. One way of achieving improved angular resolution is being pursued by Cabrera (1982b). His group is developing a new detector, in which a thin cylindrical sheet of superconducting material is used to record the track of a monopole. A monopole passing twice through the sheet leaves in its wake two vortices, each carrying two flux quanta; the vortices remain pinned, and determine the trajectory. The angular resolution is intrinsically limited only by the size of a vortex, about 10^{-6}m, divided by the distance between the vortices. Two such detectors, with a magnet in between, could serve as a monopole mass spectrometer, provided that the velocity is also measured.

Studying the monopoles that fall out of the sky is inconvenient. It would be much easier to measure the mass of a stationary monopole, or one moving at a velocity much less than 10^{-4}c. Although the abundance of monopoles in the earth's crust is probably extremely small, the advantages of finding a monopole stuck in a rock are so great that a careful search is appropriate. One such search, which will process 10^6 tons of iron ore per year, is planned (Cline 1982). To hold on to a monopole once you find it is problematic, but one should at least have the opportunity to make some measurements as the monopole begins its descent toward the center of the earth.

With a stationary or very slowly moving monopole, another extremely interesting measurement could be made - a measurement of the electric charge of the monopole. We have noted that the electric charge of a monopole with the Dirac magnetic charge is $Q = e\theta/2\pi$, where the angle θ is a parameter which characterizes the CP-nonconservation of electrodynamics (Witten 1979b). The only way to determine θ is by measuring the electric charge of a monopole.

I have saved for last some comments about the most intriguing recent development in monopole theory, which concerns the catalysis by monopoles of baryon number nonconserving processes. There is no doubt that the monopoles of a typical grand unified theory are capable of catalyzing baryon number nonconserving processes, such as nucleon decay (Dokos and Tomaras 1980; Blaer *et al.* 1981; Wilczek 1982). The remarkable new suggestion of Rubakov (1981; 1982) and Callan (1982) is that the cross sections for such processes are independent of the size of the monopole core, and hence much larger than previously believed. In fact, Rubakov and Callan claim that monopoles catalyze nucleon decay at a rate typical of the strong interactions. In my opinion, this claim has not yet been convincingly demonstrated, but we may nevertheless consider its implications.

The most interesting, and disturbing, implication concerns the effects of monopoles in neutron stars (Kolb *et al.* 1982; Dimopoulos *et al.* 1982b). Surrounded by matter at nuclear density, monopoles behaving as suggested by Rubakov and Callan catalyze nucleon decay at a furious rate. A modest number of monopoles in a neutron star would raise the luminosity of the star to the point where it emits a substantial flux of ultraviolet or x-ray photons. Since neutron stars capture any monopoles incident on them, one obtains from observational limits on the diffuse ultraviolet and x-ray spectrum a limit on the product of the cross section σ for monopole-catalyzed nucleon decay and the interstellar monopole flux nv, namely

$$\left(\frac{\sigma}{10^{-27}\,\mathrm{cm}^2}\right) nv \leqslant 10^{-12}\ \mathrm{m}^{-2}\,\mathrm{yr}^{-1} \ . \qquad (37)$$

If σ is really a strong-interaction cross section, we infer that the interstellar monopole flux must be smaller than the Parker limit, Eq. (25), by at least eight orders of magnitude. Apparently, if the cross section for monopole-catalyzed baryon number nonconservation is as large as Rubakov and Callan claim, monopoles are so rare that there is little hope of observing them directly, even if the local flux is substantially enhanced. The best way to find evidence for the existence of monopoles would then be by observing their effect on the luminosity distribution of neutron stars.

On the other hand, if the Parker limit on the interstellar flux is nearly saturated, we conclude that the cross section σ for monopole-catalyzed nucleon decay is smaller by at least eight orders of magnitude than estimated by Rubakov and Callan. This discrepancy could arise because their estimates are wrong, or because the correct unified theory is one to which their analysis does not apply - for example, one in which

"X-boson" exchange is baryon number conserving.

8 *CONCLUSIONS*

The discovery of a magnetic monopole, if genuine, has many implications. Among them are the following:

The monopole must be superheavy. If the monopole mass m is less than 10^{17} GeV, there is no reason to doubt the validity of the Parker limit, Eq. (25). Therefore, the only credible explanation for the apparently large local flux is the solar cloud scenario. But the sun can act as an efficient monopole collector only if the typical velocity of a monopole incident on the solar system is comparable to or smaller than the solar escape velocity, about $2\times10^{-3}c$. In view of Eq. (20), this means that m cannot be much less than 10^{16} GeV.

Since no astrophysical process is energetic enough to produce superheavy monopoles, any monopoles present today must have been produced in the very early universe. That any exist at all is evidence that the universe was once *extremely* hot.

The discovery of a monopole will have an important impact on cosmology. The monopole abundance will join the helium abundance and the baryon abundance as a number which must be explained by a realistic cosmological scenario. It appears that this will be a very severe constraint.

If the local monopole flux is as large as suggested by the Cabrera event, then either monopoles play an unexpected active role in the dynamics of the galactic magnetic field, or the monopole flux in the solar system is much larger than that in the galaxy at large. The astrophysical consequences of either alternative will be interesting to explore.

The existence of monopoles poses a welcome challenge to experimenters, for to further elucidate the properties of monopoles will continue to require ingenious new experimental methods.

Finally, from the perspective of particle physics, the mere existence of a magnetic monopole is enormously significant, because it confirms a fundamental prediction of grand unification. If the mass of the monopole can be measured, we will learn, at least in order of magnitude, the fundamental symmetry breaking scale M_{GUT} at which electrodynamics becomes truly united with the other particle interactions. Possibly, we will also be able to learn this mass scale by measuring the proton lifetime, but if M_{GUT} is only slightly larger than suggested by the standard estimates based on the desert hypothesis, proton decay will be extremely difficult to observe. Then a measurement of the monopole mass may be the only way to determine M_{GUT} by doing a low energy experiment.

Whether a monopole has in fact been discovered remains to be verified. If so, then for the above reasons, and more, we are all very lucky indeed.

ACKNOWLEDGEMENTS

It is a great pleasure to extend thanks to the organizers, Gary Gibbons and Stephen Hawking, for the opportunity to attend the Nuffield Workshop, and to all the participants who helped to make it such a spectacular success.

This work was supported in part by the National Science Foundation under Contract Number PHY77-22864, and by the Alfred P. Sloan Foundation.

REFERENCES

Ahlen, S. P. and Kinoshita, K. (1982), Phys. Rev. D 26, 2347.
Albrecht, A. and Steinhardt, P. J. (1982), Phys. Rev. Lett. 48, 1220.
Alvarez, L. W. *et al.* (1971), Phys. Rev. D 4, 3260.
Bais, F. A. and Langacker, P. (1982), Nucl. Phys. B197, 520.
Bais, F. A. and Rudaz, S. (1980), Nucl. Phys. B170, 507.
Barr, S. M., Reiss, D. B., and Zee, A. (1982), Seattle preprint 40048-31 P2.
Blaer, A., Christ, N., and Tang, J. (1981), Phys. Rev. Lett. 47, 1364.
Cabrera, B. (1982a), Phys. Rev. Lett. 48, 1378.
Cabrera, B. (1982b), in *Third Workshop on Grand Unification*, eds. P. H. Frampton, S. L. Glashow, and H. van Dam. Boston: Birkhäuser.
Callan, C. (1982), Phys. Rev. D 26, 2058; and (1982), Princeton preprint.
Carrigan, R. A. and Trower, W. P., ed. (1982). *Proceedings of the Magnetic Monopole Workshop*. New York: Plenum.
Cline, D. (1982), in *Proceedings of the Magnetic Monopole Workshop*, eds. R. A. Carrigan and W. P. Trower. New York: Plenum.
Coleman, S. (1977), Phys. Rev. D 15, 2929.
Coleman, S. (1982), Harvard preprint HUTP-82/A032, to be published in *Proceedings of the International School of Subnuclear Physics*, Ettore Majorana, Erice, 1981, ed. A. Zichichi. London: Plenum.
Coleman, S. and Weinberg, E. J. (1973), Phys. Rev. D 7, 1888.
Dicus, D. A., Page, D. N., and Teplitz, V. L. (1982), Phys. Rev. D 26, 1306.
Dimopoulos, S., Glashow, S. L., Purcell, E. M., and Wilckek, F. (1982a), Nature 298, 824.
Dimopoulos, S., Preskill, J., and Wilczek, F. (1982b), Phys. Lett., 119B, 320.
Dirac, P. A. M. (1931), Proc. Roy. Soc. (London) Ser. A, 133, 60.
Dokos, C. and Tomaras, T. (1980), Phys. Rev. D 21, 2940.
Drell, S. D. *et al.* (1982), SLAC-PUB-3012.
Drukiev, A. K. and Nussinov, S. (1982), Phys. Rev. Lett. 49, 102.
Einhorn, M. B., Stein, D. L. and Toussaint, D. (1980), Phys. Rev. D 21, 3295.
Fleischer, R. L. *et al.* (1969), Phys. Rev. 184, 1393, 1398.
Fry, J. N. (1981), Ap. J. (Letters) 246, L93.

Fry, J. N. (1982), Enrico Fermi Inst. preprint No. 82-17.
Fry, J. N. Olive, K. A., and Turner, M. S. (1980), Phys. Rev. D 22, 2953, 2977.
Georgi, H. and Glashow, S. L. (1974), Phys. Rev. Lett. 32, 438.
Georgi, H., Quinn, H. R., and Weinberg, S. (1974), Phys. Rev. Lett. 33, 451.
Giacomelli, G. (1982), in *Proceedings of the Magnetic Monopole Workshop*, eds. R. A. Carrigan and W. P. Trower. New York: Plenum.
Goldman, T., Kolb, E. W., and Toussaint, D. (1981), Phys. Rev. D 23, 867.
Guth, A. H. (1981), Phys. Rev. D 23, 347.
Guth, A. H. (1982), these Proceedings.
Guth, A. H. and Tye, S.-H. (1980), Phys. Rev. Lett. 44, 631, 963(E).
Guth, A. H. and Weinberg, E. J. (1983), Nucl. Phys. B212, 321.
Kibble, T. W. B. (1976), J. Phys. A9, 1387.
Kolb, E. W. and Wolfram, S. (1980), Nucl. Phys. B172, 244.
Kolb, E. W., Colgate, S. and Harvey, J. A. (1982), Phys. Rev. Lett. 49, 1373.
Kolm, H. H. *et al.* (1971), Phys. Rev. D 4, 1285.
Langacker, P. (1981), Phys. Rep. 72C, 185.
Langacker, P. and Pi, S.-Y. (1980), Phys. Rev. Lett. 45, 1.
LaRue, G. S., Fairbank, W. M., and Hebard, A. F. (1977), Phys. Rev. Lett. 38, 1011.
LaRue, G. S., Fairbank, W. M., and Phillips, J. D. (1979), Phys. Rev. Lett. 42, 142, 1019(E); (1981) Phys. Rev. Lett. 46, 967.
Lazarides, G. and Shafi, Q. (1980), Phys. Lett. 94B, 149.
Lazarides, G., Shafi, Q., and Walsh, T. F. (1981), Phys. Lett. 100B, 21.
Lazarides, G., Shafi, Q., and Trower, W. P. (1982), Phys. Rev. Lett. 49, 1756.
Linde, A. D. (1980), Phys. Lett. 96B, 293.
Linde, A. D. (1982a), Phys. Lett. 108B, 389; 114B, 431.
Linde, A. D. (1982b), these Proceedings.
Pantalone, J. (1982), Cornell preprint.
Parker, E. N. (1970), Ap. J. 160, 383.
Parker, E. N. (1979), *Cosmical Magnetic Fields*. Oxford: Clarendon.
Peebles, P. J. E. (1971), *Physical Cosmology*. Princeton: Princeton University Press.
Polyakov, A. M. (1974), JETP Lett. 20, 194.
Preskill, J. (1979), Phys. Rev. Lett. 43, 1365.
Preskill, J. (1982), Harvard preprint HUTP-82/A059. To appear in *Proceedings of the Magnetic Monopole Workshop*, eds. R. A. Carrigan and W. P. Trower. New York: Plenum.
Price, P. B. *et al.* (1978), Phys. Rev. D 18, 1382.
Rubakov, V. (1981), JETP Lett. 33, 644; (1982), Nucl. Phys. B203, 311.
Salpeter, E. E., Shapiro, S. L., and Wasserman, I. (1982), Phys. Rev. Lett. 49, 1114.
Schwinger, J. (1966), Phys. Rev. 144, 1087.
Steinhardt, P. J. (1982), these Proceedings.
Strominger, A. E. (1982), IAS preprint.
't Hooft, G. (1974), Nucl. Phys. B79, 276.
Turner, M. S. (1982), Phys. Lett. 115B, 95.
Turner, M. S., Parker, E. N. and Bogdan, T. J. (1982), Phys. Rev. D 26, 1296.
Weinberg, S. (1972), *Gravitation and Cosmology*. New York: Wiley.
Weinberg, E. J. (1982), in *Particles and Fields - 1981: Testing the Standard Model*, eds. C. A. Heusch and W. T. Kirk. New York: A.I.P.
Wilczek, F. (1982), Phys. Rev. Lett. 48, 1146.
Witten, E. (1979a), Nucl. Phys. B160, 57.

Witten, E. (1979b), Phys. Lett. 86B, 283.
Zeldovich, Ya. B. and Khlopov, M. Y. (1978), Phys. Lett. 79B, 239.
Zwanziger, D. (1968), Phys. Rev. 176, 1480, 1489.

STRINGS IN THE VERY EARLY UNIVERSE

Q. SHAFI

International Centre for Theoretical Physics, Trieste, Italy.

Abstract. Topologically stable strings with mass per unit length characterized by a superheavy scale are predicted by some grand unified theories. The presence of these strings in the very early universe would cause density perturbations $\frac{\delta\rho}{\rho} \sim 10^{-3} - 10^{-4}$. Cosmic strings therefore might form the basis of a theory of galaxy formation.

Introduction.

About two years ago Zeldovich put forward the suggestion that the presence of stringlike structures in the very early universe might provide the basis of a theory of galaxy formation. The basic idea underlying Zeldovich's suggestion may be understood from the following simple argument. Consider a region of size t^3 where t denotes the cosmic time, say in the radiation dominated era, through which passes a string. Let $\sigma \sim M^2$ denote the string mass per unit length and ρ_s/ρ_r the ratio of the energy densities of the string and the radiation. Then,

$$\left(\frac{\rho_s}{\rho_r}\right)_t \sim \frac{o(10)t^{-2}\sigma}{\frac{1}{Gt}2} \sim o(10) \left(\frac{M}{M_{PL}}\right)^2 \qquad (1)$$

where $G = M_{PL}^{-2}$ is Newton's constant. For $M \sim 10^{17}$GeV, $(\rho_s/\rho_r)_t \sim 10^{-3}$, which is about the right order of magnitude for the density perturbation $\delta\rho/\rho$ that is needed to start the condensation process.

In Zeldovich's scenario the effect of closed loops is assumed to be unimportant. According to Vilenkin, the presence of long lived closed loops produces a significant density contrast. For instance, in the radiation dominated era one finds (Vilenkin 1981 a),

$$\left(\frac{\rho_{loops}}{\rho_r}\right)_t \sim (G\sigma)^{\frac{1}{2}} \sim \left(\frac{M}{M_{PL}}\right) \sim 10^{-3} \text{ for } M \sim 10^{16} \text{ GeV} \qquad (2)$$

Whichever of these two scenarios is assumed, the desired mass scale for the strings (10^{17} or 10^{16}GeV) is not available in the minimal grand unified (GUT) model based on SU(5) (Georgi & Glashow 1974). This is just as well (!) since the SU(5) model anyway cannot deliver the stringlike structures. The simplest GUT model which produces topologically stable strings that might form the basis of a theory of galaxy formation is based on Spin (10) (Georgi 1975; Fritzsch & Minkowski 1975). Before proceeding to a detailed discussion of stringlike structures in Spin (10) (sometimes denoted as SO(10)), let us briefly state the criterion (Kibble 1976) which establishes whether or not topologically stable strings appear in GUTs.

Suppose that a GUT symmetry G is spontaneously broken to its subgroup H by the expectation value $<\phi>$ of an Higgs field. The vacuum manifold is then identified with the quotient space G/H of the gauge group G by its unbroken subgroup H. The existence of topologically stable stringlike structures is determined by the first homotopy group $\pi_1(G/H)$. The elements of this group are homotopy classes of closed curves that start and finish at some designated point $<\Phi>_0$ in G/H. If G is simply connected [Spin (10) is],

$$\Pi_1(G/H) = \Pi_0(H)$$

Thus, topologically stable strings appear if H has disconnected pieces.

SO(10) GUT Model

The fermions (quarks and leptons) of a single family are assigned to the 16 dimensional spinor represention of SO(10). The breaking of SO(10) to the low energy symmetry SU(3) x SU(2) x U(1) can proceed in various ways. We shall consider in detail the following symmetry breaking patterns:

(i) SO(10) $\longrightarrow$ SU(4) x SU(2) x SU(2) $\longrightarrow$ SU(3) x SU(2) x U(1)

(ii) SO(10) $\longrightarrow$ SU(5) $\longrightarrow$ SU(3) x SU(2) x U(1)

The predictions for $\sin^2\theta_W$ and proton lifetime that follow from these two symmetry breaking chains are in accord with present day experiments. Both chains also predict the existence of topologically stable strings (Kibble et. al., 1982 a,b). However, as we shall see, only the second chain leads to strings that have all the required properties.

Symmetry breaking via SU(4) x SU(2) x SU(2) $\simeq$ SO(6) x SO(4)

$$SO(10) \xrightarrow[M_x \sim 10^{15}-10^{16}\,GeV]{\underset{\sim}{54}} SO(6)\times SO(4) \xrightarrow[M \sim 10^{13}\,GeV]{\underset{\sim}{126}} SU(3)\times SU(2)\times U(1)$$

The expectation value of the Higgs 54-plet (second rank symmetric traceless tensor) takes the form

$$\langle\phi_{54}\rangle = y\ \mathrm{diag}\ (\underbrace{2, \ldots, 2}_{6\ \text{times}}, \underbrace{-3, \ldots, -3}_{4\ \text{times}}) \qquad (3)$$

It is certainly left invariant by the subgroup H_o = Spin (6) x Spin (4) of Spin (10). Let $T_{ij} = \sigma_{ij}/2 = [\Gamma_i, \Gamma_j]/4i$ $(i,j = 1, \ldots, 10)$ be the generators of Spin (10), where Γ_i $(i = 1, \ldots, 10)$ are the generalized Dirac matrices in 10 dimensions. The subgroup H_o of Spin (10) is then generated by T_{ab} $(1 \leq a,b \leq 6)$ and $T_{\alpha\beta}$ $(7 \leq \alpha,\beta \leq 10)$. Note that H_o is not a global direct product of Spin(6) and Spin (4) since these groups have the following non trivial element in common

$$\exp\,(2\pi i\, T_{ab}) = \exp\,(2\pi i\, T_{\alpha\beta}) = -1$$
$$(1 \leq a\,,b \leq 6,\ 7 \leq \alpha,\ \ \beta \leq 10) \qquad (4)$$

This leads to the appearance of topologically stable Z_2 magnetic monopoles (Lazarides et. al. 1980).

A closer examination reveals that the stability group of $\langle\phi_{54}\rangle$ is bigger than Spin (6) x Spin (4). This may be seen as follows. Consider the following element of Spin (10)

$$\exp\,(i\theta\, T_{a\alpha}),\ 1 \leq a \leq 6,\ \ 7 \leq \alpha \leq 10, \qquad (5)$$

which represents a rotation by angle θ in the $a\alpha$ plane. The $a\alpha$ submatrix of $\langle\phi_{54}\rangle$ transforms under this rotation as follows:

$$\begin{pmatrix} 2 & 0 \\ 0 & -3 \end{pmatrix} \longrightarrow \begin{pmatrix} -\frac{1}{2} + \frac{5}{2}\cos 2\theta & -\frac{5}{2}\sin 2\theta \\ -\frac{5}{2}\sin 2\theta & -\frac{1}{2} - \frac{5}{2}\cos 2\theta \end{pmatrix} \qquad (6)$$

We see that $\langle\phi_{54}\rangle$ is left invariant under

$$\exp(i\pi n T_{a\alpha}),\ n \in Z. \qquad (7)$$

It is easy to verify that successive rotations in various $a\alpha$ planes that leave $\langle\phi_{54}\rangle$ invariant can always be written as products of group elements from (7). Since $\exp\,(2\pi i T_{a\alpha}) = -1 \in H_o$ = Spin (6) x Spin (4) and

$$\exp(i\pi T_{a\alpha})\exp(i\pi T_{\alpha\beta}) = \exp(i\pi T_{a\beta}),\ \alpha \neq \beta,$$

$$\exp(i\pi T_{ba})\exp(i\pi T_{a\alpha}) = \exp(i\pi T_{b\alpha}),\ a \neq b, \tag{8}$$

all the group elements in (7) can be reduced, by transformations that belong to H_o, to the single group element

$$\exp(i\pi T_{67}) = i\sigma_{67} \notin H_o. \tag{9}$$

Note that $(i\sigma_{67})(i\sigma_{23}) = C$, the charge conjugation operator. Now $i\sigma_{67}$ anticommutes with $i\sigma_{a6}$ ($a = 1, 2, \ldots, 5$) and with $i\sigma_{7\alpha}$ ($\alpha = 8,9,10$), and it commutes with all other $i\sigma$'s. It then follows that, for every $g \varepsilon H_o$, there exists a $g' \varepsilon H_o$ such that

$$g\,(i\sigma_{67}) = (i\sigma_{67})\,g'. \tag{10}$$

Consequently, the stability group H of $\langle\phi_{54}\rangle$ contains not only the elements of H_o but also the set of elements

$$K = [g(i\sigma_{67}),\ g\ \varepsilon\ H_o] \tag{11}$$

Since H_o and K have no elements in common, H is not a connected group. It consists of two connected components, H_o and K.

The fact that H is not connected implies the appearance of topologically stable Z_2 strings after the first step in the symmetry breaking of Spin (10). This is because

$$\Pi_1\,(\text{Spin}\,(10)/H) = \Pi_o\,(H) = Z_2. \tag{12}$$

Note that these strings are unrelated to the Z_2 monopoles that are also produced in the first stage of the symmetry breaking.

Consider an open string along the z-axis. Its thickness is of order M_x^{-1} and it has mass per unit length σ of order M_x^2/α_G. The expectation value $\langle\phi_{54}\rangle$ vanishes along the z-axis. For $r = (x^2+y^2)^{1/2} \gg M_x^{-1}$, the Higgs 54 lies in the vacuum manifold and may be taken in the form

$$\langle\phi_{54}\rangle\,(\psi) = \exp\left[\frac{i\psi}{2}\,(T_{23} + T_{67})\right]\langle\phi_{54}\rangle\,(\psi = 0) \tag{13}$$

where ψ is the azimuthal angle. As one moves around the string, the expectation value of the Higgs 54-plet gets transformed by a gauge transformation which continuously interpolates between the identity element and the charge conjugation operator C.

At the second stage of symmetry breaking the 126 (which is a fifth rank antisymmetric tensor) acquires a vacuum expectation value which breaks H down to SU(3) x SU(2) x U(1) (A Z_2 subgroup of H_o is also left unbroken by $<\theta_{126}>$. This leads to the formation of new Z_2 strings which are unrelated to the Z_2 strings that we previously considered. These new strings are too light to cause significant density perturbations). This transition also breaks the discrete charge conjugation symmetry and has a profound effect on the Z_2 strings that we previously discussed. The strings become boundaries of domain walls that separate vacua related by charge conjugation. This may be seen as follows. Consider a string that lies along the z-axis. Assume that $<\theta_{126}>$ lies along the $(\overline{10}, 1, 3)$ direction on the semi-infinite plane $y = 0^+$, $x \gtrsim 0$ ($\psi = 0$). Then

$$<\theta_{126}>(\psi) = \exp\left[\frac{i\psi}{2}(T_{23} + T_{67})\right]<\theta_{126}>(\psi=0),\quad 0 \lesssim \psi \lesssim 2\pi \qquad (14)$$

which means that $<\theta_{126}>$ ($\psi = 2\pi$) lies along the (10, 3, 1) direction, which is the charge conjugate of the $(\overline{10},1,3)$ direction. Thus in contrast to $<\phi_{54}>$, the expectation value of the Higgs <u>126</u> does not return to its original value after a full rotation around the string. This implies the existence of a physical domain wall along the $y = 0$, $x \gtrsim 0$ semi-infinite plane. The wall is bounded by the string along the z-axis and separates vacua related by charge conjugation. Its thickness is of order M_R^{-1} and its mass per unit area μ is of order M_R^3/α_G. Note that $<\theta_{126}> = 0$ on the mid plane of the wall $y = 0$, $x \gtrsim 0$ (It should be kept in mind that the breaking of the discrete charge conjugation symmetry may also produce domain walls without boundaries).

The domain walls, with or without boundaries, are not topologically stable. This is because charge conjugation symmetry is part of Spin (10) and can be continuously connected to the identity in the full Spin (10) space. Quantum and/or thermal fluctuations can destroy the walls locally through the creation of holes bounded by Z_2 strings. However, we now show that the probability of hole creation is utterly negligible so that the walls can be taken to be locally stable.

Consider a plane wall along the xy plane at zero temperature. The probability per unit area per unit time to create a hole (bounded by string) on the wall due to quantum fluctuations is given (Coleman, 1977) in the semi-classical approximation by

$$\frac{dP}{dAdt} \sim \frac{M_R^3}{\alpha_G} \exp(-S_o) \qquad (15)$$

Here S_o is the Euclidean action of the bounce solution which corresponds to this tunnelling process. The creation and subsequent disappearance of the hole causes the action to change by

$$S = \Sigma\sigma - V\mu$$

where Σ is the invariant area of the closed world surface described by the string and V is the invariant three volume described by the hole. Minimization of S can be achieved by a spherical world surface of invariant radius $R_o = \frac{2\sigma}{\mu}$. The corresponding change in the action is then

$$S_o = \frac{16\pi\sigma^3}{3\mu^2} \sim \left(\frac{M_x}{M_R}\right)^6 \sim 10^{10} \tag{17}$$

Thus,

$$\frac{dP}{dAdt} \sim \frac{M_R^3}{\alpha_G} \exp(-10^{10}), \tag{18}$$

which shows that the probability to create a hole, even on a wall the size of our present Universe, within the age of our Universe is utterly negligible.

The bounce solution considered above also describes the hole creation process at non zero temperatures, provided $T \lesssim R^{-1}$. For $T \gg R_o^{-1}$ (relevant just after the wall producing transition at $T_c \sim M_R/g$), one has (Linde 1977)

$$\frac{dP}{dAdt} \sim \frac{M_R^3}{\alpha_G} \exp(-E_o/T), \tag{19}$$

where E_o, in our case, is the change in energy due to the presence of a static hole of radius $(1/2)R_o$ bounded by a Z_2 string. One finds that

$$\frac{E_o}{T} \sim \left(\frac{M_x}{M_R}\right)^3 \frac{M_x}{T} \gtrsim 10^6 \tag{20}$$

This shows that the probability for hole creation, just after the phase transition that produces the walls, is extremely small. For all practical purposes, the walls are locally stable.

The extended structures we have been considering presumably can be produced in the very early Universe. As the Universe expands and cools after the Big Bang, it undergoes a series of phase transitions. For $T \gtrsim M_x/g$ the Spin (10) gauge symmetry is unbroken. For $T \sim M_x/g$, Spin (10) breaks down to H and a network of topologically stable Z_2 strings is produced. The scale of the string network grows (Kibble 1976) with time and becomes of order t at $t \sim t_*^s$, where

$$t_*^s \simeq \frac{\alpha_G \pi^2}{333} \frac{M_p^2}{M_x^3} \simeq 7.7 \times 10^{-36} \text{ sec.} \tag{21}$$

Here $M_P = 1.2 \times 10^{19}$ GeV is the Planck mass. For $t \gtrsim t_*^s$, friction due to the relativistic gas of particles can be neglected, the strings move with relativistic velocities and the scale of the string network is essentially given by the size of the horizon (also see Everett, 1981).

Let us next discuss the phase transition from H to SU(3) x SU(2) x U(1). Suppose that the phase transition takes place without supercooling at some critical temperature $T_c \sim M_R/g$. For definiteness, we take $T_c \simeq 5 \times 10^{13}$ GeV which corresponds to cosmic time $t_c \sim 6 \times 10^{-35}$ sec. Since $t_c > t_*^s$, friction effects on the strings are negligible and the scale of the string network at this phase transition is of order t_c. The phase transition leads to the production of domain walls which terminate on the strings.

The most optimistic assumption one can make is that the production of the walls at t_c forces the string network to rearrange itself in such a way that one ends up with one wall per horizon volume t_c^3, bounded by a closed string of length $\sim t_c$ (closed walls of size $\lesssim t_c$ may also be produced). The walls at t_c are to be thought of as random surfaces with persistence area $\ell_c^2 \gtrsim (g M_R)^{-2}$. The total area of a typical wall is $\sim t_c^3/\ell_c$, so that the energy density in the walls at t_c is given by

$$\rho_w (t_c) \simeq \frac{M_R^3}{\alpha_G \ell_c} \tag{22}$$

This is smaller than the radiation density $\rho_r (t_c)$.

The persistence area ℓ^2 grows rapidly immediately after the phase transition and ρ_w/ρ_r drops sharply. For $t >> t_c$, the growth of ℓ^2 is given by (Kibble 1976)

$$\ell^2 \simeq t^2 \left(\frac{t}{t_*^W}\right) \tag{23}$$

where

$$t_*^W \simeq \frac{\alpha_G \pi^2 \eta_o}{333\, \eta_3} \frac{M_P^2}{M_R^3} \simeq 7.2 \times 10^{-32} \text{ sec} \tag{24}$$

$\eta_o = 11.25$ is the effective number of massless degrees of freedom which are reflected by the walls and $\eta_3 = 106.75$ is the effective number of massless degrees of freedom in the SU(3) x SU(2) x U(1) phase. The size of the wall boundary d grows as $t^{1/2}$ while the persistence area ℓ^2 becomes of order d^2 at $t = t_f \underset{-}{\sim} (t_*^W\, t_c)^{1/2} \simeq 2 \times 10^{-33}$ sec. For $t_c << t \lesssim t_f$,

$$\frac{\rho_w}{\rho_r}(t) \underset{-}{\sim} \frac{333}{\alpha_G \pi^2} \left(\frac{M_R}{M_P}\right)^3 \left(\frac{t_*^W\, t}{t_P^2}\right)^{\frac{1}{2}} \tag{25}$$

At $t = t_f$, the size of a typical wall is equal to $d_f = (t_f t_c)^{\frac{1}{2}} \simeq 3.5 \times 10^{-34}$ sec and the ratio of the wall mass to the string mass is given by

$$\left(\frac{M_R^3}{M_X^2}\right) d_f \simeq 12 \tag{26}$$

Clearly, the wall mass dominates over the string mass.

Due to friction from the surrounding medium, walls of size d and curvature $\sim d^{-1}$ acquire a limiting velocity $v_\ell \tilde{}\, t_d/d$, where $t_d \simeq t^2/t_*^W$ is the viscous dissipation time. At $t = t_f$, $v_\ell t_f \sim d_f$ and the walls can contract to a point within one expansion time. The energy loss due to friction during such a contraction of a typical wall is $\Delta E \sim F d_f$, where

$$F \sim v_\ell\, \rho_o\, d_f^2 \tag{27}$$

is the force of friction on the wall. Here

$$\rho_o = \frac{\pi^2}{30} \eta_o T^4 \tag{28}$$

is the radiation energy density of the massless particle species which are reflected by the walls. It is easy to see, from Eqs.(24),(27) and (28), that

$$\Delta E \sim \frac{M_R^3}{\alpha_G} d_f^2 \tag{29}$$

This means that the walls lose their energy by friction and disappear at a cosmic time $t \sim t_f$. From Eq. (25), we see that the Universe was never wall dominated. Therefore, in the case where the transition from H to SU(3) x SU(2) x U(1) takes place without supercooling, the presence of the walls bounded by strings does not seem to affect standard cosmology in any essential way. (See also Everett & Vilenkin, 1982 a).

Matters may become worse if the phase transition from H to SU(3) x SU(2) x U(1) proceeds only after supercooling by two or more orders of magnitude. The walls in this case never enter the horizon. They become gravitationally unstable and collapse into long lived black holes that dominate the energy density of the Universe, thereby destroying standard cosmology. In the model we have been discussing, this scenario occurs for $T_c' \lesssim 1.4 \times 10^{11}$ GeV, where T_c' is the temperature at which the transition from H to SU(3) x SU(2) x U(1) takes place. This temperature corresponds to cosmic time $t_c' \gtrsim 1.7 \times 10^{-33}$ sec. For $T < T_E \simeq 3.2 \times 10^{13}$ GeV ($t > t_E \simeq 1.5 \times 10^{-34}$ sec.), the vacuum energy density of the phase in which H is unbroken dominates over the radiation energy density and the Universe undergoes an exponential expansion. As a consequence one finds that the scale d_c' of the string network at t_c' is much greater than t_c'. In particular, for $T_c' \lesssim 1.4 \times 10^{11}$ GeV, $d_c' \gtrsim 3.4 \times 10^{-32}$ sec. We assume that the walls produced at t_c' terminate on closed strings of length d_c' and therefore extend over several t_c' horizons. For $t > t_c'$, the length d of the boundary of a wall follows the cosmological expansion. Thus,

$$d \simeq d_c' \left(\frac{t}{t_c'}\right)^{1/2} \tag{30}$$

The walls do not enter the horizon at least till the cosmic time $t_H \simeq 6.8 \times 10^{-31}$ sec. Since $t_H > t_*^W$, the persistence area of the walls at t_H is $\sim t_H^2$, and there is effectively one wall piece per horizon. This gives

$$\frac{\rho_W}{\rho_r}(t_H) = \frac{333}{\alpha_G \pi^2}\left(\frac{M_R}{M_P}\right)^3 \frac{t_H}{t_P} = 1. \tag{31}$$

For $t > t_H$, the Universe becomes wall dominated. The walls are conformally stretched and one finds that

$$\rho_W \propto a^{-1}, \quad a \propto t^2, \tag{32}$$

where a is the cosmic scale factor. The horizon grows as

$$a(t)\int_{t_H}^{t} \frac{dt'}{a(t')} \sim t^2/t_H. \tag{33}$$

Since the cosmic expansion keeps up with the horizon, the walls never enter the horizon and the Universe remains wall dominated. At $t \simeq 4\times10^{-30}$ sec. the wall piece within the horizon enters its Schwarzschild radius and collapse into black holes starts (Vilenkin 1981 b) . These black holes have mass greater than 5.6×10^{33} GeV and their lifetime $\tau \gtrsim 10^{4}$ sec. Clearly, they lead to a cosmology totally different from ours.

To summarize, the breaking of SO(10) via SU(4) x SU(2) x SU(2) produces extended structures consisting of walls bounded by strings. It also produces Z_2 strings whose mass scale is characterized by 10^{13} GeV. None of these structures possess the properties needed to produce $\delta\rho/\rho \sim 10^{-3}$ in the very early universe and so we turn next to the discussion of the breaking of SO(10) via SU(5).

SYMMETRY BREAKING VIA SU(5)

Consider the symmetry breaking pattern

$$SO(10) \xrightarrow[\underset{\sim}{126}]{M_S} H_1 \xrightarrow[\underset{\sim}{45}]{M_x} H_2$$

where H_1 and H_2 are locally isomorphic to SU(5) and SU(3) x SU(2) x U(1) respectively. The scale M_s can be of order 10^{16} to 10^{17} GeV, and $M_x \sim 10^{14}$ GeV. Almost the same symmetry breaking scheme can be achieved in a more economical way using a Higgs field in the 16 representation in the first stage. In that case, H_1 = SU(5) globally, so that neither strings nor monopoles are produced in the first transition.

However, in the scheme shown H_1 is not isomorphic to SU(5). Instead (Kibble et al. 1982a,b)

$$H_1 = SU(5) \times Z_2$$

where $Z_2 = [1, -1]$. This easily follows if one notes that the rotation through 2π is not contained in SU(5). Thus

$$\Pi_o\ (H_1) = Z_2$$

so that topologically stable strings with mass scale characterized by M_s are formed in the phase transition from SO(10) to SU(5).

The major difference between this and the previous case is that this Z_2 symmetry remains unbroken in the two subsequent transitions. This is because all the Higgs fields belong to single-valued representations of SO(10). Thus the strings remain stable and survive the sequence of transitions. Moreover no domain walls appear.

This is a particularly attractive scenario. It allows for the appearance of stable strings which can generate density perturbations $\delta\rho/\rho \sim 10^{-3} - 10^{-4}$ required for galaxy formation, without at the same time leading to domain walls whose presence would be liable to conflict with standard cosmology. The evolution of such strings in the very early universe is discussed in Vilenkin's talk.

More on Strings

We wish to furnish some more examples of string producing theories in order to display the variety of strings that can occur.

(a) Strings connected to Monopoles

Consider the following breaking of SO(10) to SU(5) (Lazarides et. al. 1982 a)

$$SO(10) \xrightarrow[M_1]{\underset{\sim}{45}} SU(5) \times U(1) \xrightarrow[M_2]{\underset{\sim}{126}} SU(5)$$

Monopoles will be produced at the first stage of symmetry breaking because an U(1) factor appears. The monopoles carry SU(5) and U(1) magnetic fields. At the second stage the U(1) symmetry is broken. The U(1) magnetic flux is confined to strings of radius $\sim M_2^{-1}$. The flux strings usually run from a monopole to an antimonopole. However, it is also possible to form closed U(1) flux strings from fluctuations of the Higgs $\underset{\sim}{126}$. The presence of monopoles leads to a cutoff in the length of the strings. Preliminary investigations (Lazarides et al 1982 a; Vilenkin 1981 c) suggest that these strings live too short a time to be of any relevance to galaxy formation.

(b) Strings from the Spontaneous Breaking of Global Symmetry:

Consider a Goldstone model described by the Lagrangian

$$L = \partial_\mu \phi^+ \partial_\mu \phi - \frac{h^2}{2} (|\phi|^2 - \eta^2)^2$$

where ϕ is a complex scalar field. The model possesses a global U(1) symmetry which is spontaneously broken. The manifold of degenerate vacuum states is a circle and hence topologically stable strings can occur. If the scale of the string network is of order ξ, the energy per unit length μ of a single string is approximately

$$\mu \sim \eta^2 \ \ell n \ (\xi/\lambda)$$

where $\lambda \sim (h\eta)^{-1}$. With $\eta \sim 10^{15}$ GeV, the presence of such strings in the very early universe would cause density perturbations $\delta\rho/\rho$ of order $10^{-3} - 10^{-4}$ (Everett & Vilenkin 1982 b).

(c) Hybrid Strings

An example of a theory that produces hybrid strings is provided by the SO(10) x U(1) model, in which the centre Z_4 of the SO(10) gauge symmetry is contained in the global U(1) symmetry. Thus, SO(10) x U(1) is defined as the usual direct product of the two groups, but with the proviso that the three pairs of SO(10) x U(1) transformations $[(i\Gamma^o)^{-k}, e^{i\pi k/2}]$ (k=1,2,3) are to be identified with the identity transformation (1, 1). Consider the symmetry breaking

$$SO(10) \xrightarrow{\underset{\sim}{16},\ \underset{\sim}{45}} SU(5)$$

Strings are expected to occur since (Lazarides & Shafi 1982)

$$\Pi_1 \ (SO(10) \times U(1) \ / \ SU(5)) = Z, \ \text{the set of integers.}$$

The generator of Z is represented by the homotopy class of closed curves (g(t), h(t)) ($o \leq t \leq 1$, $g(t) \in SO(10)$, $h(t) \in U(1)$) in SO(10)xU(1) with $g(0) = 1$, $h(0) = 1$, $g(1) = (i\Gamma^o)^{-1} = -i\Gamma^o$, $h(1) = \exp(i\pi/2)$ and

$0 \leq \arg h(t) \leq \pi/2$.

The minimal hybrid string has the following structure. As one moves around a string the expectation values of ϕ_{16} and ϕ_{45} get transformed by a pair of transformations. One of them belongs to the gauge group SU(5) and continuously interpolates between 1 and $-i\Gamma^0$. The other belongs to the global group U(1) and continously interpolates between 1 and $\exp(i\pi/2)$, with its phase ranging between 0 and $\pi/2$.

(d) Supersymmetric Strings

A simple example of this is provided by the supersymmetric SO(10) GUT in which the SO(10) gauge symmetry breaks to SU(3) x SU(2) x U(1) with supersymmetry unbroken. Topologically stable supersymmetric strings will be produced. The mass scale associated with these strings can be of order $10^{16} - 10^{-17}$ GeV as required.

Remarks on Closed Loops

We mentioned earlier on that the presence of closed loops with long lifetimes can produce density perturbations $\delta\rho/_{\rho} \sim 10^{-3} - 10^{-4}$. Thus, it is important to study the dynamics of closed loops in order to see whether or not they can play an important role in galaxy formation. A start in this direction has recently been made (Kibble & Turok 1982) . If one assumes that the only significant energy loss mechanism is gravitational radiation the closed loops can survive for a long time. Typically, a loop of length ℓ will have a lifetime $\tau \sim 10^8 \ell$. Thus, at some cosmic time t there will be a certain number of loops surviving from earlier epochs. (It is possible, although not very likely since they do not carry any net charges, that loops may disappear more quickly by emitting other types of radiation).

A closed loop may from time to time intersect itself and break into smaller loops, whose lifetime would be correspondingly shorter. This could reduce, in principle, the number of closed loops to a level well below that required for significant density perturbations. The complete answer to this problem is not known. However, the following result has recently been obtained (Kibble & Turok 1982). Any initially static string will collapse after half an oscillation period to a linear configuration. On the other hand there exist solutions, differing infinitesimally from these, in which the strings never self-intersect. What needs to be shown now is that a reasonable proportion of loops formed by self-intersection of long strings can be in a self-avoiding configuration. If this can be achieved then it seems plausible that closed loops will play an essential role in the string theory of galaxy formation.

Summary and Conclusions

Spontaneously broken gauge theories can produce a variety of stringlike structures. Thus, one could have open strings, closed strings, strings connected to domain walls, strings connected to monopoles, etc. Superheavy strings occur in realistic GUT models of which the SO(10) model provides the simplest example. The presence of open and possibly also closed strings in the very early universe would cause density perturbations $\delta\rho/\rho \sim 10^{-3} - 10^{-4}$ in an otherwise homogeneous and isotropic universe. It seems plausible that strings might form the basis of the theory of galaxy formulation.

Acknowledgement

I thank Gary Gibbons, Steve Hawking, Steven Siklos and the Nuffield Foundation for offering me the kind hospitality of the Nuffield Workshop on the Very Early Universe.

References

Coleman, S. (1977), Phys. Rev. D15, 2929

Everett, A. (1981), Phys. Rev. D24, 858

Everett, A. & Vilenkin, A. (1982 a), Nucl. Phys.B207, 43

Everett, A. & Vilenkin, A. (1982 b), Tufts University Preprint

Fritzsch, H. & Minkowski, P. (1975), Ann Phys. 93, 193

Georgi, H. & Glashow, S.L. (1974), Phys. Rev. Lett. 32, 438

Georgi, H. Particles and Fields (1975), C.E. Carlson,(Ed.,ATP) (for earlier work on grand unification, see J. Pati & A.Salam)

Kibble, T. (1976), J. Phys. A9, 1387

Kibble, T., Lazarides, G. & Shafi, Q. (1982 a), Phys. Lett. B

Kibble, T., Lazarides, G. & Shafi, Q. (1982 b), Phys. Rev.D26 435

Kibble, T. & Turok, N. (1982), Imperial College Preprint

Lazarides, G., Magg, M. & Shafi, Q. (1980), Phys. Lett.97B, 87

Lazarides, G., Shafi, Q. & Walsh, T. (1982 a), Nucl. Phys.B195 157

Lazarides, G. & Shafi, Q. (1982), Phys. Lett. 115B, 21

Linde, A. (1977), Phys. Lett. 70B, 306

Pati, J. & Abdus Salam, (1973), Phys. Rev. Lett. 31, 661

Vilenkin, A. (1981 a), Phys. Rev. Lett. 46, 17

Vilenkin, A. (1981 b), Phys. Rev. D23, 852

Vilenkin, A. (1981 c), Tufts University Preprint

Zeldovich, Y. (1980), Mon. Not. R. Astr. Soc. 192, 663

COSMIC STRINGS

A. Vilenkin
Physics Department, Tufts University
Medford, MA 02155

In this talk I would like to discuss the evolution and the cosmological consequences of vacuum strings which could be produced at a phase transition in the early universe. Qaisar Shafi has discussed various types of strings which can occur in grand unified models. Here we shall be concerned only with topologically stable strings; that is, it will be assumed that the strings cannot break producing monopole-antimonopole pairs at the free ends. Such strings can be either infinite or closed.

Suppose that the strings are formed at a phase transition when a Higgs field ϕ develops a vacuum expectation value, $\langle\phi\rangle = \eta$. Then the linear mass density of the strings is $\mu \sim \eta^2$. The force of tension (per unit length) in a curved string of curvature radius R is $F \sim \mu/R$. At the time of formation ($t = t_o$) the strings have the shape of Brownian trajectories with a persistence length not exceeding t_o. The dynamics of convoluted strings in expanding universes has been discussed by Kibble (1976, 1980, 1981) and Vilenkin (1981b,c). Let us consider a straight string with perturbations of amplitude A and wavelength λ. In flat space-time such perturbations result in oscillations of constant amplitude and period $\sim \lambda$. In expanding universe λ grows proportionally to the scale factor, $\lambda \propto a(t)$. Perturbations of wavelength greater than the horizon ($\lambda > t$) have not started oscillating yet. The amplitude of such perturbations also grows like a(t), so that the whole string is just stretched by the expansion, its shape remaining unchanged. For $\lambda << t$ the amplitude of the oscillations remains constant (Vilenkin 1981b), the curvature radius $R \sim \lambda^2/A$ grows like $a^2(t)$, and thus the perturbations are smoothed out. (This effect is similar to the red shift of photons.) If we assume that large perturbations behave in a similar manner, then the strings are conformally stretched by the expansion on scales greater than the horizon and are straightened out on scales smaller than the horizon. The persistence length of the strings at time t is $\sim t$. The typical distance between adjacent strings is of the same order. (Here we disregard closed loops smaller than the horizon; see below.)

So far, we have neglected the interaction of strings with one another and with surrounding matter. It can be shown that friction due to interaction of strings with particles becomes negligible soon after the string formation (Kibble 1976, 1980; Everett 1981). A potentially important process is the intercommuting of intersecting strings which can lead to closed loop formation (Kibble 1976). The intercommuting probability, p, is unknown, and we shall consider various possibilities.

Strings move with relativistic speeds, and there should be about one intersection per horizon volume ($\sim t^3$) per expansion time ($\sim t$). If $p \sim 1$, then we expect about one closed loop of size $\sim t$ to be formed per horizon volume per horizon time. At first sight it appears that for $p \ll 1$ the loop formation is suppressed. However, it will be argued that the number of closed loops produced is independent of p. Suppose we have two string segments of length $\sim t$ intersecting at time $t \gg t_o$. These segments can be parts of infinite strings or of large closed loops; they can also belong to the same string. We first note that our segments are likely to have had a large number of intersections at earlier times. At time t_1 such that $t_o < t_1 \ll t$, our segments had the shape of Brownian trajectories with persistence length $\sim t_1$, end-to-end distance $d \sim (t_1 t)^{\frac{1}{2}}$ and total length $\ell \sim d^2/t_1 \sim t$, and were confined to a volume $\sim d^3$. If we divide this volume into cubes of size $\sim t_1$, then an intersection of the two strings occurs at time $\sim t_1$ with probability ~ 0.5 whenever the two strings pass through the same cube. Each of the two segments belongs to $\sim t/t_1$ cubes, and we expect $\sim(t/t_1)^{\frac{1}{2}} \gg 1$ intersections. If $p(t/t_1)^{\frac{1}{2}} \gtrsim 1$, the probability of intercommuting is large. Even if $p(t/t_1)^{\frac{1}{2}} \ll 1$, we can go back to times $\sim t_o$ when the strings are "soft", fluctuations of the Higgs expectation value $\langle\phi\rangle$ are large and intercommutings between neighboring strings are probable. These considerations lead us to conjecture that, even if the intercommuting is totally suppressed at $t \gg t_o$, the distribution of closed loops at the time of formation is such that at any time $t \gg t_o$ about one loop comes within the horizon per horizon volume per expansion time. Then a simple calculation shows that the loop distribution at $t \sim t_o$ is

$$n_d \, \Delta d \sim d^{-4} \, \Delta d, \tag{1}$$

where $n_d \, \Delta d$ is the number of loops of size $d \pm \Delta d$ per unit volume. This equation applies only for values of d greater than the persistence length of the strings, ξ. (Note that the size of a loop, d, is different from its length. d can be defined as the diameter of the smallest sphere containing the loop.) A nice property of the distribution (1) is its scale invariance. If we smooth out all irregularities of the strings on scales smaller than $R > \xi$ and remove all closed loops of size smaller than R, we would get a system of strings statistically identical to the original one, up to a scale transformation $\xi \rightarrow R$.

The cosmological evolution of closed loops has been discussed by Vilenkin (1981b,c). Loops greater than the horizon are conformally stretched by the expansion, while their small-scale irregularities are smoothed out. When a loop comes within the horizon at time t, it has length $R \sim t$ and starts oscillating with a typical frequency $\omega \sim R^{-1}$. If the intercommuting probability p is large, the loop can rapidly decay as a result of multiple self-intersections, the typical lifetime being $\tau \sim R/pp'$, where p' is the probability of self-intersection per oscillation. The value of p' is unknown, and one could think that p' should be ~ 1. However, Kibble and Turok (1982) have found a large class of loop trajectories which never self-intersect. This suggests that p' may in fact be small and the loops may have long lifetimes even for $p \sim 1$.

For sufficiently small pp' intercommuting is unimportant and the loops can lose energy only by radiation of one kind or another. Loops oscillating with a frequency $\omega \sim R^{-1}$ cannot emit particles of mass greater than R^{-1}. If neutrinos have nonzero masses, then loops greater than, say, 1 cm can emit only electromagnetic or gravitational waves. Strings do not carry any net charges, so the emission of photons must be rather indirect. On the other hand, the mass density of the strings can be very large and the coupling to gravity quite strong. Here we shall assume that the gravitational radiation is dominant. Then (Vilenkin 1981c)

$$dM/dt \sim -GM^2 R^4 \omega^6 \sim - G\mu^2, \tag{2}$$

where $M \sim \mu R$ is the mass of the loop. The lifetime of the loop is

$$\tau \sim (G\mu)^{-1} R. \tag{3}$$

$G\mu$ is an important dimensionless parameter, $G\mu \sim (\eta/m_p)^2 << 1$, where m_p is the Planck mass.

We now turn to the cosmological consequences of the strings. It has been suggested (Zel'dovich 1980, Vilenkin 1981b,c) that the strings can produce density fluctuations sufficient to explain the galaxy formation. On scales greater than the horizon the density fluctuations due to strings are balanced by the corresponding variations of the radiation density; so at any time t the fluctuations can only be produced on scales smaller than t. If $pp' \sim 1$, closed loops are unimportant, and the density of strings is $\rho_s \sim \mu t/t^3 \sim \mu t^{-2}$. The density fluctuations produced on the horizon scale are of the order $(\delta\rho)_H \sim \rho_s \sim \mu t^{-2}$, so that

$$(\delta\rho/\rho)_H \sim 30G\mu, \tag{4}$$

where $\rho = 3/32\pi G t^2$ is the radiation density. The fluctuations are adiabatic and their amplitude is the same for all scales at the time when the corresponding scale comes within the horizon. This is the so-called Zel'dovich spectrum.

Adiabatic fluctuations do not grow until recombination time, which roughly coincides with the time of equal matter and radiation densities, $t_{eq} \sim 10^{13}$ s. For $t > t_{eq}$ the density contrast grows like $t^{2/3}$. For the galaxy formation we need $(\delta\rho/\rho)_H \sim 10^{-3}$, so that $G\mu \sim 3 \times 10^{-5}$ and $\eta \sim 5 \times 10^{16}$ GeV (Zel'dovich 1980). If the universe is dominated by heavy neutrinos, then the fluctuations can start to grow earlier (see, e.g., Silk 1981) and one can have $(\delta\rho/\rho)_H \sim 10^{-4}$, $G\mu \sim 3 \times 10^{-6}$ and $\eta \sim 2 \times 10^{16}$ GeV.

If pp' is small, closed loops are important and can serve as seeds for galaxy formation (Vilenkin 1981c). Galactic scale comes within the horizon at $t_G \sim 10^9$ s. At that time $\delta\rho/\rho \sim 30\ G\mu$ on galactic scales. (Loops smaller than the horizon have a larger total density than loops of size $\sim t$, but give a smaller contribution to the density fluctuations.) The density contrast due to the loops grows like $t^{\frac{1}{2}}$ for $t < t_{eq}$ and like $t^{2/3}$ for $t > t_{eq}$. Galaxies form at $t \sim t_*$ when $30\ G\mu\ (t_{eq}/t_G)^{\frac{1}{2}} (t_*/t_{eq})^{2/3} \sim 1$. Taking $t_* \sim 3 \times 10^{16}$ s we find $G\mu \sim 10^{-6}$, $\eta \sim 10^{16}$ GeV. (This estimation is not very sensitive to the presence of heavy neutrinos;

it seems hard to push $G\mu$ below $\sim 10^{-7}$.) We conclude that strings can be important for galaxy formation only if the corresponding energy scale is $\eta \gtrsim 10^{16}$ GeV. The constraint is weaker for global symmetry strings (Vilenkin & Everett 1982) which have $\mu \sim 2\pi\eta^2 \ell n(\eta t_G)$. With $G\mu \sim 10^{-6}$ this gives $\eta \gtrsim 4 \times 10^{14}$ GeV.

The string theory of galaxy formation predicts that substantial density fluctuations should be absent on scales greater than the horizon. (In fact, there may still be small fluctuations due to surface effects; see Carr 1975, Zel'dovich 1980, and references therein.) This prediction can be tested experimentally by studying the fluctuations in the temperature of the cosmic background radiation on various angular scales. Angular separations greater than a few degrees correspond to scales greater than the horizon at the time of matter-radiation decoupling (see, e.g., Silk 1981).

Is it possible to see the strings? If $pp' \sim 1$, there can only be a few strings in the visible universe. One expects that the strings will gravitationally attract surrounding galaxies and will look like garlands in the sky. However, this is not the case. Neglecting $(G\mu)^2$ the gravitational field of a straight string parallel to the z-axis is described by the metric (Vilenkin 1981a)

$$ds^2 = dt^2 - dz^2 - dr^2 - (1 - 8G\mu)\, r^2 d\phi^2. \tag{5}$$

A coordinate transformation $\phi' = (1 - 4G\mu)\phi$ brings the metric to a locally Minkowskian form. It is clear that massive objects located near the string do not feel its presence at all. However, the space-time around the string is not globally Minkowskian, since the angle ϕ' changes from 0 to $(1 - 4G\mu)2\pi$. Light propagating in the xy-plane is deflected by the angle $4\pi G\mu$, independently of the impact parameter, and we can see double images of objects located behind the string with an angular separation $\delta\phi \leqq 8\pi G\mu$. This is analogous to the Aharonov-Bohm effect (Ford & Vilenkin 1981). With $G\mu \sim 10^{-6}$, $\delta\phi \lesssim 6$ arc sec. Note that the angular separation in double quasars has the same order of magnitude.

If $pp' \ll 1$, then there should be a large number of closed loops in the visible universe. With $pp' < G\mu$, the smallest loops have size $R \sim G\mu t_{pres} \sim 10^{22}$ cm and mass $M \sim \mu R \sim 10^{11} M_\odot$. (Here, $t_{pres} \sim 3 \times 10^{17}$ s is the present cosmic time and I have used $G\mu \sim 10^{-6}$.)

The typical distance between the loops is $\sim 10^{26}$ cm $\sim$ 30 Mpc, and there are about 10^6 such loops in the visible universe. At distances >> R the loops produce a regular Schwarzschild field and can attract surrounding matter.

Another observational consequence of strings is the existence of very long wavelength gravitational radiation emitted by oscillating loops. For $pp' \lesssim G\mu$, the energy spectrum of gravitational waves in the interval $10^{-11}\ s^{-1} \lesssim \omega \lesssim 10^{8}\ s^{-1}$ is (Vilenkin 1981d)

$$d\varepsilon_g \sim 30(G\mu)^{\frac{1}{2}}\ \varepsilon_\gamma\ d\omega/\omega,$$

where ε_γ is the energy density of the electromagnetic background radiation. This is beyond the present experimental capabilities, but not hopelessly so.

This work was supported in part by the National Science Foundation under Grant No. PHY-8206202.

REFERENCES

Carr, B.J. (1975). The primordial black hole mass spectrum. Astrophys. J. 201, 1-19.
Everett, A.E. (1981). Cosmic strings in unified gauge theories. Phys. Rev. D24, 858-68.
Ford, L.H. & Vilenkin, A. (1981). A gravitational analogue of the Aharonov-Bohm effect. J. Phys. A14, 2353-7.
Kibble, T.W.B. (1976). Topology of cosmic domains and strings. J. Phys. A9, 1387-98.
Kibble, T.W.B. (1980). Some implications of a cosmological phase transition. Phys. Rep. 67, 183-99.
Kibble, T.W.B. (1981). Phase transitions in the early universe. In Proceedings of the Workshop on Quantum Gravity, ed. C.J. Isham and M.J. Duff. Cambridge: Cambridge University Press.
Kibble, T.W.B. & Turok, N. (1982). Self-intersection of cosmic strings. Phys. Lett. 116B, 141.
Silk, J. (1981). Fundamental tests of galaxy formation theory. University of California preprint.
Vilenkin, A. (1981a). Gravitational field of vacuum domain walls and strings. Phys. Rev. D23, 852-7.
Vilenkin, A. (1981b). Cosmic strings. Phys. Rev. D24, 2082-9.
Vilenkin, A. (1981c). Cosmological density fluctuations produced by vacuum strings. Phys. Rev. Lett. 46, 1169-72, 1496 (E).
Vilenkin, A. (1981d). Gravitational radiation from cosmic strings. Phys. Lett. 107B, 47-50.
Vilenkin, A. & Everett, A.E. (1982). Cosmic strings and domain walls in models with Goldstone and pseudo-Goldstone bosons. Phys. Rev. Lett. 48, 1867-70.

Zel'dovich, Ya. B. (1980). Cosmological fluctuations produced near a singularity. Mon. Not. R. Astr. Soc. 192, 663-7.

PHASE TRANSITIONS IN THE VERY EARLY UNIVERSE

Alan H. Guth
Massachusetts Institute of Technology
Cambridge, Massachusetts 02139 USA

I INTRODUCTION

Since this talk is supposed to encompass the general subject of phase transitions in the early universe, I have shown in Figure 1 all of the phase transitions which are generally thought to have taken place. The figure shows kT (Boltzmann constant times temperature) on a logarithmic scale, running from the Planck energy ($E_P = (\hbar c^5/G)^{1/2} = 1.22 \times 10^{19}$ GeV) to the current temperature ($k(2.7^\circ K) = 2.3 \times 10^{-13}$ GeV).

Figure 1. Phase transitions in the very early universe

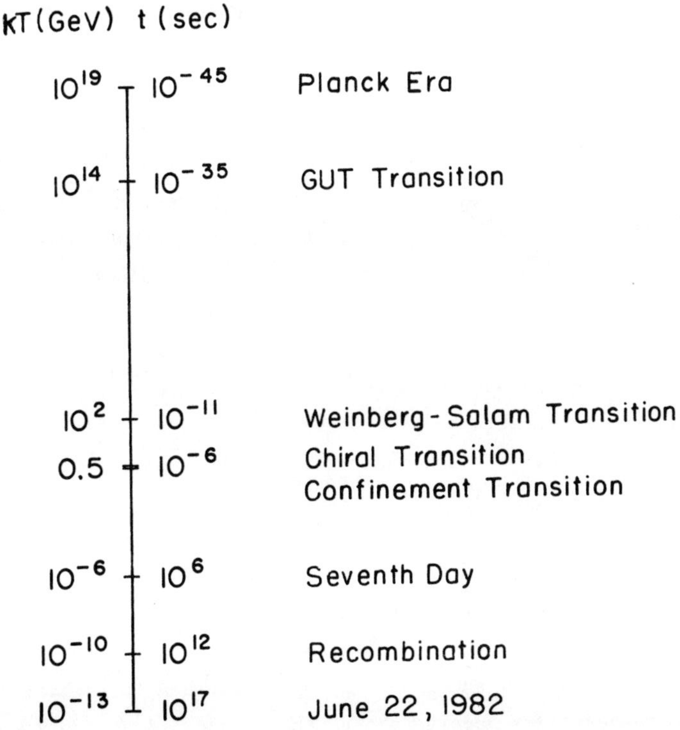

The figure also shows the time t at which these temperatures were reached, as calculated in the standard cosmological scenario.

The Planck era is shrouded by our uncertainties about quantum gravity: there has been discussion of a phase transition in this era (Krauss, 1982), but the subject is highly speculative. In any case, the quantum effects of gravity are expected to be proportional to higher powers of G. By dimensional analysis, these effects would be suppressed by powers of the dimensionless factor $GE^2/\hbar c^5 = (E/E_P)^2$, where E is an energy characteristic of the process under consideration. Thus, once the temperature has fallen several orders of magnitude below the Planck scale, one expects the corrections due to quantum gravity to be negligible.

If grand unified theories (GUTs) provide an accurate picture of physics at energies below the Planck scale, then there will be a phase transition (or possibly several) at $kT \approx 10^{14}$GeV. This appears to be the most interesting of the phase transitions, and it will be the subject of most of my talk.

At kT of order 10^2GeV, one expects a phase transition associated with the Weinberg-Salam-Glashow model of weak and electromagnetic interactions. I will have a few words to say about this transition at the very end of Section 4.

Finally, at kT the order of a few hundred MeV, there are two closely spaced phase transitions associated with quantum chromodynamics (QCD). (For a review of finite temperature QCD, see Gross, Pisarski, & Yaffe, 1981.) The first transition occurs at the onset of the spontaneous breaking of chiral symmetry. Since the chiral symmetry was already weakly broken by the "current algebra" masses of the quarks (which are responsible for giving the pion a nonzero mass), this transition is presumably "rounded" – i.e., thermodynamic quantities will vary rapidly as a function of temperature, but they will not have the mathematical singularities characteristic of true phase transitions. This phase transition has recently been studied numerically in the context of lattice QCD (Kogut et.al., 1982). At a slightly lower temperature is the confinement phase transition: the quarks, which at higher temperature could exist as free particles, now become absolutely confined into colorless bound states. (Note, however, that the particle spectrum in the unconfined phase may contain quark bound states in addition to free quarks.) If quarks are omitted from the theory altogether, then it is believed that the QCD gauge fields undergo a true phase transition (Polyakov,

1978; Susskind, 1979); this transition has been studied numerically in lattice QCD using the Wilson criterion for confinement (Kuti et.al., 1981; McLerran & Svetitsky, 1981; Engels et.al., 1982). However, when dynamical quarks are incorporated into the theory, a precise definition of confinement is lacking. It is therefore quite conceivable that this phase transition is also rounded.

As reference markers, I have included two other events on the timeline. I have marked the beginning of the seventh day, which according to the biblical account was the time of the completion of the world. According to current cosmology, the universe at this time was filled with a plasma with $kT \simeq 1.6$ KeV. I have also marked the time of "recombination", when the plasma neutralized into atoms of hydrogen and helium. It was at this time that the cosmic background radiation decoupled from the matter. Optical observations are capable of probing the universe only at times later than that of recombination.

The timeline ends at June 22, 1982 — the date this talk was given. I terminated it at this time because I did not wish to engage in unnecessary speculation; Frank Wilczek (Wilczek, 1982; Turner & Wilczek, 1982) has informed us that our vacuum might be metastable, and time as we know it might come to an abrupt halt at any moment.

The body of this talk will deal almost exclusively with the GUT phase transition. In Section 2 I review the essential properties of GUTs, placing particular emphasis on those properties which are relevant for cosmology. In Section 3 I describe what may be called the standard scenario of the very early universe; the monopole, horizon, and flatness problems are discussed. The original inflationary universe scenario, which offers a possible solution to these three problems, is described in Section 4. In Section 5 I explain the problems of this scenario. These problems are avoided by the new inflationary scenario, proposed by Linde (1982a) and Albrecht & Steinhardt (1982). This new scenario is discussed in Section 6. I include also a summary of the calculation of density fluctuations in the new inflationary universe. Section 7 contains concluding remarks.

Before going on I would like to simplify my system of units. I will set $\hbar = c = k = 1$, and will take the GeV as my fundamental unit. Note then that 1 GeV = 1.16×10^{13}°K = 1.78×10^{-24} g, and 1 GeV^{-1} = 1.97×10^{-14} cm = 6.58×10^{-25} sec.

2 GRAND UNIFIED THEORIES

In this section I will outline the basic properties of GUTs, placing particular emphasis on those properties which are most relevant for cosmology. For a thorough review of GUTs, see Langacker, 1981.

GUTs unite the SU_3 and $SU_2 \times U_1$ gauge fields associated with the standard low energy theory of strong and electroweak interactions into a single gauge group. The simplest possible unifying group is SU_5 (Georgi & Glashow, 1974). By computing the variation of the three gauge coupling constants with energy (Georgi, Quinn, & Weinberg, 1974; Goldman & Ross, 1979 & 1980), one finds that all three interactions have equal strength at $E \approx 10^{15}$ GeV. Thus, the unifying gauge symmetry must be spontaneously broken (by the Higgs mechanism) at energies of this order.

There are two important predictions of GUTs which lead me to believe that they have a reasonable chance of being valid. The first is the prediction of the equality in magnitude of the electron and proton charges. This equality is known to hold to an accuracy of one part in 10^{20} (King, 1960), but in the standard $SU_3 \times SU_2 \times U_1$ model it is achieved only by an ad hoc fine-tuning of parameters. (Of course, one might easily imagine that other types of theories, such as composite models, would also require this equality.) The second prediction is the value of $\sin^2\Theta_W$ (where Θ_W is the Weinberg angle), which is calculated from the requirement that the three gauge coupling constants are all equal at some energy. The predicted value (see Gasiorowicz & Rosner, 1981, and references therein) is 0.209 ± 0.005, while the experimental value is 0.215 ± 0.012.

There are three consequences of GUTs which are of crucial cosmological importance, and I would like to discuss these one at a time. The first is the nonconservation of baryon number. This means that the proton has a finite lifetime, estimated in the simplest models to lie in the range of 10^{30} to 10^{33} years. It also means that the net baryon number of the universe need not be taken as an initial condition, but can instead be generated dynamically (for a review, see Yoshimura, 1981). Possibly even more important, it means that there may be no conserved quantity which distinguishes the observed universe from the vacuum. (This last point will be discussed further in Section 4.)

The second consequence is the existence of one or more phase transitions when T is of order 10^{14} GeV. To illustrate the nature of these phase transitions, I will use the minimal SU_5 model of Georgi &

Glashow (1974); the main points, however, will be fairly model-independent.

In the SU_5 model, the full gauge symmetry is broken to the subgroup $SU_3 \times SU_2 \times U_1$ by a set of Higgs fields Φ which transform according to the adjoint representation of SU_5. That is, Φ represents a traceless, hermitian 5×5 matrix of fields, with 24 independent components. The symmetry breaking is accomplished by the fields Φ acquiring a vacuum expectation value of the form

$$<\Phi> = \sqrt{\frac{2}{15}}\, \phi_v \,\mathrm{diag}[1,1,1,-3/2,-3/2], \tag{1}$$

or a form which is gauge-equivalent to this. ($\mathrm{diag}[\lambda_1,\lambda_2,\ldots]$ denotes the diagonal matrix with entries $\lambda_1,\lambda_2,\ldots$. The normalization has been fixed so that $\mathrm{Tr}\Phi^2 = \phi_v^2$.) The generators of the unbroken subgroup are the hermitian traceless matrices which commute with the matrix displayed in (1); the matrices with non-zero entries in the upper-left 3×3 block generate an SU_3 subgroup, those with non-zero entries in the lower-right 2×2 block generate an SU_2 subgroup, and the matrix in (1) generates a U_1 subgroup.

The phase transition occurs because this non-zero expectation value is destroyed by thermal fluctuations at high temperatures (Kirzhnits & Linde, 1972; S. Weinberg, 1974; Dolan & Jackiw, 1974; for a review, see Linde, 1979) just as thermal fluctuations destroy the expectation value of the magnetic spin of a ferromagnet at temperatures above the Curie point. Thus, at the highest temperatures,

$$<\Phi> = 0 \,. \tag{2}$$

For a wide range of parameters, there is an intermediate phase (Guth & Tye, 1980) in which

$$<\Phi> = \frac{1}{\sqrt{20}}\, \phi'_v \,\mathrm{diag}[1,1,1,1,-4] \tag{3}$$

Before going on, let me point out a defect in this description: the expectation value of Φ is a gauge-dependent quantity, and it is not clear that the value which it takes when computed with any particular gauge condition has any real physical meaning. It is clear that these phase transitions would take place if the gauge coupling constant were set equal to zero. With a weak gauge coupling (as is the case in GUTs), it is conceivable that the discontinuities that one expects in a phase transition are in fact rounded by the gauge interactions.

The third consequence is the existence of magnetic monopoles in the particle spectrum. Any GUT in which a simple group is broken eventually to $SU_3 \times U_1$ (i.e., QCD and electromagnetism) will necessarily (see for example the review by Goddard & Olive, 1978) contain magnetic monopoles of the 't Hooft-Polyakov type ('t Hooft, 1974; Polyakov, 1974). These monopoles are topologically stable knots in the Higgs field expectation value. Their masses are generically of order M_X/α, where M_X is the mass of the superheavy vector bosons and α is the fine-structure constant of the grand unified gauge interactions. In the minimal SU_5 GUT, $M_X \simeq 4 \times 10^{14}$ GeV and $\alpha \simeq 1/45$, so the monopole mass is of order 10^{16} GeV.

To see how these topological knots are formed in the SU_5 example, define the 5×5 matrices

$$M_0 \equiv \mathrm{diag}[1,1,1,-3/2,-3/2]$$

$$M_1 \equiv \mathrm{diag}[1,1,-1/4,-1/4,-3/2]$$

$$\Sigma_i \equiv \begin{bmatrix} 0 & 0 & 0 & 0 & 0 \\ 0 & 0 & 0 & 0 & 0 \\ 0 & 0 & \multicolumn{2}{c}{\boxed{\sigma_i}} & 0 \\ 0 & 0 & & & 0 \\ 0 & 0 & 0 & 0 & 0 \end{bmatrix} \qquad (4)$$

where σ_i denotes the 2×2 Pauli spin matrices. Note then that $M_0 = M_1 + (5/4)\Sigma_3$. The SU_5 gauge group contains an SU_2 subgroup which operates only on the 3 and 4 directions of the 5-dimensional space. Under this SU_2 subgroup, M_1 is invariant and the Σ_i rotate as a 3-vector. Now consider a (classical) Higgs field configuration which at large distances from the origin has the behavior

$$\phi(\vec{r}) \xrightarrow[r\to\infty]{} \sqrt{\frac{2}{15}}\, \phi_v \{M_1 + (5/4)\hat{r}_i \Sigma_i\}. \qquad (5)$$

In any small region the above expression can be gauge transformed to match the vacuum expectation value of Eq. (1), but it is topologically impossible to extend this gauge transformation nonsingularly to a region which surrounds the origin. For essentially the same reason, it is impossible to nonsingularly continue the Higgs configuration of Eq.(5) into the central region while maintaining the property that at each point it is a gauge transformation of Eq.(1). Thus, any configuration with the asymptotic form of Eq.(5) will necessarily have a positive energy. The configuration with the lowest possible energy is the minimal SU_5 monopole. The gauge fields will arrange themselves to cancel the gauge-covariant derivative of Φ at large distances, and these long range gauge fields constitute the long range magnetic fields. The twisting of the Higgs field in the 3-4 direction results in long range magnetic fields in all three pieces of the unbroken subgroup. The SU_3 magnetic fields are screened by quantum effects at distances larger than about 1 GeV^{-1} (a typical QCD scale). The SU_2 and U_1 magnetic fields will combine in just the right combination to form an ordinary electromagnetic magnetic field; the magnetic charge will have the Dirac magnitude of $(1/2\alpha)e$, where e is the charge of an electron and $\alpha = e^2/4\pi \approx 1/137$ is the electromagnetic fine-structure constant.

For more details on magnetic monopoles, see Preskill(1982); Coleman(1982); or Goddard & Olive(1978). For more details on the SU_5 magnetic monopole, see Dokos and Tomaras (1980).

3 THE STANDARD SCENARIO

Now I will discuss what may be called the standard scenario of the very early universe. (For a general background in cosmology, see S. Weinberg, 1972 or Silk, 1980. At a less technical level, see S. Weinberg, 1977).

The universe is described by a Robertson-Walker metric. Whether it is closed, open, or flat, the curvature is negligible at very early times. Thus,

$$ds^2 = -dt^2 + R^2(t)d\vec{x}^2. \tag{6}$$

The expansion of the universe is described by the scale factor R(t) which is governed by the Einstein field equation

$$\left(\frac{\dot{R}}{R}\right)^2 = \frac{8\pi}{3}G\rho, \tag{7}$$

where ρ is the mass density and the dot denotes differentiation with respect to t. (The above equation would have an additional term if the curvature were not negligible.) The universe expands adiabatically, with a mass density which is dominated by the radiation of effectively massless (i.e., $M \ll T$) particles at temperature T:

$$\rho = cT^4, \tag{8}$$

where

$$c = \frac{\pi^2}{30}(N_b + \frac{7}{8}N_f). \tag{9}$$

Here N_b denotes the number of effectively massless bosonic spin degrees of freedom (e.g., the photon contributes two units), and N_f denotes the corresponding number for fermions (e.g., electrons and positrons together contribute four units). For the minimal SU_5 theory, $c \approx 52.9$ at the highest temperatures ($T > 10^{14}$ GeV), while $c \approx 1.11$ at the lowest temperatures ($T < \frac{1}{2}$ MeV). (At the lowest temperatures I take $T \equiv T_\gamma$ (the photon temperature), and take into account the lower neutrino temperature $T_\nu = (4/11)^{1/3}T_\gamma$. For a discussion, see Guth, 1983.) The expansion is assumed to be adiabatic, which means that R^3s is conserved, where the entropy density s is given by

$$s = \frac{2\pi^2}{45}(N_b + \frac{7}{8}N_f)T^3. \tag{10}$$

Thus, as long as the number of effectively massless particle species does not change, one has RT = constant. The temperature then varies with the time t according to

$$T^2 = \frac{E_P}{2\gamma t}, \tag{11}$$

where $\gamma \equiv (8\pi c/3)^{1/2}$ ranges from 21.0 at the highest temperatures to 3.04 at the lowest temperatures, and $E_P \approx 1.2 \times 10^{19}$ GeV is the Planck energy. Since RT = constant, it follows that $R \propto \sqrt{t}$. (When T was 10^{14} GeV, these equations give $t \approx 1.9 \times 10^{-35}$ sec, and $\rho \approx 1.2 \times 10^{75}$ g cm^{-3}. Objects which are separated by 10^{10} light-years today were separated by only 6 cm at that time.)

In the standard scenario one also assumes that any phase transitions which may occur are inconsequential. That is, they occur quickly when the temperature falls to the critical temperature, and the release of latent heat is negligible.

The standard scenario suffers from three problems, which I will now discuss. The first is the monopole problem. As described in the previous section, magnetic monopoles are topologically stable knots which can be tied in the Higgs expectation value. When the universe cools below the critical temperature T_c of the GUT transition, it becomes thermodynamically favorable for the Higgs field to align uniformly over large distances. However, it takes time for these correlations to be established. Note that the horizon length, defined as the distance which a light pulse could have travelled since the initial singularity, is given by

$$\ell_H = R(t)\int_o^t dt'\ R^{-1}(t'). \tag{12}$$

Thus, in the standard scenario $\ell_H = 2t$. Causality implies that the Higgs field correlation length ξ must be less than ℓ_H. Since the monopoles are knots in the Higgs expectation value, their density is given roughly by (Kibble, 1976)

$$n_M \approx \xi^{-3} \geq 1/8t^3, \tag{13}$$

where t is the time at which the universe cools to the critical temperature T_c of the phase transition. (If there is more than one GUT phase transition, T_c refers to the critical temperature of the final one, which leaves the universe in the $SU_3 \times SU_2 \times U_1$ phase.) Using Eqs.(10) and (11), it follows that

$$n_M/s \geq 10.43(N_b + \tfrac{7}{8}N_f)^{1/2}(T_c/E_P)^3. \tag{14}$$

This ratio is unchanged by the expansion of the universe, and Preskill (1979) has shown that monopole-antimonopole annihilation is totally ineffective at these densities (see also Zeldovich & Khlopov, 1978, and Goldman, Kolb & Toussaint, 1981). Entropy is essentially conserved in this model, so the ratio n_M/s should be about the same today. The entropy density of the universe today (taking $T_\gamma = 2.7^\circ K$) is

$$s \approx 2.8 \times 10^{3}\ cm^{-3}, \tag{16}$$

and for 10^{16} GeV monopoles this would give an energy density

$$\rho_M \gtrsim 5 \times 10^{-18}\ g\ cm^{-3}. \tag{17}$$

Taking an estimate of the current Hubble "constant" $H \equiv (\dot{R}/R) \approx (10^{10}\ yr)^{-1}$, the critical mass density (which gives a precisely flat (k=0) universe) is given by

$$\rho_c = 3H^2/8\pi G \approx 1.8 \times 10^{-29}\ g\ cm^{-3}. \tag{18}$$

Thus, this density of monopoles would require

$$\Omega \equiv \rho/\rho_c \gtrsim 3 \times 10^{11}, \tag{19}$$

which is totally incompatible with observations. (In particular, the age of such a universe would be less than 30,000 yr.) Thus, if GUTs are to be compatible with cosmology, one must understand how the monopole production was suppressed. (One possible solution is the proposal of Langacker & Pi, 1980. See also Bais & Langacker, 1982.)

The second problem is the horizon problem, first pointed out by Rindler (1956). The observational basis for this problem is the uniformity of the cosmic background radiation, which is known to be isotropic to about one part in 10^3. This fact is particularly difficult to understand when one considers the existence of the horizon length, the maximum distance that light could have travelled since the beginning of time. Consider two microwave antennas pointed in opposite directions. Each is receiving radiation which is believed to have been emitted (or "decoupled") at the time of hydrogen recombination, at $t \approx 10^5$ yr. At the time of emission, these two sources were separated from each other by over 90 horizon lengths (Guth, 1983). The problem is to understand how two regions over 90 horizon lengths apart came to be at the same temperature at the same time. Within the standard model this large scale homogeneity is simple assumed as an initial condition.

The third problem is the flatness problem, which was first

pointed out by Dicke & Peebles (1979). To describe this problem I must first write down the general Robertson-Walker metric:

$$ds^2 = -dt^2 + R^2(t)\left[\frac{dr^2}{1-kr^2} + r^2(d\Theta^2 + \sin^2\Theta d\phi^2)\right], \tag{20}$$

where k = +1, −1, or 0 for a closed, open, or flat universe, respectively. It should be emphasized that any value of k is possible, but by convention r and R(t) are rescaled so that k takes on one of the three discrete values. Eq.(7) generalizes to

$$H^2 = \left(\frac{\dot{R}}{R}\right)^2 = \frac{8\pi}{3} G\rho - \frac{k}{R^2} . \tag{21}$$

For the remainder of this discussion I will take $k = \pm 1$; the special case k=0 is still included as the limit $R \to \infty$. The flatness problem is then associated with the value of the dimensionless quantity

$$S \equiv R^3 s , \tag{22}$$

which I call the characteristic entropy of the universe. To determine S from current observations, note that s is given by Eq.(16) and that Eqs.(21) and (18) imply that

$$R = H^{-1}|1-\Omega|^{-1/2} . \tag{23}$$

If we assume that today $\Omega < 2$, then $R > H^{-1} \simeq 10^{10}$ yr. It then follows that

$$S > 2 \times 10^{87} . \tag{24}$$

The flatness problem is essentially the problem of understanding why S is such a large number. Since it is a dimensionless constant fixed by the initial conditions of the model, it seems more reasonable to expect it to be of order unity.

There are at least two other ways of phrasing the flatness problem which should also be described. One way involves looking at the time evolution of Ω. Eq.(21) can be rewritten (with the help of Eqs.(18)

and (22)) as

$$\frac{\Omega-1}{\Omega} = \frac{3}{8\pi G}\frac{ks^{2/3}}{S^{2/3}\rho} \tag{25}$$

Numerically, one finds that $\Omega = 1 \pm \mathcal{O}(10^{-15})$ when T = 1 MeV, and $\Omega = 1 \pm \mathcal{O}(10^{-49})$ when $T = 10^{14}$ GeV. The flatness problem is then a fine-tuning problem: can one explain, in terms of physical processes occurring at these early times, how the value of Ω became so extraordinarily close to unity. Finally, the flatness problem can be expressed in terms of the age of the universe. In the standard scenario the value $\Omega = 1$ is an unstable equilibrium point. Furthermore, the only time scale which appears in the equations describing the early universe is the Planck time, $E_P^{-1} = 5.4 \times 10^{-44}$ sec. Since the universe is about 10^{60} Planck times old, how do we explain why Ω is still of order unity?

The horizon and flatness problems are both statements about the initial conditions required by the standard scenario. It is therefore possible that the solution to these problems lies in the era of quantum gravity, at times so early that the assumptions of the standard model are not yet applicable. However, I will show in the next section how these problems might be solved at the level of GUT phase transitions, using the inflationary universe scenario. Even if the solution really occurs at the level of quantum gravity, it is still conceivable that the mechanism closely resembles the inflationary scenario. (See, for example, Starobinsky, 1980.)

4 THE (ORIGINAL) INFLATIONARY UNIVERSE

In this section I will describe the inflationary universe scenario in its original form (Guth, 1981), and show how it can possibly solve the three cosmological problems of the standard scenario that were discussed in the previous section.

The inflationary scenario requires a Higgs field effective potential $V(\phi)$ similar to the form shown schematically in Figure 2. (Note that ϕ is really a multicomponent field.) At zero temperature, the global minimum of the effective potential is called the true vacuum, with $\phi = \phi_{true}$; the universe which we observe today consists of small perturbations about this true vacuum. The value of $V(\phi_{true})$ is related to the cosmological constant Λ by

$$\Lambda = 8\pi G V(\phi_{true}). \tag{26}$$

For reasons which we do not understand, the value of the cosmological constant (and hence $V(\phi_{true})$) is negligibly small, and I will take it to be zero. The zero temperature effective potential is required to also have a secondary local minimum at $\phi \equiv \phi_{false}$, which is called a false vacuum; the energy density of this state will be called ρ_o. Furthermore, I assume that there is a critical temperature T_c above which the finite temperature effective potential has a lower value near the false vacuum than it has near the true vacuum. Thus, T_c is the critical temperature for a first order phase transition.

For a wide range of parameters (Guth & Tye, 1980) the SU_5 Higgs effective potential has precisely these properties. The false vacuum is the $SU_4 \times U_1$ configuration shown in Eq.(3).

The starting point of a cosmological scenario is somewhat a matter of taste and philosophical prejudice. Some physicists find it plausible that the universe began in some highly symmetrical state, such as a de Sitter space. I prefer to believe that the universe began in a highly chaotic state; one advantage of the inflationary scenario, from my point of view, is that it appears to allow a wide variety of starting configurations. I require only that the initial universe is hot ($T > 10^{14}$ GeV) in at least some places, and that at least some of these regions are expanding rapidly enough so that they will cool to T_c before gravitational effects reverse the expansion.

Figure 2. A schematic Higgs effective potential appropriate for the original inflationary scenario.

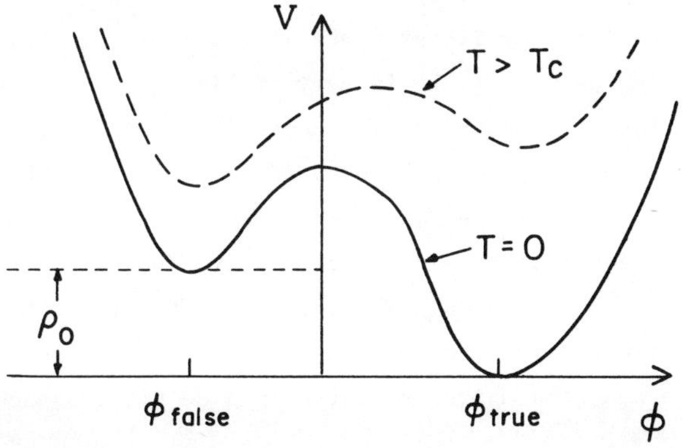

In these hot regions, thermal equilibrium would imply $\langle\phi\rangle \approx \phi_{false}$. (Actually, though, the universe has not had time at this point to thermalize (Steigman, 1982). Thus, I need to assume that there are some regions of high energy density with $\langle\phi\rangle \approx \phi_{false}$.) Such a region will cool to T_c, and it will then start to supercool below T_c. At this point the phase transition will begin, occurring through the spontaneous nucleation of bubbles of the new phase. However, I will assume that the nucleation rate λ ($\lambda \equiv$ the expected number of nucleations per volume per time) is small compared to H^4. (For the SU_5 model, $\lambda << H^4$ for a wide range of parameters (Guth & Weinberg, 1981).) The phase transition could also be driven by the nucleating effect of impurities such as magnetic monopoles (Steinhardt, 1981a; Guth & Weinberg, 1981); however, I will assume that the monopoles of the false vacuum phase are stable against nucleation. (For the SU_5 model, this assumption also holds for a wide range of parameters (Steinhardt, 1981a).) Thus, the region of space will supercool toward the false vacuum.

Assuming that this local region is sufficiently homogeneous and isotropic to be approximated by a Robertson-Walker metric, then the constant energy density ρ_o leads to exponential expansion

$$R(t) \propto e^{\chi t} , \tag{27}$$

where

$$\chi = \left(\frac{8\pi}{3} G\rho_o\right)^{1/2} . \tag{28}$$

For $\rho_o \sim (10^{14} \text{ GeV})^4$, $\chi \sim 10^{10}$ GeV. Soon R(t) will become so large that the Robertson-Walker metric of Eq.(20) can be approximated, so far as any local measurements are concerned, by the $k = 0$ form of Eq.(6). The resulting space is called a de Sitter space (see for example Hawking & Ellis, 1973). As long as $\rho = \rho_o$, it is very stable against any small perturbation — any locally measurable inhomogeneity or anisotropy is damped exponentially on the time scale of χ^{-1}(Guven, 1982; Gibbons, 1982; Frieman & Will, 1982). Hawking & Moss (1982) conjecture that this situation is in fact governed by a non-perturbative "no-hair" theorem, similar to the one for black holes. Thus, a smooth de Sitter metric arises naturally, without any need to fine-tune the initial conditions. (Those who advocate that the universe begins in a de Sitter space are welcome to join the scenario at this point.)

Let me now assume that the exponential expansion continues for a time interval Δt, and that the phase transition then occurs instantaneously, with rapid thermalization of the latent heat. (This assumption, however, is very unrealistic; in the next section I will discuss what would actually happen.)

During the exponential expansion, the value of $R(t)$ would increase by a factor

$$Z = e^{\chi \Delta t} \quad . \tag{29}$$

In order to solve the cosmological problems, one must insist that $Z > 10^{29}$. It is very easy to find parameters (in the SU_5 model) which give a nucleation rate which is low enough to yield this value of Z. In fact, reasonable parameters can lead (Guth & Weinberg, 1981) to values of Z as large as $10^{10^{10}}$, which is also quite acceptable. Nucleation rates will be discussed further in the next section.

When the phase transition takes place, the energy density ρ_o of the false vacuum will be released as latent heat. If this energy is rapidly thermalized, then the temperature will again rise to the order of 10^{14}GeV. From here on, the inflationary scenario rejoins the standard scenario. (It is essential that baryon number production occurs after the exponential expansion, so that the baryon number density is not diluted.) Thus, the temperature immediately after the phase transition (and hence also the entropy density s) will be comparable in order of magnitude to the value it had at the onset of exponential expansion. Since $R(t)$ increases by a factor of Z, it follows that $S \equiv R^3 s$ increases by a factor of order Z^3. Thus if S were initially of order unity, it would be greater than 10^{87} after the phase transition. The flatness problem is solved.

Note that since Z is typically much larger than 10^{29}, the scenario leads one to expect that $S \gg 10^{87}$. Thus, one would expect that today Ω should be equal to one to a very high degree of accuracy.

The horizon problem disappears because the early universe was much smaller than in the standard scenario. After the reheating (at a temperature of order 10^{14} GeV), the region which evolves to become our observed universe has a size of order 10 cm. Before the exponential expansion it was smaller by a factor of Z, and thus its size was less than

10^{-14} GeV^{-1}; this is $\sim 10^4$ times smaller than the horizon length.

As for the monopole problem, the argument which led to the overproduction of monopoles in the standard scenario has clearly been obviated. However, it is not clear in the inflationary scenario how many magnetic monopoles would be produced in the phase transition. Thus, a large discrepancy has been replaced by an open question. The answer to this open question would depend on the detailed mechanism of the phase transition, and no workable mechanism has yet been found (see the next section).

Before going on to discuss the problems of the inflationary universe, let me discuss its virtues a bit longer. (The new inflationary scenario will preserve these virtues, while avoiding the problems.) If it is valid, the inflationary universe scenario explains the source of essentially all the matter, energy, and entropy in the universe: it all came from the expansion of the false vacuum.

The above statement seems to violate our naive notions of conservation laws, but in fact it violates no established laws of physics. As the universe expands in the false vacuum state, the energy density is fixed at ρ_o; thus, the total matter energy (i.e., all energy other than gravitational) is increasing! However, one must remember that the gravitational field can exchange energy with the matter fields. The energy-momentum tensor for matter obeys a covariant conservation equation, which in the Robertson-Walker metric reduces to

$$\frac{d}{dt}(R^3\rho) = -p\,\frac{d}{dt}(R^3) , \tag{30}$$

where p is the pressure. Since the false vacuum is completely Lorentz invariant, its energy-momentum tensor must have the form

$$T_{\mu\nu} = \rho_o g_{\mu\nu} . \tag{31}$$

(In curved space $T_{\mu\nu}$ could also have a term proportional to the covariantly conserved tensor, $R_{\mu\nu} - \frac{1}{2}g_{\mu\nu}R$, but such a term could be absorbed in the Einstein field equations by a redefinition of G.) It follows that $p = -\rho_o$; the false vacuum has a large and negative pressure! Eq.(30) is then satisfied identically, with the energy of the expanding system increasing due to the negative pressure. If the space were asymptotically Minkowskian it would be possible to define a conserved total energy (matter plus gravitational). (See, for example, Witten, 1981a.) However, in the Robertson-Walker metric, a global conservation law of this type does not exist.

Let me discuss briefly the question of whether or not our universe is vacuous — that is, does it possess any conserved quantities which distinguish it from the vacuum? The electric charge, color (i.e., QCD) charge, and the angular momentum of the universe are apparently all consistent with zero. The matter energy is clearly much larger than zero; but as I have just explained, it is not conserved. The baryon number of the universe is apparently $\gtrsim 10^{78}$, but according to GUTs baryon number is also non-conserved. Thus, as far as we know the universe may be devoid of all conserved quantities. In that case it is very tempting to believe that the universe began from nothing, or from almost nothing. The inflationary universe illustrates the latter possibility. (Ideas along these lines have also been discussed by Tryon, 1973; Brout, Englert, & Spindel, 1979; Gott, 1982; Atkatz & Pagels, 1982). It is often said that there is no such thing as a free lunch. However, it now appears quite possible that the universe itself is a free lunch.

Before closing the section, I want to say a few words about the impossibility of inflation at the Weinberg-Salam phase transition. The important point is that baryon number is essentially conserved at the temperature of this transition and below. Thus, if inflation by a factor of Z takes place at this phase transition, then the baryon-to-entropy ratio will be reduced by a factor of order Z^3. Given our current understanding of baryon number production, it is hard to see how a Z greater than about 10^2 could be tolerated.

The Weinberg-Salam-Glashow theory contains one undetermined parameter — the mass m_H of the Higgs boson. The behavior of the phase transition depends crucially on how m_H compares to m_{CW}, the Coleman-Weinberg value for the mass. (The Coleman-Weinberg value corresponds to an effective potential which has a vanishing second derivative at the origin. See Coleman & E. Weinberg, 1973.) The value of m_{CW} depends sensitively on Θ_W, but it is about 10 GeV (Mahanthappa & Sher, 1980). The possibility that m_H could be less than m_{CW} is excluded by the fact that Z would then be enormous (Linde, 1977 & 1980; Guth & E. Weinberg(1980); see also Sher, 1980; Steinhardt, 1981b). The case of $m_H \equiv m_{CW}$ has been studied by Witten (1981b) and found to be marginally acceptable ($Z \approx 10^2$).

For $m > m_{CW}$, there is negligible supercooling.

5 PROBLEMS OF THE (ORIGINAL) INFLATIONARY UNIVERSE

In the previous section I assumed, for the sake of discussion, that the phase transition occurred instantaneously, with rapid thermalization of the latent heat. In this section I will discuss what actually happens when a slow first order phase transition occurs in an exponentially expanding space. It turns out that the randomness of the bubble nucleation process leads to gross inhomogeneities, rendering the original inflationary scenario untenable. The problems discussed here are avoided by the new inflationary scenario, which will be discussed in the following section.

The process of bubble nucleation and growth in a false vacuum has been analyzed by Coleman (1977) and Callan & Coleman (1977) (see also Coleman, 1979), and the description has been extended to curved space-time by Coleman & De Luccia (1980). Here I will only summarize the most basic facts about the nucleation process. The nucleation rate λ (defined) in the previous section) can be written as

$$\lambda = Ae^{-B} . \tag{32}$$

Here B is the classical action associated with an O(4)-invariant (see Coleman, Glaser, & Martin, 1978) bubble solution to the Euclidean field equations. The quantity A is calculated by Callan & Coleman (1977); it has the dimension of mass^4, and for our purposes it will be sufficient to estimate it as the fourth power of some mass scale M characteristic of the phase transition. Thus for the GUT phase transition, I take $A \simeq (10^{14}\ \text{GeV})^4$.

The bubble wall expands at a speed which rapidly approaches that of light. As energy is released by the conversion of false vacuum to true, this energy is transferred to the accelerating bubble wall. The interior of the bubble approaches the true vacuum. Note that bubble formation does not affect the metric outside the bubbles; since they expand at essentially the speed of light, the region outside is causally disconnected. Thus, the outside region is described by the flat Robertson-Walker metric (Eq.(6)), with $R(t) = e^{\chi t}$.

For simplicity, I will assume that bubbles nucleate with zero size and then grow precisely at the speed of light. Light trajectories obey $ds^2 = 0$, so the coordinate velocity of light is $1/R(t)$. Then the coordinate radius at time t of a bubble which forms at time t_o is given by

$$r(t,t_o) = \int_{t_o}^{t} \frac{dt'}{R(t')} = \chi^{-1}(e^{-\chi t_o} - e^{-\chi t}) . \tag{33}$$

Note that as $t \to \infty$,

$$r(t,t_o) \to \chi^{-1} e^{-\chi t_o} . \tag{34}$$

The finite limit of $r(t,t_o)$ constitutes an event horizon.

Let us first ask if these bubbles will fill the space. The answer will turn out to be yes. To explain this, I will calculate p(t), the probability that an arbitrary point remains in the old phase (i.e., outside of all bubbles) at time t. I will use two simplifications. First, I will ignore the fact that bubbles cannot form inside of bubbles; instead I will assume that bubbles nucleate at a constant rate λ per physical volume at all points in space. This simplification will produce no error in p(t) whatever, since any bubble which forms inside of another bubble will remain forever inside it. Second, I will pretend that each bubble appears instantaneously with its asymptotic coordinate radius of Eq.(34). Since χ^{-1} will be a short time for our purposes, this will be a good approximation.

Now consider some arbitrary coordinate volume V, and the time interval between t and t + dt. The physical volume is $e^{3\chi t}V$, so the number of bubbles expected to nucleate is $\lambda e^{3\chi t}Vdt$. The coordinate volume of each bubble is $(4\pi/3)\chi^{-3}e^{-3\chi t}$, so the expected total volume of the bubbles formed is $(4\pi/3)\lambda\chi^{-3}Vdt$. Thus a fraction $(4\pi/3)\lambda\chi^{-3}dt$ of the volume is covered with new bubbles. Since the new bubbles are uncorrelated with any preexisting bubbles, it follows that this same fraction of the previously uncovered region becomes covered during the time interval dt. This leads to

$$p(t) = \exp\{-\tfrac{4\pi}{3}\, \varepsilon\chi(t-t_B)\} , \tag{35}$$

where t_B is the time at which bubble formation begins, and

$$\varepsilon \equiv \lambda/\chi^4 . \tag{36}$$

Thus, as $t \to \infty$, $p(t) \to 0$. The fraction of space which remains in the old phase goes exponentially to zero. (Note, however, that for $\varepsilon << 1$ the physical volume of the region which remains in the old phase increases exponentially with time.) For more details about $p(t)$, see Guth & Tye (1980) and Guth & E. Weinberg (1981, 1982).

To give you some feeling of the range of values of ε which might be considered reasonable, I have shown in Figure 3 (Guth & E. Weinberg, 1982) the results of numerical calculations of the bubble action in the simplest SU_5 model. The nucleation rate is a function of the Higgs potential parameters β, γ, and b, but the definition of these parameters will not be important for present purposes. The figure shows ε as a function of β/γ in the limit of $\gamma \to \infty$, for several values of b. In order to have exponential expansion by a factor $Z \sim 10^{29}$ before the phase transition is "completed" (i.e., $p(t) << 1$), one needs $\varepsilon \lesssim 10^{-2}$. However, note that values like $\varepsilon = 10^{-100}$ or even smaller are very reasonable. In what follows I will assume that $\varepsilon << 1$.

Figure 3. Values of ε for a phase transition in a simple SU_5 model.

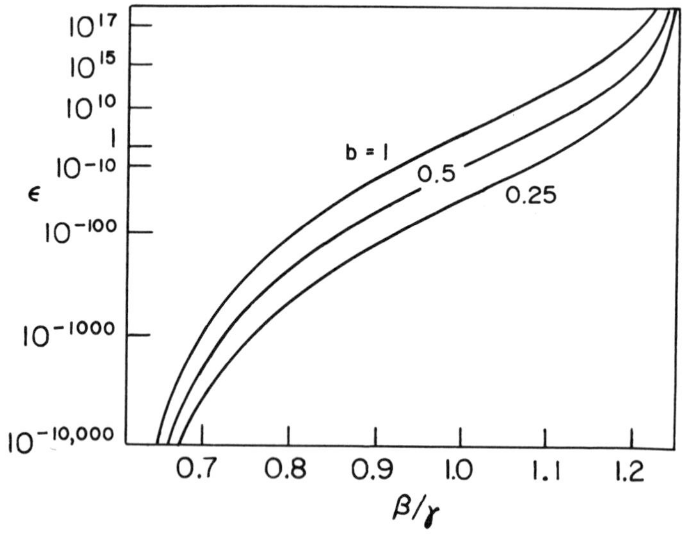

Thus, the instantaneous phase transition assumed in the previous section must be replaced by one in which p(t) approaches zero with an exponential time constant which is very long compared to the expansion time χ^{-1}.

The real problems appear when one looks at the properties of bubble clusters (Guth & E. Weinberg, 1982). It can be proven rigorously that if $\varepsilon < 10^{-6}$, then the bubbles will form finite size clusters only, no matter how long one waits, even though $p(t) \to 0$. (In mathematical terminology, the bubbles do not "percolate". It can also be shown, by the way, that the bubbles do percolate if $\varepsilon \gtrsim 0.24$.)

Since the regions which are in the new phase exist as isolated clusters only, one must look in more detail at these clusters. It can be shown that a typical cluster is dominated by its largest bubble, with the second largest bubble being smaller by a factor of order $\sqrt{\varepsilon}$. For $\varepsilon \ll 1$, this means that a typical cluster would appear as a single dominant bubble, with an insignificant fuzz of smaller bubbles around its boundary. This structure clearly bears no resemblance to the observed universe.

The problems of the original inflationary universe have also been analyzed by Hawking, Moss, & Stewart (1982). Taking $\varepsilon \simeq 1/200$ (which is roughly the maximal value consistent with the inflationary scenario) these authors calculate that about 1/3 of all bubbles will grow to a galactic size (i.e., a size which will evolve to galactic proportions in the present universe) before undergoing their first collision. They then point out that, in the standard model, a region of galactic size does not come within the horizon until $t \approx 10^8$ sec, when $T \approx 10^{6\,\circ}$K. Thus, they argue, it is implausible that the energy in the wall of such a bubble could thermalize before T fell to $\sim 10^{6\,\circ}$K, a value too low for baryon production or even nucleosynthesis. Their conclusion then is the same as ours: the scenario is unworkable, unless one finds a more graceful end to the period of inflation.

6 THE NEW INFLATIONARY UNIVERSE

In the previous section I have summarized the problems of the original inflationary universe scenario. This scenario illustrated how inflation could potentially solve the horizon and flatness problems, and maybe the monopole problem as well; but the description of the ending of the inflationary period was fatally flawed. These problems, however, have been overcome by a modified scenario, known as the new inflationary universe, which was proposed by Linde (1982a,b) and Albrecht & Steinhardt

(1982)). The new inflationary universe is also described in an article in this volume by Linde (1982d), but I would like to briefly summarize here my view of this scenario.

The key feature of this scenario is the realization that, under special circumstances, it is possible for the observed universe to have evolved from a single bubble or fluctuation region. This kind of evolution is not possible for a generic Higgs potential. It can be shown (Guth & E. Weinberg, 1982) that if all the mass parameters in the Higgs potential are of the same order, then the total entropy produced within the bubble is always of order one, which is far short of the required entropy of 10^{87} or greater. This problem is evaded in the new inflationary scenario by the use of a Coleman-Weinberg (Coleman & E. Weinberg, 1973) potential for the Higgs field; i.e., the second derivative of the effective potential is required to vanish at the origin. More precisely, the scenario requires that this derivative be less than about $(10^9\ \text{GeV})^2$, which is lower than the characteristic GUT value of $\sim 10^{29}\ \text{GeV}^2$ by 11 orders of magnitude.

This kind of fine-tuning is generally regarded as distasteful, and one would presumably not consider it if the original scenario had worked. However, many of us are willing to temporarily turn a blind eye to this fine-tuning problem, and go on to pursue the consequences. After all, at least one has a vague idea of where to look for a mechanism to enforce the Coleman-Weinberg condition: if GUTs have any claim to truth, they are presumably embedded in some more sophisticated theory (perhaps supergravity) which fixes most or all of the >20 free parameters. GUT enthusiasts are already depending on some such mechanism to solve the hierarchy problem (i.e., to explain why $(M_W/M_X)^2 \approx 10^{-25}$). Since the new inflationary universe maintains all the successes of the original scenario while avoiding all of its problems, it is a highly attractive proposal. (Elsewhere in this volume, Steinhardt (1982) discusses a variant which avoids this fine-tuning problem.)

Including the one-loop gauge field quantum corrections (calculated in the flat-space approximation), the Coleman-Weinberg potential for the SU_5 Higgs field is given by

$$V(\phi) = \frac{25}{16}\,\alpha^2[\phi^4 \ln(\phi^2/\sigma^2) + \tfrac{1}{2}(\sigma^4 - \phi^4)] , \tag{37}$$

where

$$\Phi = \sqrt{\frac{2}{15}}\,\phi\,\mathrm{diag}[\,1,1,1,-3/2,-3/2],$$

$\alpha \equiv g^2/4\pi \approx 1/45$ is the gauge coupling, and $\sigma \approx 1.2 \times 10^{15}$ GeV. The form of this potential is shown in Figure 4. The minimum lies at $\phi = \sigma$, corresponding to the true vacuum. The point $\phi = 0$ is an equilibrium point which is just barely unstable. I will refer to the field configuration $\phi = 0$ as the false vacuum, even though this term is traditionally reserved for configurations which are classically stable. For any small temperature $T > 0$, the false vacuum is stabilized by a bump in the finite temperature effective potential. The height of this bump is of order T^4, and the width is of order T. There is a phase transition with critical temperature $T_c \sim 10^{14}$ GeV, above which the $\phi = 0$ configuration is the thermal equilibrium state.

As in the original scenario, the universe can begin in any of a wide variety of starting configurations. We again must assume that the initial universe contains some hot ($T > 10^{14}$ GeV) regions, and that at least some of these regions are expanding fast enough to cool to T_c before

Figure 4. The Coleman-Weinberg potential for the SU_5 adjoint Higgs field.

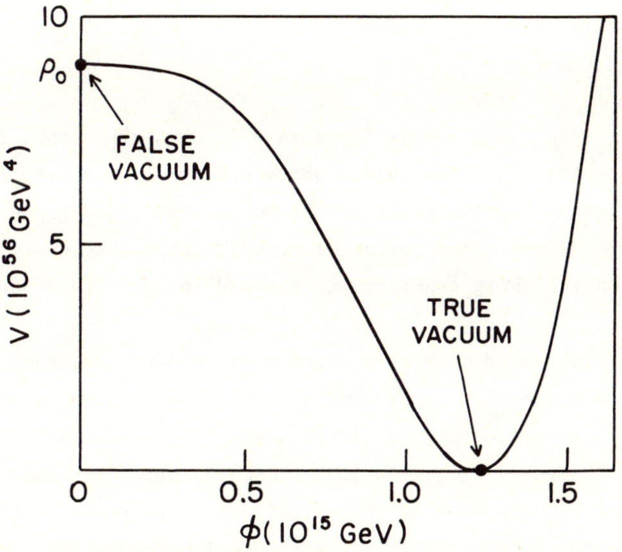

gravitational effects reverse the expansion. Such a region will cool to T_c, and then it will supercool. Flat space calculations of the finite temperature nucleation rate (Sher, 1981; see also Abbott, 1981, Billiore & Tamvakis, 1982, and Cook & Mahanthappa, 1982) are reliable for $T \gtrsim 10^{10}$ GeV, and these calculations show that the nucleation rate is negligible — $\lambda/H^4 << 1$. When these calculations are continued to lower temperatures (where their validity is dubious — see below), one finds (Albrecht & Steinhardt, 1982) that the calculations break down at $T \approx 10^{7-8}$ GeV, at which point the energy of a critical size bubble becomes comparable to T, resulting in a failure of the steepest descent approximation. The situation becomes even more difficult to analyze when T falls to $\sim 10^6$ GeV, at which point the SU_5 gauge coupling constant becomes strong (Sher, 1981). While these calculations are not reliable, it is still interesting to note that the calculated nucleation rates are always negligibly small; the false vacuum appears surprisingly stable.

More realistic attempts to understand the details of the initial stage of the phase transition have been made by Hawking & Moss (1982), Vilenkin (1982), Vilenkin & Ford (1982), and Linde (1982c), but no definitive conclusions have been reached. These studies confront three significant complications.

First, there are curved space effects: quantum fluctuations in de Sitter space mimic thermal fluctuations at a Hawking temperature (Gibbons & Hawking, 1977)

$$T_H = \chi/2\pi. \tag{38}$$

In our case, $T_H \approx 10^9$ GeV. It is impossible in de Sitter space to have a temperature less than T_H. The curved space effects also modify the effective potential (Shore, 1980; Allen, 1982; Moss, 1982), and Linde (1982a&b) emphasizes that one must speak of an "improved" Coleman-Weinberg theory, in which the Coleman-Weinberg condition is imposed on the curved space effective potential.

The problem of bubble nucleation in de Sitter space has been worked out, in semiclassical approximation, by Coleman & De Luccia (1980). They analytically continue the de Sitter space to S^4 (the surface of a sphere in five Euclidean dimensions), a procedure which automatically incorporates the Gibbons-Hawking thermal effects. However, when Hawking & Moss (1982) studied these equations with a Coleman-Weinberg (or nearly Coleman-Weinberg) potential, they found that the only solutions are homogeneous — the whole universe apparently tunnels at once. But they also encountered the second of these complications: the classical action of their solution is small compared to one. Thus the semiclassical (or steepest descent) approximation seems to break down, and it is hard to know if these mathematical solutions have any physical implications.

The third complication is the unusual sensitivity of the problem. While order of magnitude estimates are good enough for many questions in cosmology, in this calculation a factor of two could make the difference of whether the scenario works or not.

While the problem has the complications mentioned above, in at least two ways it is simpler than one might have guessed. First, the existence of the Hawking temperature $T_H \approx 10^9$ GeV saves us from having to deal with the SU_5 gauge theory at energy scales of order 10^6 GeV, when the gauge coupling becomes strong. At 10^9 GeV the value of the SU_5 fine-structure constant α is about 0.057, which indicates that perturbation theory in α is presumably reliable. The second simplification is that gravity can apparently be treated as a fixed classical background field. The crucial stage of the phase transition is when the Higgs field is just beginning to fluctuate away from the false vacuum. At this time the energy density of the quantum fluctuations in the Higgs field is negligible compared to the false vacuum energy density ρ_o, and so the quantum fluctuations in the gravitational field are presumably negligible.

While the details of the initial stage of the phase transition are not well understood, one expects that the Higgs field ϕ will undergo fluctuations due to thermal and/or quantum effects. Some fluctuations would begin to grow, and at some point these fluctuations will become large enough so that their subsequent evolution can be described by the classical equations of motion. I will use the term "fluctuation region" (abbreviated FR) to denote any region in which this has occurred. The shape and size of an FR are quite uncertain. Linde refers to bubbles, suggesting that the regions are spherical. Hawking & Moss suggest that the entire universe forms a single FR. However, Albrecht & Steinhardt point out that the breakdown of the semiclassical approximation probably means that the FR's are irregular; their size would presumably be characterized by χ^{-1}.

There is one crucial requirement for the scenario to work: the initial fluctuation in ϕ , which I will call ϕ_i, must be small. (By "initial", I refer to the time at which the classical equations of motion become applicable.) The field will then "roll" down the potential of

Figure 4, but on a time scale which is slow compared to the exponential expansion rate. In this way a single FR can grow to be larger than (or much larger than) the observed universe, which is assumed to lie within one such region. This slow "rolling" is the key ingredient of the new inflationary universe.

It is presumably safe to assume that the initial FR is homogeneous on a length scale of order χ^{-1}. For the scenario to work, it is necessary for the size of this region of homogeneity to be stretched to about 10 cm. (After the phase transition the universe will reheat to $T \approx 10^{14}$ GeV, and it will rejoin the standard scenario. A region of 10 cm size will expand to $\sim 10^{10}$ light-years by the time T has fallen to 2.7°K.) Thus, the exponential expansion must continue for at least 58 time constants, in order to give an expansion greater than $10\ \mathrm{cm}/\chi^{-1} \approx 10^{25}$.

To see how this expansion can be obtained, we must analyze the classical evolution of a homogeneous scalar field in de Sitter space, which is described by

$$\ddot{\phi} + 3\chi\dot{\phi} = -\frac{\partial V}{\partial \phi} . \tag{39}$$

For the potential of Eq.(37),

$$\frac{\partial V}{\partial \phi} = \frac{25}{4}\alpha^2\phi^3 \ln(\phi^2/\sigma^2) . \tag{40}$$

For α one should use the running coupling constant of unbroken SU_5 (Sher, 1981), given by

$$\alpha(Q^2) = \frac{4\pi}{\frac{40}{3}\ln(Q^2/\Lambda^2)} \tag{41}$$

where Q^2 should presumably be taken of order ϕ^2, and $\Lambda \approx 2.5 \times 10^5$ GeV. (I have chosen the parameters to correspond to $M_X \approx 4 \times 10^{14}$ GeV, and $\alpha(M_X^2) \approx 1/45$.)

It will be useful to have an approximate solution to these classical equations. We will be interested in ϕ of order $\chi/4$, and Eq.(40) can then be approximated by

$$\frac{\partial V}{\partial \phi} = -\lambda\phi^3 , \tag{42}$$

where $\lambda \approx 1/2$. By neglecting the $\ddot{\phi}$ term in Eq.(39), one finds

$$\phi^2(t) \approx -3\chi/2\lambda t , \tag{43}$$

where I have chosen the time at which $\phi_o \to \infty$ to be $t = 0$. (The neglecting of the $\ddot{\phi}$ term is a self-consistent approximation for $-\chi t >> 1$.) The behavior of this equation is shown in Figure 5. Note that the Higgs field varies very slowly for many time constants, and then rapidly shoots off to large values. Note also that if Eq.(43) is to accurately describe the evolution for the required 58 time constants of exponential expansion, then the classical equations must be valid for $\phi \geq \phi_i = \phi(t = -58\chi^{-1}) \approx 0.23\chi \approx 1.4T_H$. Thus, the scenario can tolerate initial fluctuations as large as $1.4T_H$, but larger fluctuations would cause it to fail. On dimensional grounds one would estimate the initial fluctuations to be of order T_H, so clearly a more detailed understanding is necessary.

When the ϕ field reaches the steep part of the potential, it

Figure 5. The time evolution of ϕ as it "rolls" in the Coleman-Weinberg potential.

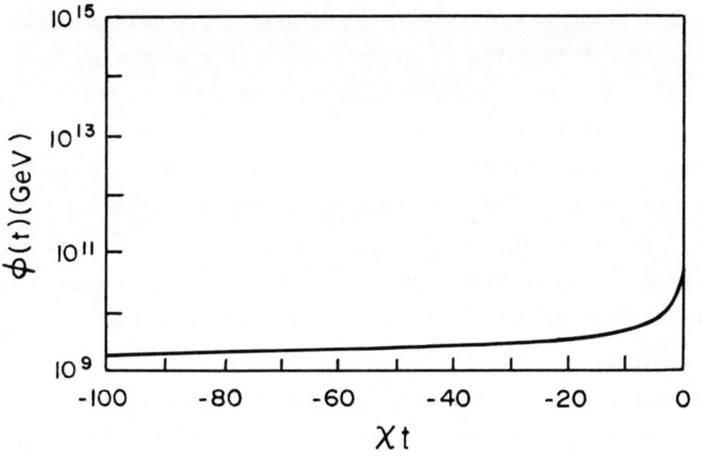

falls quickly to the bottom and oscillates about the minimum. The time scale of this motion is a typical GUT time of $(10^{14}\ \mathrm{GeV})^{-1}$, which is very fast compared to the expansion rate. The Higgs field oscillations are then quickly dámped by the couplings to the other fields, and the energy is rapidly thermalized (Albrecht, Steinhardt, Turner, & Wilczek, 1982). (The Higgs field oscillations correspond to a coherent state of Higgs particles; the damping is simply the decay into other species.) The release of this energy (which is just the latent heat of the phase transition) raises the temperature to the order of 10^{14} GeV. (More precisely, the reheating temperature is about one sixth of the heavy vector boson mass.) From here on the standard scenario ensues, including the production of a net baryon number (Abbott, Farhi, & Wise, 1982; Dolgov & Linde, 1982).

In the new inflationary universe scenario, the horizon and flatness problems disappear for the same reasons as in the original inflationary scenario. In the new scenario, the monopole problem is clearly solved — the Higgs field would be correlated in the initial FR, and inflation would give rise to a Higgs field which is correlated throughout the entire observed universe. Some monopoles would still be produced by thermal fluctuations after the reheating, but this number would be negligible in the SU_5 model (Lazarides, Shafi, & Trower, 1982) and presumably also in most other models.

Because the expansion takes place after the Higgs field has chosen a direction to roll, the new inflationary scenario allows one to consider particle theories in which a discrete symmetry is spontaneously broken. The universe would then contain domain walls, but inflation can stretch the typical domain size to be much greater than 10^{10} light-years. (Note that the SU_5 Coleman-Weinberg potential in fact has the discrete symmetry $\phi \rightarrow -\phi$, which is broken spontaneously.)

Returning to the flatness problem, one should note that the inflationary mechanism will typically suppress the curvature by much more than the required amount. For example, we have seen that the known fact that Ω is within an order of magnitude of unity today can be explained if the initial fluctuation of the Higgs field ϕ_i is about 0.23χ. Suppose, however, that ϕ_i is just 10% smaller. Using Eq.(43) with the new value of λ (evaluated by using Eq.(41) at the new value of ϕ_i), one finds that the exponential expansion factor increases by $\sim e^{11}$. The characteristic entropy

S then increases by $\sim e^{33}$, and according to Eq.(25) the quantity $\Omega-1$ is suppressed by an additional factor of $\sim e^{22} \approx 10^{10}$. This leads to the prediction that the value of Ω today is expected to be equal to one to within an extraordinary degree of accuracy.

Probably the most exciting development to come out of the Nuffield Workshop was the intense study of the density fluctuations produced in the new inflationary universe. Results on this question were announced by Aleksei Starobinsky (1982), myself (reporting on work done with So-Young Pi (Guth & Pi, 1982)), Stephen Hawking (1982), and Jim Bardeen, Paul Steinhardt, & Mike Turner (1982).

In general, one finds that a density fluctuation $\delta\rho/\rho$ on any given comoving length scale ℓ will begin to oscillate acoustically with a constant amplitude when the Hubble length H^{-1} exceeds ℓ. The goal of our calculations was to estimate the amplitude of these acoustic waves. To account for the mass distribution in the observed universe, one would like $\delta\rho/\rho$ to be roughly independent of ℓ (Harrison, 1970), and to have a value of maybe 10^{-4} (Zeldovich, 1972) or 10^{-2} (Press & Vishniac, 1980). In our first rounds of discussion, all of the people involved agreed that the new inflationary universe leads to a $\delta\rho/\rho$ which is roughly independent of ℓ. However, the initial estimates of the amplitude ranged from 10^{-16} to 10^{2}. By the end of the workshop, all of the groups converged to a value at or near the top of this range.

A variety of methods were used by the different groups, so I will summarize only the calculations by Pi and myself. We worked in the approximation in which the Higgs field rolls slowly across the plateau of the potential in Figure 4, producing only a negligible perturbation in the de Sitter metric; it then falls off the plateau instantaneously. From Figure 5 one can see that this is not such a bad approximation. We further assumed that the phase transition is completed instantaneously, with all of the energy converted to an ideal radiation fluid. (In a subsequent paper we intend to show that these approximations, although they greatly simplify the analysis, have only a negligible effect on the answer.) The fluctuations arise from the quantum fluctuations in the Higgs field ϕ, which cause it to fall off the plateau and trigger the phase transition at a time which varies from one position to another. (This statement depends of course on one's choice of coordinates; it is valid in the flat Robertson-Walker coordinates which describe our idealized system until the time of the phase transition.) We use the quantum theory of free scalar fields in de Sitter space to estimate the initial fluctuations in ϕ, conservation of energy and momentum to determine the conditions immediately following the instantaneous phase transition, and the linearized perturbation equations of Olson (1976) to follow the evolution of the system until $H^{-1} >> \ell$. Our final result is

$$\delta\rho/\rho \approx 0.2\lambda^{1/2}\ell n^{3/2}(\ell_{present}/\ell_o) , \qquad (44)$$

where $\lambda \approx 1/2$, $\ell_{present}$ denotes the length scale as measured in the present universe, and $\ell_o \approx 10^{-15}$ light-yr. A typical galaxy condenses from a region of size $\ell_{present} \approx 10^6$ light-yr, which gives $\delta\rho/\rho \approx 50$ on this scale. On a scale of 10^{10} light-yr, one finds $\delta\rho/\rho \approx 65$.

Thus, it appears that the new inflationary universe, attractive though it is, has encountered a clear-cut failure. Apparently it still requires modification. The estimates used in the fluctuation calculation are somewhat model-dependent, and it is conceivable that the desired amplitude can be obtained by modifying the scenario and/or the underlying particle theory. Some work along these lines is described by Steinhardt (1982) in this volume.

7 CONCLUSION

The new inflationary universe scenario had been studied enthusiastically by myself and many others for about six months prior to the Nuffield Workshop. It was therefore, I think, very disappointing to all of us to learn that the scenario appears to have a fatal flaw: the prediction it gives for density fluctuations is about 10^5 times too large.

I would still like to argue, however, that this discrepancy should be viewed not as a discouraging failure, but rather as an encouraging near miss. First, one should remember that this is the first time we have ever had a scenario with enough detail to even allow an estimate of $\delta\rho/\rho$ from first principles. Traditionally, an *ad hoc* initial spectrum of fluctuations has simply been assumed. Calculations based on phase transitions which produce cosmic strings (Vilenkin, 1981) have given reasonable spectra of fluctuations, but these models require the *ad hoc* assumption of an initial space which is perfectly homogeneous. Thus, the mere fact that $\delta\rho/\rho$ is calculable can be considered progress.

Furthermore, a prediction which is 10^5 times too large is not nearly as bad as it could have been. To see how bad an answer we could have gotten, consider the following version of the standard scenario. (This model has been considered by Peebles (1968), who had a very different motivation.) I will assume here that quantum gravity "turns off" at the Planck temperature, corresponding to $t \approx 10^{-45}$ sec. Since I am trying to generate a homogeneous universe, I will assume that the mysterious effects of quantum gravity have left us with a universe filled with matter in thermal equilibrium at a perfectly uniform temperature T. I will assume, however, that the matter distribution has the statistical fluctuations which are characteristic of a gas of random particles: in any volume which contains N particles, the statistical fluctuations are given by $\sqrt{N}$. (This assumption could be avoided if we assumed instead that quantum gravity left the particles arranged in some sort of a uniform lattice — but such an assumption would seem wildly implausible.) A galaxy might contain $\sim 10^{68}$ nucleons today, and hence it would have contained $N \approx 10^{78}$ massless particles at these very early times. Thus, on galactic scales at $t \approx 10^{-45}$ sec, one has $\delta\rho/\rho \approx N^{-1/2} \approx 10^{-39}$. Now note that $\delta\rho/\rho$ can be decomposed into two modes (see, for example, Olson, 1976) which grow as t and $t^{1/2}$. I will make the reasonable assumption that some significant fraction of the initial $\delta\rho/\rho$ is contained in the first mode. Then, by the time the galactic scale comes within the Hubble radius at $\sim 10^8$ sec, the fluctuation has grown to $\delta\rho/\rho \sim 10^{14}$, which is $\sim 10^{17}$ times too large.

Thus, inflation does serve its intended purpose of creating homogeneity and surpressing fluctuations. On galactic scales, it does about 10^{12} times better than the assumption of thermal equilibrium at the Planck scale. However, this is still not quite good enough. Some modification of the new inflationary scenario must be sought.

In conclusion, I want to say that I began working on cosmology only because I was pressured into it by Henry Tye. He had to do a lot of arm-twisting, because at that time I very strongly believed that cosmology was the kind of field in which a person could say anything he wanted, and no one could ever prove him wrong. I am sure that truth does not change with time, but after three years of working in cosmology my prejudices about the subject have completely reversed. It now appears that it is very easy to show that a cosmological scenario is wrong, and far more difficult than I had ever imagined to develop a totally consistent picture.

ACKNOWLEDGEMENTS

I would like to acknowledge the hospitality of the Nuffield Workshop on the Very Early Universe, and I would like to particularly thank its organizers, Gary Gibbons and Stephen Hawking. I greatly enjoyed interacting with all of the participants at the Workshop.

I would also like to acknowledge the support of the U.S. Department of Energy (DOE) under contract number DE-AC02-76ER03069, and an Alfred P. Sloan Fellowship.

REFERENCES

Abbott, L.F. (1981). Nucl. Phys. B185, 233.

Abbott, L.F., Farhi, E., & Wise, M.B. (1982). Phys. Lett. 117B, 29.

Albrecht, A. & Steinhardt, P.J. (1982). Phys. Rev. Lett. 48, 1220.

Albrecht, A., Steinhardt, P.J., Turner, M.S., & Wilczek, F. (1982). Phys. Rev. Lett. 48, 1437.

Allen, B. (1982). Phase transitions in deSitter space. To be published in Nucl. Phys., B

Atkatz, D. & Pagels, H. (1982). Phys. Rev. D25, 2065.

Bais, F.A. & Langacker, P. (1982). Nucl. Phys. B197, 520.

Bardeen, J.M., Steinhardt, P.J., & Turner, M.S. (1982). Seminar presented at the Nuffield Workshop on the Very Early Universe. Manuscript in preparation.

Billiore, A. & Tamvakis, K. (1982). Nucl. Phys. B200, 329.

Brout, R., Englert, F., & Spindel, P. (1979) Phys. Rev. Lett. 43, 417.

Callan, C.G. & Coleman, S. (1977). Phys. Rev. D16, 1762.

Coleman, S. (1977). Phys. Rev. D15, 2929.

Coleman, S. (1979). In The Whys of Subnuclear Physics, Proc. of the International School of Subnuclear Physics, Ettore Majorana, Erice, 1977; ed. A. Zichichi. New York: Plenum.

Coleman, S. (1982). The magnetic monopole fifty years later. Harvard University preprint, to be published in the Proc. of the International School of Subnuclear Physics, Ettore Majorana, Erice 1981; ed. A. Zichichi. New York: Plenum.

Coleman, S. & De Luccia, F. (1980). Phys. Rev. D21, 3305.

Coleman, S., Glaser, V., & Martin, A. (1978). Commun. Math. Phys. 58, 211.

Coleman, S. & Weinberg, E.J. (1973). Phys. Rev. D7, 1888.

Cook, G.P. & Mahanthappa, K.T. (1982). Phys. Rev. D25, 1154.

Dicke, R.H. & Peebles, P.J.E. (1979). In General Relativity: An Einstein Centenary Survey; eds. S.W. Hawking & W. Israel. Cambridge, England: Cambridge Univ. Press.

Dolan, L. & Jackiw, R. (1974). Phys. Rev. D9, 3320.

Dokos, C.P. & Tomaras, T.N. (1980). Phys. Rev. D21, 2940.

Dolgov, A.D. & Linde, A.D. (1982). Baryon asymmetry in inflationary universe. Phys. Lett.116B, 329.

Engels, J., Karsch, F., Satz, H. & Montvay, I. (1982). To be published.

Frieman, J.A. & Will, C.M. (1982). Ap. J. 259, 437.

Gasiorowicz, S. & Rosner, J.L. (1981). Am. J. Phys. 49, 954.

Georgi, H. & Glashow, S.L. (1974). Phys. Rev. Lett. 32, 438.

Georgi, H., Quinn, H.R., & Weinberg, S. (1974). Phys. Rev. Lett. 33, 451.
Gibbons, G.W. (1982). Seminar presented at the Nuffield Workshop on the Very Early Universe.
Gibbons, G.W. & Hawking, S.W. (1977). Phys. Rev. D15, 2738.
Goddard, P. & Olive, D.I. (1978). Rep. Prog. Phys. 41, 1357.
Goldman, T., Kolb, E.W., & Toussaint, D. (1981). Phys. Rev. D23, 867.
Goldman, T.J. & Ross, D.A. (1979). Phys. Lett. 84B, 208.
Goldman, T.J. & Ross, D.A. (1980). Nucl. Phys. B171, 273.
Gott, J.R. (1982). Nature 295, 304.
Gross, D.J., Pisarski, R.D., & Yaffe, L.G. (1981). Rev. Mod. Phys. 53, 43.
Guth, A.H. (1981). Phys. Rev. D23, 347.
Guth, A.H. (1983). To be published in Asymptotic Realms of Physics: Essays in Honor of Francis Low; eds. A.H. Guth, K. Huang, R.L. Jaffe. Cambridge, Mass.: M.I.T. Press.
Guth, A.H. & Pi, S.-Y. (1982). Fluctuations in the new inflationary universe. Phys. Rev. Lett. 49 , 1110.
Guth, A.H. & Tye, S.-H. (1980). Phys. Rev. Lett. 44, 631, 963(E).
Guth, A.H. & Weinberg, E.J. (1980). Phys. Rev. Lett. 45, 1131.
Guth, A.H. & Weinberg, E.J. (1981). Phys. Rev. D23, 876.
Guth, A.H. & Weinberg, E.J. (1982). Could the universe have recovered from a slow first order phase transition. To be published in Nucl. Phys. B212, 321.
Guven, J. (1982). Manuscript in preparation.
Harrison, E.R. (1970). Phys. Rev. D1, 2726.
Hawking, S.W. (1982). Seminar presented at the Nuffield Workshop on the Very Early Universe; Phys. Lett. 115B, 295.
Hawking, S.W. & Ellis, G.F.R. (1973). The Large Scale Structure of Space-Time. Cambridge, England: Cambridge Univ. Press.
Hawking, S.W. & Moss, I.G. (1982). Phys. Lett. 110B, 35.
Hawking, S.W., Moss, I.G., & Stewart, J.M. (1982). Bubble collisions in the very early universe. Phys. Rev. D26, 2681.
Kibble, T.W.B. (1976). J. Phys. A9, 1387.
King, J.G. (1960). Phys. Rev. Lett. 5, 562.
Kirzhnits, D.A. & Linde, A.D. (1972). Phys. Lett. 42B, 471.
Kogut, J., Stone, M., Wyld, H.W., Shigemitsu, J., Shenker, S.H., & Sinclair, D.K. (1982). Phys. Rev. Lett. 48, 1140.
Krauss, L.M. (1982). Gravitation, phase transitions, and the big bang. Phys. Rev. Lett. 49, 1459.
Kuti, J., Polonyi, J., & Szlachany, K. (1981). Phys. Lett. 98B, 199.
Langacker, P. (1981). Phys. Rep. 72C, 185.
Langacker, P. & Pi, S.-Y. (1980). Phys. Rev. Lett. 45, 1.
Lazarides, G., Shafi, Q., & Trower, W.P. (1982). Consequences of a monopole with Dirac magnetic charge. Virginia Polytechnic Institute preprint VPI-EPP-82-2.
Linde, A.D. (1977). Phys. Lett. 70B, 306.
Linde, A.D. (1979). Rep. Prog. Phys. 42, 389.
Linde, A.D. (1980). Phys. Lett. 92B, 119.
Linde, A.D. (1982a). Phys. Lett. 108B, 389.
Linde, A.D. (1982b). Phys. Lett. 114B, 431.
Linde, A.D. (1982c). Scalar field fluctuations in expanding universe and the new inflationary universe scenario. Phys. Lett. 116B, 335.
Linde, A.D. (1982d). Seminar presented at the Nuffield Workshop on the Very Early Universe.
Mahanthappa, K.T. & Sher, M. (1980). Phys. Rev. D22, 1711.
McLerran, L. & Svetitsky, B. (1981). Phys. Lett. 98B, 195.

Moss, I. (1982). Seminar presented at the Nuffield Workshop on the Very Early Universe.
Olson, D.W. (1976). Phys. Rev. D14, 327.
Peebles, P.J.E. (1968). Nature 220, 237.
Polyakov, A.M. (1974). Pis'ma Zh. Eksp. Teor. Fiz. 20, 430 [JETP Lett. 20, 194].
Polyakov, A.M. (1978). Phys. Lett. 72B, 477.
Preskill, J.P. (1979). Phys. Rev. Lett. 43, 1365.
Preskill, J.P. (1982). Seminar presented at the Nuffield Workshop on the Very Early Universe.
Press, W.H. & Vishniac, E.T. (1980). Ap. J. 239, 1.
Rindler, W. (1956). Mon. Not. R. Astron. Soc. 116, 663.
Sher, M. (1980). Phys. Rev. D22, 2989.
Sher, M. (1981). Phys. Rev. D24, 1699.
Shore, G.M. (1980). Ann. of Phys. 128, 376.
Silk, J. (1980). The Big Bang. San Francisco: W.H. Freeman & Co.
Starobinsky, A.A. (1980). Phys. Lett. 91B, 99.
Starobinsky, A.A. (1982). Seminar presented at the Nuffield Workshop on the Very Early Universe; manuscript in preparation.
Steigman, G. (1982). Proceedings of the Europhysics Study Conference: Unification of the Fundamental Interactions-II, Erice, 6-14 October, 1981.
Steinhardt, P.J. (1981a). Phys. Rev. D24, 842; Nucl. Phys. B190, 583.
Steinhardt, P.J. (1981b). Nucl. Phys. B179, 492.
Steinhardt, P.J. (1982). Seminar presented at the Nuffield Workshop on the Very Early Universe.
Susskind, L. (1979). Phys. Rev. D20, 2610.
't Hooft, G. (1974). Nucl. Phys. B79, 276.
Tryon, E.P. (1973). Nature 246, 396.
Turner, M.S. & Wilczek, F. (1982). Might our vacuum be metastable? Enrico Fermi Institute (Univ. of Chicago) Nature, 298, 633.
Vilenkin, A. (1981). Phys. Rev. D24, 2082; Phys. Rev. Lett. 46, 1169, 1496(E).
Vilenkin, A. (1982). Phys. Lett. 115B, 91.
Vilenkin, A. & Ford, L.H. (1982). Gravitational effects upon cosmological phase transitions. Phys. Rev. D26, 1231.
Weinberg, S. (1972). Gravitation and Cosmology. New York: Wiley.
Weinberg, S. (1974). Phys. Rev. D9, 3357.
Weinberg, S. (1977). The First Three Minutes. New York: Bantam Books.
Wilczek, F. (1982). Seminar presented at the Nuffield Workshop on the Very Early Universe.
Witten, E. (1981a). Commun. Math. Phys. 80, 381.
Witten, E. (1981b). Nucl. Phys. B177, 477.
Yoshimura, M. (1981). In Grand Unified Theories and Related Topics: Proceedings of the 4th Kyoto Summer Institute; eds. M. Konuma & T. Maskawa. Singapore: World Scientific Publishing.
Zeldovich, Ya. B. (1972). Mon. Not. R. Astr. Soc. 160, 1P.
Zeldovich, Ya. B. & Khlopov, M.Y. (1978). Phys. Lett. 79B, 239.

THE NEW INFLATIONARY UNIVERSE SCENARIO

A.D. Linde
Lebedev Physical Institute, Moscow 117924, USSR

1 INTRODUCTION

There is considerable interest now in cosmological consequences of grand unified theories (GUTs). One of the main reasons for such an interest is that these theories may help us to solve many important cosmological problems such as the horizon, flatness, homogeneity and isotropy problems, the problem of the baryon asymmetry of the universe, the problem of the origin of fluctuations which are necessary for galaxy formation etc. Several years ago the questions why there is no antimatter in the universe, why the universe is almost homogeneous and isotropic etc. seemed almost metaphysical. Typical answers were based on the so-called Anthropic Principle: it was argued that in a matter - antimatter symmetric, anisotropic and inhomogeneous universe there would be no observers who could ask such questions (see e.g. Dicke 1961; Collins & Hawking 1973; Carr & Rees 1979). However after the discovery of a mechanism of quasi-isotropization (i.e. local isotropization) of the universe due to particle creation (Zeldovich 1970; Zeldovich & Starobinsky 1972; Hu & Parker 1978) and especially after the discovery of a possible mechanism of cosmological baryon production (Sakharov 1967; Yoshimura 1978; Weinberg 1979), it became clear that recent developments in elementary particle theory may provide us with physical answers to all these almost metaphysical questions. The situation became even more interesting after the important paper of Guth (1981), who suggested a possible way of solving the horizon, flatness and primordial monopole problem by means of exponential expansion (inflation) of the universe in a supercooled symmetric state during symmetry breaking phase transitions in GUTs.

There are two main features of the inflationary universe scenario suggested by Guth. First of all, the stress tensor $T_{\mu\nu}$ in the supercooled symmetric state $\phi = 0$ (ϕ is the Higgs scalar field which breaks the symmetry) becomes equal to the vacuum energy tensor,

$T_{\mu\nu} \approx g_{\mu\nu}V(0)$, where $V(\phi)$ is the effective potential of the theory (Kirzhnits & Linde 1976). This leads to exponentially fast expansion of the universe, $a(t) \sim a_0 e^{Ht}$. Here $a(t)$ is the scale factor, H is the Hubble constant at that time, $H = \sqrt{8\pi/3M_P^2\, V(0)}$, and M_P is the Planck mass, $M_P \approx 1.2\times10^{19}$ GeV . During the exponential expansion period (de Sitter stage) the vacuum energy density $V(0)$ remains constant. This means that the temperature of the universe, which appears after the reheating of the universe during the phase transition, does not depend on the duration of the de Sitter stage (Linde 1979). The only effect of the exponential expansion stage, therefore, is the enormously large increase of the scale factor $a(t)$ in the supercooled state, which makes the universe after the reheating very flat. Guth has pointed out that this effect may provide us with a natural explanation of why the universe at present is almost flat, and may help us to solve simultaneously the horizon problem and the primordial monopole problem if the scale factor $a(t)$ grows at least $10^{28} \sim e^{65}$ times during the exponential expansion period. This means that the duration τ of this period should exceed $65\ H^{-1}$. (For a more detailed discussion of this scenario and of the horizon and flatness problems see the paper by Guth (1981) and his paper in this volume.)

Another important feature of this scenario was an implicit assumption, that the field ϕ inside the bubbles of the new phase, formed during the phase transition, immediately grows up to its equilibrium value ϕ_0 , which corresponds to the minimum of $V(\phi)$, and the reheating of the universe occurs only after the bubble wall collisions. Unfortunately, this assumption, which is actually true for many types of GUTs, makes the inflationary universe scenario in the form suggested by Guth unacceptable. As was pointed out by Guth and by other authors who studied this problem later, this scenario leads to an extremely large inhomogeneity and anisotropy of the universe after the phase transition (Guth 1981; Hawking, Moss and Stewart 1981; Guth & Weinberg 1982).

Recently a new inflationary universe scenario was suggested which was based on the theory of phase transitions in GUTs with the Coleman-Weinberg mechanism of symmetry breaking (Linde 1982). This scenario provides a possible solution of the horizon, flatness, homogeneity, isotropy, primordial monopole and domain wall problems, and may help us also to solve the baryon asymmetry problem (Dolgov & Linde 1982), the galaxy formation problem (Mukhanov & Chibisov 1981, 1982; Hawking 1982; Starobinsky 1982; Guth & Pi 1982; Bardeen et al. 1982), and the gravitino mass problem (Ellis et al. 1982).

The new inflationary universe scenario was further developed by many authors (Hawking & Moss 1982; Albrecht & Steinhardt 1982; Albrecht et al. 1982; Linde 1982 a,b,c; Vilenkin 1982; Ford & Vilenkin 1982; Starobinsky 1982; Parke 1982; Moss 1982; Albrecht et al. 1982; Ellis et al. 1982). The simplest versions of this scenario suggested so far are not devoid of their own difficulties. However these difficulties do not seem unavoidable. The aim of this paper is to discuss the present status of the new inflationary universe scenario and to outline some possibilities related to this scenario which deserve further investigation.

2 COLEMAN-WEINBERG THEORY

Since the new inflationary universe scenario is based essentially on the theory of the symmetry breaking phase transition in the Coleman-Weinberg theory, we shall start with the brief discussion of this theory (Coleman-Weinberg 1973). As an example let us consider first the ordinary scalar electrodynamics, which describes the complex scalar field $\chi = 1/\sqrt{2}\,(\chi_1 + i\chi_2)$, interacting with the Abelian vector field A_μ. The Lagrangian of this theory looks as follows

$$L = -\frac{1}{4}(\partial_\mu A_\nu - \partial_\nu A_\mu)^2 + (\partial_\mu + ieA_\mu)\chi^*(\partial_\mu - ieA_\mu)\chi + \mu^2\chi^*\chi - \lambda(\chi^*\chi)^2 . \tag{2.1}$$

The scalar field χ in this theory may acquire a classical part ϕ for some values of the mass squared $-\mu^2$ (e.g. for $\mu^2 > 0$). The simplest way to describe this effect is to use the unitary gauge, or equivalently, to make the change of variables

$$\chi(x) \longrightarrow \frac{1}{\sqrt{2}}(\chi(x) + \phi)\exp\left(\frac{i\zeta(x)}{\phi}\right) , \qquad A_\mu(x) \longrightarrow A_\mu(x) - \frac{1}{e\phi}\partial_\mu\zeta(x) , \tag{2.2}$$

after which the Lagrangian (2.1) is transformed into

$$L = -\frac{1}{4}(\partial_\mu A_\nu - \partial_\nu A_\mu)^2 + \frac{e^2}{2}(\chi+\phi)^2 A_\mu^2$$
$$+ \frac{1}{2}(\partial_\mu \chi)^2 - \lambda\phi^2\chi^2 - \lambda\phi\chi^3 - \frac{\lambda}{4}\chi^4 \qquad (2.3)$$
$$- \frac{\lambda}{4}\phi^4 + \frac{\mu^2}{2}(\chi^2+\phi^2) - (\lambda\phi^2 - \mu^2)\chi\phi .$$

Note that the auxiliary field $\zeta(x)$ is completely transformed away. The energy of the constant field ϕ at the classical level, according to (2.3), is given by

$$V_o(\phi) = -\frac{\mu^2}{2}\phi^2 + \frac{\lambda}{4}\phi^4 . \qquad (2.4)$$

At $\mu^2 < 0$, $V_o(\phi)$ has only one minimum at $\phi = 0$, whereas at $\mu^2 > 0$ there is a minimum of $V_o(\phi)$ at $\phi = \phi_o = \mu/\sqrt{\lambda}$, i.e. spontaneous symmetry breaking takes place.

The situation becomes somewhat different after an account is taken of the one-loop corrections to $V_o(\phi)$. These corrections arise due to contributions of vector and scalar particles, but in the case $e >> \lambda$, which will be most important for us, the vector particles contribution dominates, and the one-loop effective potential $V(\phi)$ before renormalization is given by

$$V(\phi) = V_o(\phi) + V_1(\phi) = -\frac{\mu_1^2}{2}\phi^2 + \frac{\lambda\phi^4}{4} + \frac{3e^4\phi^4}{32\pi^2}\ln\frac{\phi}{\mu_2} , \qquad (2.5)$$

where μ_1 and μ_2 are determined by normalization conditions imposed on $V(\phi)$. Coleman and Weinberg have suggested the following normalization conditions:

$$\left.\frac{d^2V}{d\phi^2}\right|_{\phi=0} = -\mu^2 ,$$
$$\frac{1}{6}\left.\frac{d^4V}{d\phi^4}\right|_{\phi=M} = \lambda , \qquad (2.6)$$

where M is some normalization point ($M \neq 0$ since $\left.\frac{d^4V}{d\phi^4}\right|_{\phi=M} \sim e^4 \ln\frac{\phi}{M} \to \infty$ at $M \to 0$). In this case

$$V(\phi) = -\frac{\mu^2}{2}\phi^2 + \frac{\lambda\phi^4}{4} + \frac{3e^4\phi^4}{32\pi^2}\left(\ln\frac{\phi}{M} - \frac{25}{12}\right) . \qquad (2.7)$$

From (2.7) it follows that symmetry can be broken even at $\mu^2 \geq 0$. In the most interesting case $\mu^2 = 0$ (massless scalar electrodynamics) it follows from (2.7) that $V(\phi)$ has a minimum not at $\phi = 0$, but at some $\phi = \phi_o(\lambda, M)$, and $V(\phi)$ looks as follows

$$V(\phi) = \frac{3e^4\phi^4}{32\pi^2}\left(\ell n\,\frac{\phi}{\phi_o} - \frac{1}{4}\right) . \tag{2.8}$$

This theory, in which $V''(0) = 0$, is called the Coleman-Weinberg theory.

Two comments are in order.

1. Since the parameter μ_1 in (2.5) is fixed by the normalization conditions, the same theory (2.8) can be obtained from theories with different bare mass μ and different normalization conditions. In particular, one can consider the Higgs model, which is just the theory (2.1) with $\mu^2 > 0$ and with spontaneous symmetry breaking $\phi_o = \mu/\sqrt{\lambda}$, and impose the following normalization conditions (Linde 1976, 1979):

$$\left.\frac{dV}{d\phi}\right|_{\phi=\phi_o} = 0 , \qquad \left.\frac{d^2V}{d\phi^2}\right|_{\phi=\phi_o} = 2\mu^2 . \tag{2.9}$$

These normalization conditions imply that quantum corrections do not change the value of the classical field $\phi = \phi_o = \mu/\sqrt{\lambda}$ and the mass of the Higgs bosons in the minimum of $V(\phi)$. The effective potential in such a theory is given by

$$V(\phi) = -\frac{\mu^2\phi^2}{2}\left(1 - \frac{3e^4}{16\pi^2\lambda}\right) + \frac{\lambda\phi^4}{4}\left(1 - \frac{9e^4}{32\pi^2\lambda}\right) + \frac{3e^4\phi^4}{64\pi^2}\,\ell n\,\frac{\lambda\phi^2}{\mu^2} . \tag{2.10}$$

It can be easily seen that at $\lambda = 3e^4/16\pi^2$ the effective potentials (2.8) and (2.10) coincide with each other. This means that the Coleman-Weinberg model (2.8) is equivalent to the Higgs model (2.10) with the normalization conditions (2.9) and $\lambda = 3e^4/16\pi^2$ (Linde 1976, 1979). One can draw an important lesson from this result. The Coleman-Weinberg theory can be formulated starting from the true vacuum state of this theory, $\phi = \phi_o$. In this case the theory would seem very unnatural, $\lambda = 3e^4/16\pi^2$. (This is not surprising since all spontaneously broken gauge theories look very unnatural <u>after</u> symmetry breaking.) However, being formulated before the symmetry breaking takes place, the Coleman-Weinberg theory looks very natural. It is just massless scalar electrodynamics.

2. Effective coupling constant $\lambda(\phi) = \frac{1}{6}\frac{d^4V}{d\phi^4}$ in the Coleman-Weinberg theory grows in magnitude as $\phi \to 0$. According to (2.8),

$$\lambda(\phi) = \frac{3e^4}{8\pi^2} \ln \frac{\phi}{\phi_o} + \frac{11e^4}{16\pi^2} . \tag{2.11}$$

Note that $\lambda(\phi)$ is large and <u>negative</u> at $\phi \to 0$. In non-Abelian gauge theories the effective gauge coupling $g(\phi)$ also rapidly increases as $\phi \to 0$. As a result, it is impossible to obtain a reliable expression for $V(\phi)$ at small ϕ by means of perturbation theory and the renormalization group equation. Therefore, strictly speaking, it is unknown whether a theory in which $V''(0) = 0$ and the absolute minimum of $V(\phi)$ occurs at $\phi = \phi_o \neq 0$ actually exists. However such a possibility seems plausible, since according to eq.(2.10) the value of $V''(0)$ can be arbitrarily varied if one varies the value of μ^2 in the corresponding Higgs model, and it is very likely that there exists a value of μ^2 for which $V''(0) = 0$. Therefore it is commonly believed that the Coleman-Weinberg theory with $V''(0) = 0$ and with the dynamical symmetry breaking $\phi = \phi_o$ does actually exist.

In conclusion let us consider the SU(5) theory. The effective potential in this theory in the tree approximation is given by

$$V(\Phi) = \frac{1}{4} a(\mathrm{Tr}\ \Phi^2)^2 + \frac{1}{2} b\ \mathrm{Tr}\ \Phi^4 , \tag{2.12}$$

where Φ is the Higgs scalar field in the adjoint representation. For simplicity we shall consider here symmetry breaking $SU(5) \to SU(3) \times SU(2) \times U(1)$ due to the appearance of the classical field

$$\Phi = \phi \sqrt{\frac{2}{15}}\ \mathrm{diag}(1, 1, 1, -\frac{3}{2}, -\frac{3}{2}) . \tag{2.13}$$

The effective potential of the SU(5) Coleman-Weinberg theory is

$$V(\phi) = \frac{25g^4}{128\pi^2}\left(\phi^4 \ln \frac{\phi}{\phi_o} - \frac{\phi^4}{4} + \frac{\phi_o^4}{4}\right) , \tag{2.14}$$

where g is the SU(5) gauge coupling constant. The last term is added in order to ensure that $V(\phi_o) = 0$. Since the vector meson mass M_X in this theory is equal to $\sqrt{\frac{5}{3}}\frac{g\phi_o}{2}$, the effective potential (2.14) can be also represented as

$$V(\phi) = \frac{25g^4\phi^4}{128\pi^2}(\ln\frac{\phi}{\phi_o} - \frac{1}{4}) + \frac{gM_x^4}{32\pi^2} . \qquad (2.15)$$

3 THE NEW INFLATIONARY UNIVERSE SCENARIO: A GENERAL OUTLOOK

The theory of phase transitions in GUTs with the Coleman-Weinberg mechanism of symmetry breaking is very complicated. Therefore we would like first to give a somewhat simplified description of the phase transition in the SU(5) Coleman-Weinberg theory, which will make it possible to outline the main idea of the new inflationary universe scenario (Linde 1982). A more detailed discussion of this phase transition will be contained in Sections 4-6.

The effective potential in the SU(5) Coleman-Weinberg theory at finite temperature is given by (Kirzhnits 1972; Kirzhnits & Linde 1972, 1974, 1976; Weinberg 1974; Dolan & Jackiw 1974):

$$V(\phi,T) = \frac{18T^4}{\pi^2}\int_0^\infty dx\, x^2 \ln(1 - \exp(-\sqrt{x^2 + \frac{5g^2\phi^2}{12T}})) + \frac{25g^4}{128\pi^2}\phi^4(\ln\frac{\phi}{\phi_o} - \frac{1}{4}) + \frac{9}{32\pi^2}M_x^4 , \qquad (3.1)$$

where $M_x = \sqrt{\frac{5}{3}}\frac{g\phi_o}{2} \approx 5\times10^{14}$ GeV , $g^2 \approx 0.3$, $\phi_o \approx 1.4\times10^{15}$ GeV . At $T \gg \phi_o$ the only minimum of $V(\phi)$ is that at $\phi = 0$, and the symmetry is restored. With a decrease of temperature the absolute minimum of $V(\phi)$ appears at $\phi \approx \phi_o$. However, according to (3.1), at any $T \neq 0$ the point $\phi = 0$ remains a local minimum of $V(\phi)$:

$$m^2(T) = \left.\frac{d^2V}{d\phi^2}\right|_{\phi=0} = \frac{5}{4}g^2T^2 . \qquad (3.2)$$

To be more precise, one should also take into account the scalar field contribution to $V(\phi,T)$, which changes the curvature of $V(\phi,T)$ near $\phi = 0$,

$$\left.\frac{d^2V}{d\phi^2}\right|_{\phi=0} = (\frac{5}{4}g^2 + \frac{65a + 47b}{30})T^2 . \qquad (3.3)$$

We shall forget for a while about the difference between (3.2) and (3.3) since usually $a,b \ll g^2$ (see however the next Section). In that case the phase transition from the local minimum of $V(\phi,T)$ at $\phi = 0$ to the global minimum at $\phi \sim \phi_o$ proceeds by the formation and subsequent expansion of the bubbles of the field ϕ. According to Sher (1981) and Billoire & Tamvakis (1982), the probability of this process becomes significant only at a very small temperature, $T_c \leq 10^6$ GeV (this statement is not quite correct, see Sections 4,5, but we shall assume for a moment that it is true). In that case from our investigation of the bubble formation at a finite temperature (Linde 1981) it follows that the bubble of the field ϕ at the moment of its formation has a size $O(T_c^{-1})$, and the field inside the bubble is much smaller than ϕ_o (see Fig.1):

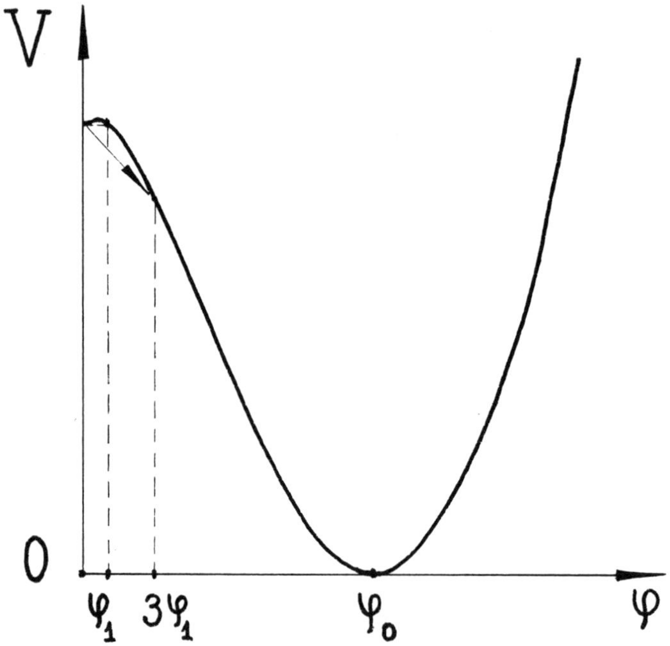

Fig.1 Effective potential in the Coleman-Weinberg theory for $T \ll \phi_o$. The arrow indicates the direction of the tunnelling with bubble formation. The field inside the bubble $\phi \leq 3\phi_1$, where $V(\phi_1) = V(o)$.

$$\phi \le \frac{12\pi T_c}{g}\left(5 \ln \frac{M_x}{T_c}\right)^{-1/2} \ll \phi_o \ . \tag{3.4}$$

This means that the mass squared of the field ϕ inside the bubble is

$$m^2 = \frac{d^2V}{d\phi^2} \le 75\ g^2 T_c^2 \sim 25\ T_c^2 \ . \tag{3.5}$$

It is clear that the field ϕ inside the bubble will grow up to its equilibrium value $\phi \sim \phi_o$ during the time interval τ, which is of the order or greater than $m^{-1} \approx 0.2\ T_c^{-1}$. (As will be shown later τ is even much greater than m^{-1}.) During the most part of this interval the field ϕ remains much smaller than ϕ_o. This means that during some time of the order of $0.2\ T_c^{-1}$ the vacuum energy density $V(\phi)$ remains almost equal to $V(0)$, and the part of the universe inside the bubble expands exponentially just as it expanded before the bubble creation. The value of the Hubble constant H in this theory is given by

$$H = \sqrt{\frac{8\pi}{3M_p^2} V(0)} = \frac{M_x^2}{2M_p}\sqrt{\frac{3}{\pi}} \approx 10^{10}\ \text{GeV} \ . \tag{3.6}$$

During the time $\tau \sim 0.2\ T_c^{-1}$ the universe grows $e^{H\tau}$ times, where

$$e^{H\tau} \sim e^{0.2 \frac{H}{T_c}} \sim e^{2000} \sim 10^{800} \ . \tag{3.7}$$

A typical size of the bubble at the moment of its creation is $O(T_c^{-1}) \sim 10^{-20}$ cm. After the period of exponential expansion this bubble has a size of the order of $10^{-20}\ e^{H\tau}$ cm $\sim 10^{800}$ cm, which is much greater than the size of the observable part of the universe at present, $\ell \sim 10^{28}$ cm. Therefore the whole observable part of the universe is contained inside one bubble, so we see no inhomogeneities caused by the bubble wall collisions.

When the field ϕ grows sufficiently large, the rate of its growth increases, and finally the field ϕ becomes convergently oscillating near its equilibrium state $\phi = \phi_o$ with frequency equal to the Higgs meson mass m at $\phi = \phi_o$, $m \approx M_x/4 \approx 10^{14}$ GeV. Note that this frequency is much larger than $H \sim 10^{10}$ GeV. Therefore the whole process of symmetry breaking can be approximately divided into two parts:

1. The field ϕ grows very slowly, whereas the size of the bubble increases exponentially, and all the observable part of the universe becomes filled with the almost homogeneous field $\phi \ll \phi_o$.

2. This almost homogeneous field rapidly grows. During this time

the universe expansion rate can be neglected as compared with the rate of the growth and subsequent oscillations of the field ϕ . The oscillating field ϕ creates Higgs bosons and X,Y bosons, which rapidly decay and reheat the universe up to the temperature $T_1 \sim M_X/6 \sim 10^{14}$ GeV (Linde 1982).

The process of the baryon asymmetry generation in the inflationary universe somewhat differs from that in the standard baryosynthesis scenario. In the first papers discussing this question it was assumed that the baryon asymmetry in the inflationary universe is generated either directly by the oscillating field ϕ (Linde 1982; Hawking & Moss 1982; Abbott et al. 1982), or by the decay of the Higgs bosons after the reheating of the universe (Albrecht et al. 1982). However, actually the process of the baryon production proves to be much more complicated and in the most interesting cases proceeds due to decay of Higgs bosons and/or X,Y bosons before the final stage of the reheating of the universe (Dolgov & Linde 1982). It appears that the baryon asymmetry in an inflationary universe under certain constraints on the masses of superheavy bosons can be one or two orders of magnitude greater than that in the standard scenario. It is important also that any initial baryon asymmetry of the universe vanishes after the inflation. Therefore the final baryon asymmetry in the inflationary universe does not depend on the initial baryon asymmetry of the universe in contrast to what occurs in the standard baryosynthesis scenario based on B-L conserving GUTs.

One of the most important problems of GUTs is the problem of primordial monopoles which are copiously produced during the phase transitions in GUTs. A large amount of monopoles should survive after the monopole-antimonopole annihilation in the very early universe (Zeldovich & Khlopov 1978; Preskill 1979), which would drastically contradict cosmological data. However in our scenario no monopole problem arises. Indeed, promordial monopoles in GUTs are created only in the points in which bubbles with different types of the Higgs field ϕ collide (Kibble 1976). Therefore no monopoles are created in the observable part of the universe in our scenario. For the same reason there will be no domain walls in the theories with broken discrete symmetries (Zeldovich et al. 1975). This solves the domain wall problem in the theories in which CP invariance is spontaneously broken before (or simultaneously with) the phase transition in the Coleman-Weinberg theory. In particular, inflation saves the SU(5) Coleman-Weinberg theory itself, since in this theory there exists a *discrete* symmetry $\phi \to -\phi$ (Lazarides *et al*. 1981; Tamvakis & Vayonakis 1981; Parke & Pi 1981).

Inflation of the universe dilutes the density of all objects which existed before the exponential expansion stage. In particular, it dilutes the density of gravitini, which have been created near the Big Bang. Thus one can avoid some stringent cosmological constraints on the gravitino mass (Ellis *et al*. 1982).

Now what can be said about the horizon problem, the flatness problem etc.? In the scenario under consideration the observable part of the universe is created from a domain with size of the order of 10^{-800} cm. This domain at the time of the phase transition ($t \sim 10^{-35}$ sec. after the Big Bang) was causally connected. After the exponential expansion period this causally connected domain covers the whole observable part of the universe, which solves the horizon problem.

From out results it follows that the size ℓ_1 of the universe after the phase transition should exceed 10^{800} cm, and the temperature T_1 is of the order of 10^{14} GeV. Therefore the total entropy of the universe after the reheating exceeds $(\ell_1 T_1)^3 \sim 10^{2400}$. This solves the flatness problem, which is equivalent to the problem why the total entropy of the universe exceeds 10^{85}. However, to be more precise, inflation of the universe completely solves the flatness problem only if our universe is open. If the universe is closed, its typical lifetime is $t_p \sim 10^{-43}$ sec, which is much smaller than the time of the beginning of the phase transitions in GUTs ($t \sim 10^{-35}$ sec in the minimal SU(5) theory). Therefore the inflationary universe scenario based on the theory of phase transitions in GUTs suggests that our universe is *open*, since otherwise the universe should be very flat *before* the phase transition in order to have a lifetime exceeding $t \sim 10^{-35}$ sec (Linde 1982a). Another important prediction of the inflationary universe scenario is that the universe at present should be almost completely flat (Guth 1982). Indeed, whereas it is possible to explain why the total entropy is much greater than 10^{85}, it is very difficult to obtain the total entropy exactly equal to 10^{85}.

Now let us discuss the problems of isotropy and homogeneity of the universe. As it is known, particle creation in the very early universe in general cannot make the universe completely isotropic, but can make it quasi-isotropic (Zeldovich & Novikov 1975), i.e. locally isotropic in small domains of space of the size of the same order or greater than the Planck length $\ell_P \sim 10^{-33}$ cm $\sim M_P^{-1}$ at the Planck time $t_P \sim 10^{-43}$ sec, when the temperature was $T_p \sim M_p \sim 10^{19}$ GeV . Since before the phase transition the quantity $a(t)T$ was constant inside each isotropic domain of the universe, at the moment of the phase transition the typical size of the isotropic domain exceeds the bubble size $O(T_c^{-1})$. Therefore the space-time inside the bubble was isotropic and the exponential expansion extends this isotropy to all the observable part of the universe. (Moreover, the remaining small anisotropy inside the bubble decreases rapidly during the exponential expansion period (Barrow & Turner 1981).) This may solve the long-standing problem of the space-time isotropy in our universe.

Density fluctuations inside the bubble immediately after its formation are negligibly small as compared with $V(0)$, i.e. the space inside the bubble is almost homogeneous. Then the exponential expansion extends this homogeneity to all the observable part of the universe, which may explain the large-scale homogeneity of the universe (Linde 1982).

However the universe after the phase transition does not become absolutely homogeneous. Indeed, during the phase transition long-wave quantum fluctuations of curvature and of the scalar field ϕ rapidly grow. This process gives rise to almost scale-independent perturbations of density, which may explain the origin of density perturbations necessary for galaxy formation (Mukhanov & Chibisov 1981, 1982; Hawking 1982; Starobinsky 1982; Guth & Pi 1982; Bardeen et al. 1982). The magnitude of these perturbations depends on the underlying particle theory and will be discussed in Section 7. It is worth noting also that for similar reasons the universe after the phase transition will not be absolutely isotropic; small violations of the isotropy induce an anisotropy of the microwave background radiation which could be experimentally measurable (Rubakov et al. 1982).

Now we have some idea of what the new inflationary universe scenario is and how it can help us to solve simultaneously many different cosmological problems. However the version of the inflationary universe scenario discussed above (as well as a similar version suggested later by Albrecht & Steinhardt (1982)) is oversimplified. As was noted by Linde (1982), to have a more precise theory of the Coleman-Weinberg phase transition one should take into account the effects connected with the non-vanishing curvature and rapid expansion of the universe, which become important at $T \leq H$ (Hawking & Moss 1982; Linde 1982 a,c; Starobinsky 1982), see Sections 5 and 6.

However first of all one should develop a more detailed theory of the high-temperature effects in the Coleman-Weinberg theory (Linde 1982b). This question will be discussed in the next Section.

4 TEMPERATURE DEPENDENCE OF COUPLING CONSTANTS AND THE PHASE TRANSITION IN THE COLEMAN-WEINBERG THEORY

The theory of bubble production in the Coleman-Weinberg theory discussed in the previous Section was based essentially on eq. (3.2) for the curvature of $V(\phi,T)$ near $\phi = 0$. As we have already mentioned, a more precise expression for $V''(0)$ is given by eq. (3.3), but usually the last two terms in this equation are small, and eq. (3.2) practically coincides with eq. (3.3). However, whereas for $\phi, T \sim \phi_o$ we have $a, b \sim g^4 << g^2$, at small ϕ and T the effective coupling constants a and b become large and cannot be neglected as compared with g^2. We will study the behaviour of $g^2(T)$, $a(T)$ and $b(T)$ at $\phi = 0$ by means of the renormalization group equations (Linde 1982b). In the minimal SU(5) model with three fermion generations these equations look as follows (Fradkin & Kalashnikov 1976)

$$g^2(t) = \frac{3\pi^2}{5t}, \tag{4.1}$$

$$4\pi^2 \frac{da}{dt} = 16a^2 + \frac{94}{5}ab + \frac{168}{25}b^2 - 15g^2a + \frac{9}{4}g^4, \tag{4.2}$$

$$4\pi^2 \frac{db}{dt} = \frac{32}{5}b^2 + 6ab - 15g^2b + \frac{15}{8}g^4, \tag{4.3}$$

where $t = \ln\frac{T}{\Lambda}$ (at $\phi = 0$) or $t = \ln\frac{\phi}{\Lambda}$ (at $T = 0$). Here Λ is the normalization parameter, $\Lambda \approx 4\times10^6$ GeV. Parameters a and b in the SU(5) Coleman-Weinberg theory are not uniquely fixed; we shall consider for simplicity the case $a(T=0, \phi=\phi_o) = b(T=0, \phi=\phi_o)$. In that case $a(T=\phi_o, \phi=0) \approx b(T=\phi_o, \phi=0) \approx \frac{125}{128\pi^2}g^4(T=\phi_o, \phi=0) \approx 10^{-2}$, $g^2(T=\phi_o, \phi=0) = 0.3$. Solving eqs. (4.1)-(4.3) with these boundary conditions gives the temperature dependence of $g^2(T)$, $a(T)$, $b(T)$ and

$$\frac{m^2(T)}{T^2} = \frac{1}{T^2}\frac{d^2V(\phi,T)}{d\phi^2}$$

shown in Fig.2. (Similar results are valid for the coupling constants $g^2(\phi)$, $a(\phi)$ and $b(\phi)$ at $T = 0$.) It is seen that at $T \sim \phi_o$, $|a|, |b| << g$, and eq. (3.2) is approximately valid. However at smaller values of T the parameters a and b become negative and rapidly growing

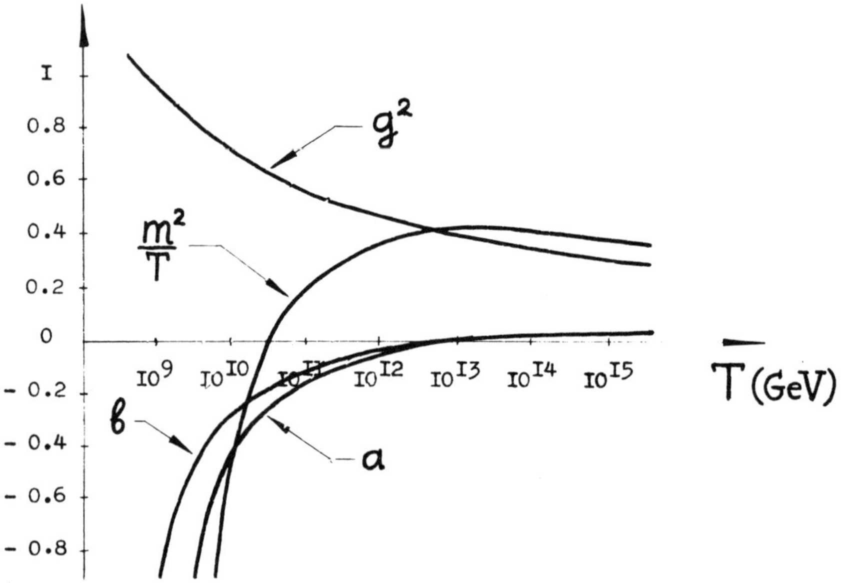

Fig.2 Temperature dependence of the effective coupling constants g^2, a and b, and of the Higgs boson mass m(T) at $\phi = 0$.

in magnitude, so that eq. (3.2) becomes inapplicable and $m^2(T) = d^2V/d\phi^2$ becomes negative at $T_c \sim 10^{-5}\ \phi_o \sim 10^{10}$ GeV. This means that the high-temperature phase transition in the SU(5) Coleman-Weinberg theory may occur not due to tunneling with bubble formation, but merely due to the "rolling down" of the scalar field from the top of the effective potential with $V''(0) < 0$. This means also, that this phase transition may start neither at $T \leq \Lambda \sim 10^6$ GeV as was assumed by Billoire & Tamvakis (1982), nor at $T \sim 10^8$ GeV as was claimed by Albrecht & Steinhardt (1982), but much earlier, at $T_c \sim 10^{10}$ GeV.

Several comments should be made in connection with this result, however.

1. The behaviour of a(T), b(T) and $m^2(T)$ depends crucially on the type of the theory and its field content. Therefore the value of T_c may be either much greater or much smaller than $H \sim 10^{10}$ GeV.

2. If $T_c >> H \sim 10^{10}$ GeV , then the phase transition occurs too early for the new inflationary universe scenario to be completely realized. However for $T_c \sim 10^{10} - 10^{11}$ GeV the scale factor of the universe increases $\sim 10^4$ times during the de Sitter stage before the phase transition. This is sufficient to solve (or at least to soften) the primordial monopole problem and the gravitino mass problem (Ellis et al. 1982), and the new scenario for baryon asymmetry generation (Dolgov & Linde 1982) remains valid.

3. At $T_c << H$ all high-temperature effects are irrelevant to the theory of the phase transition. Indeed, the time interval which is necessary for the phase transition to start due to high-temperature effects exceeds $T_c^{-1} >> H^{-1}$, and during this interval the temperature falls to $T_c\ e^{-H/T_c}$, i.e. practically vanishes. This means that at $T_c << H$ the phase transition proceeds not due to high-temperature effects considered in the first versions of the new inflationary universe scenario (Linde 1982, Albrecht & Steinhardt 1982), but due to the effects connected with the nonvanishing curvature and rapid expansion of the universe (Hawking & Moss 1982; Linde 1982 a,c; Starobinsky 1982). These effects will be considered in the Sections 5 and 6.

4. In the case $T_c \sim H$ the phase transition is started by the growth of quantum fluctuations of the field ϕ (i.e. by the growth of $<\phi^2>$) at $m^2(T) < 0$, see Section 6. However during the time $\Delta t \sim H^{-1}$ the value of $m^2(T)$ exponentially decreases and a further growth of fluctuations of the field ϕ proceeds just as at $T = 0$. The possibility of a phase transition triggered by the high-temperature effects at $T_c \sim H$ is very interesting. However the theory of this effect is rather involved and will not be considered in the present paper.

Before discussing symmetry behaviour in the Coleman-Weinberg theory at $T_c << H$ it should be noted that from the point of view of a comoving observer the temperature in the exponentially expanding universe never becomes smaller than the Hawking temperature $T_H = \frac{H}{2\pi}$ (Gibbons & Hawking 1977). This is connected with the fact that any comoving observer in the exponentially expanding universe cannot have any information about anything removed at a distance greater than the horizon radius H^{-1} from the observer. However if one tries to describe the whole universe (bearing in mind that after the phase transition the above-mentioned horizon disappears) one can use a formalism in which there is no lower bound on the temperature (in particular, we will use such a formalism in our study of the scalar field fluctuations in inflationary universe, Section 6). Just in this sense we are speaking here about temperatures $T \ll H$. For a more detailed discussion of the concept of temperature and of the definition of the vacuum state in the exponentially expanding (de Sitter) universe see e.g. Birrell & Davies (1982).

5 COLEMAN-WEINBERG THEORY IN DE SITTER SPACE

The main feature of the new inflationary universe scenario is the slow growth of the field ϕ in the exponentially expanding Friedmann universe with metric

$$ds^2 = dt^2 - e^{2Ht}(dx^2 + dy^2 + dz^2) \tag{5.1}$$

(we consider here the flat Friedmann space since during the exponential expansion before the phase transition the universe becomes almost flat and the effects connected with the spatial curvature of the universe become negligible). To have a more exact idea of what is necessary for the realization of the new inflationary universe scenario, let us consider the growth of a scalar field ϕ with mass $m^2(\phi) = V''(\phi) < 0$. The Klein-Gordon equation for the homogeneous field ϕ in the exponentially expandind universe (5.1) is

$$\ddot{\phi} + 3H\dot{\phi} + m^2(\phi)\phi = 0 . \tag{5.2}$$

In the case $|m^2| \ll H^2$ it can be shown that the first term in (5.2) can be neglected as compared with the second one, and instead of eq. (5.2) one has

$$\frac{\dot{\phi}}{\phi} = -\frac{m^2(\phi)}{3H} . \tag{5.3}$$

An approximate solution of this equation is

$$\phi \sim \phi_1 \exp(-\frac{m^2(\phi_1)}{3H} t) , \tag{5.4}$$

which is true until $m^2(\phi)$ becomes considerably different from $m^2(\phi_1)$. From (5.4) it follows that the time τ necessary for the phase transition to occur is of the order ot $\tau \geq 3H/|m^2(\phi_1)|$. As we have already mentioned, in order to obtain a solution of the horizon and flatness problems, the duration τ of the exponential expansion period should exceed approximately $65\ H^{-1}$ (Guth 1981). This leads to the following condition necessary for the realization of the new inflationary universe scenario (Linde 1982a)

$$|m(\phi_1)| \leq 0.2\ H\ . \tag{5.5}$$

The simplest possibility is that the initial point from which the field ϕ begins to grow is $\phi = 0$. In this case inequality (5.5) implies that

$$|m(0)| \leq 0.2\ H\ . \tag{5.6}$$

It is clear that theories in which inequalities (5.5) and (5.6) hold do actually exist. However these inequalities are not quite innocent. For example, in the SU(5) theory with $M_X \sim 5\times10^{14}$ GeV , $H \sim 10^{10}$ GeV , so that $|m(0)| < 0.2\ H \sim 2\times10^9$ Gev $\sim 10^{-5}\ M_X$. Therefore one may wonder whether it is possible to satisfy the hierarchy condition $|m(0)| \leq 10^{-5} M_X$ in a natural way. The simplest strategy in considering this question would be to postpone answering it until the hierarchy problem in GUTs is solved. In any case the possibility of interplay between the hierarchy problem in GUTs and the horizon, flatness, homogeneity and isotropy problems in cosmology is very intriguing.

Another possibility to have $|m(\phi_1)| \leq 0.2\ H$ is connected with the Coleman-Weinberg theory discussed in previous Sections. To analyse this question let us consider the Coleman-Weinberg theory in the exponentially expanding Friedmann universe (Linde 1982a). If the duration of the exponential expansion stage is sufficiently large, the effective potential $V(\phi)$ in the Friedmann universe becomes equal to $V(\phi)$ in de Sitter space with the same curvature scalar $R = 12\ H^2$. Therefore we will use the results of calculations of $V(\phi)$ in de Sitter space which have been obtained by Shore (1980) (see also Allen 1982, Starobinsky 1982) for the U(1) Coleman-Weinberg model (extension of these results for the SU(5) theory is trivial). Our interpretation of these results, however, will be slightly different from that of Shore.

According to Shore, an expression for $V(\phi)$ in de Sitter space with the curvature scalar R in the limit $e^2\phi^2 << R$ looks as follows

$$V(\phi,R) = \frac{1}{2}\mu_1^2\phi^2 + \frac{e^2R\phi^2}{64\pi^2}\ln\frac{R}{\mu_2^2} + \frac{3e^4\phi^4}{64\pi^2}\ln\frac{R}{\mu_3^2} + V(0,R) \ , \qquad (5.7)$$

where μ_i are some normalization constants, which are necessary for renormalization of $m^2 = \frac{d^2V}{d\phi^2}$, of the parameter $\xi = \frac{d^3V}{d\phi^2 dR}$ and of the effective coupling constant $\lambda = \frac{1}{6}\frac{d^4V}{d\phi^4}$. At $R \ll e^2\phi^2$, $V(\phi)$ coincides with the usual Coleman-Weinberg effective potential in Minkowski space.

To specify the values of μ_i one should impose normalization conditions on m^2, ξ and λ. Let us first consider the original normalization conditions (2.6) for m^2 and λ in the Coleman-Weinberg theory in Minkowski space-time:

$$\left.\frac{d^2V}{d\phi^2}\right|_{\phi=0,R=0} = 0 \ ,$$

$$\left.\frac{1}{6}\frac{d^4V}{d\phi^4}\right|_{\phi=M,R=0} = \lambda \ , \qquad (5.8)$$

where M is some normalization point, $M \neq 0$. The quantity ξ usually is taken to be equal to $1/6$ in order to maintain conformal invariance of the theory, or it is assumed that $\xi = 0$. However, the mechanism of symmetry in the Coleman-Weinberg theory is based on the breaking of conformal invariance, and therefore there is no reason to assume that $\xi = 1/6$. Moreover, such an assumption would be self-contradictory, since the value of $\xi = \frac{d^3V}{d\phi^2 dR}$ is not a constant, but depends on ϕ and R. As we have seen in Section 2, in order to formulate the theory in a natural way one should try to impose normalization conditions at the point $\phi = 0$. However, it is impossible to normalize ξ at $\phi = 0$, $R = 0$, since in this limit $\xi \to \infty$, see eq. (5.7). Therefore we suggest imposing on ξ the self-consistent normalization condition

$$\xi = \left.\frac{d^3V}{d\phi^2 dR}\right|_{\phi=0,R=32\pi GV(0,R)} = 0 \ , \qquad (5.9)$$

which means that $\xi = 0$ at $\phi = 0$ not in Minkowski space, but in de Sitter space (in the exponentially expanding Friedmann space) with scalar curvature determined by the vacuum energy $V(\phi)$ at $\phi = 0$. Normalization conditions (5.8), (5.9) lead to the following expression for $V(\phi)$ at small ϕ :

$$V(\phi,R) = -\frac{e^2\phi^2R}{64\pi^2} + \frac{e^2R\phi^2}{64\pi^2}\ln\frac{R}{32\pi GV(0,R)} + V(0,R) + O(\phi^4) \ . \qquad (5.10)$$

This means that in the theory under consideration the effective mass squared $m^2(0)$ in the exponentially expanding universe differs from zero and is given by

$$m^2(0) = \left.\frac{d^2V}{d\phi^2}\right|_{\phi=0,R=32\pi GV(0,R)} = -\frac{e^2R}{32\pi^2} = -\frac{3e^2H^2}{8\pi^2} \ . \qquad (5.11)$$

The analogous expression in the SU(5) Coleman-Weinberg theory is

$$m^2(0) = -\frac{15g^2H^2}{8\pi^2} \ . \qquad (5.12)$$

If one takes for g^2 in (5.12) the effective value of g^2 at $\phi \sim H \sim 10^{10}$ GeV , $g^2 \sim 0.65$, eq. (5.12) yields

$$|m(0)| \approx 0.35 \ H \ , \qquad (5.13)$$

which is not far from the desired value $|m(0)| \le 0.2$ H , eq. (5.6). It is worth noting that the above-mentioned result is model dependent. For example, in the U(1) theory one has a much smaller value of $|m(0)|$ (5.11).

We see therefore, that in the context of the Coleman-Weinberg theory one can obtain a natural explanation of the mass hierarchy $|m(0)| \le 0.2\ H \ll M_X$, which is necessary for the realization of the new inflationary universe scenario. One should not, however, take quite seriously the numerical estimate (5.13), since near the point $\phi = 0$ the scalar field fluctuations give an important contribution to $V(\phi)$, which has not been taken into account by Shore, see below. Moreover, the version of the Coleman-Weinberg theory discussed above does not seem to be the most natural one. Indeed, in contrast to eq. (5.9), the normalization conditions (5.8) are not self-consistent, since at $\phi = 0$ the curvature scalar does not vanish, $R = 32\pi GV(0)$. Such normalization conditions are quite legitimate like the off-shell normalization conditions in QCD . However in our opinion it is much more natural to impose self-consistent (on-shell) normalization conditions on m^2 and λ . For example, the self-consistent generalization of the normalization condition (5.8) on m^2 is

$$m^2(0) = \left.\frac{d^2V}{d\phi^2}\right|_{\phi=0,R=32\pi GV(0,R)} = 0 . \qquad (5.14)$$

We will not discuss here the normalization conditions on ξ and λ in this version of the Coleman-Weinberg theory, since they will be irrelevant to our consideration.

The effective potential $V(\phi)$ (5.7) in this theory differs from the standard Minkowski space effective potential in the Coleman-Weinberg theory only in the vicinity of the point $\phi = 0$ (at $e\phi \leq H$) and has a minimum at some $\phi = \phi_0$. At the first glance no problems with higher orders of perturbation theory appear at small ϕ , since the curvature scalar R serves here as an infrared cut off (there are no terms $\sim \ln \phi$ in (5.7)). However the situation is not so simple, since eq. (5.7) does not take into account the scalar field contribution to $V(\phi)$. Usually it is quite possible to omit the scalar field contribution since in the Coleman-Weinberg theory the scalar coupling constant is small, $\lambda \ll e^2$. However, as we have shown in Section 4, at $\phi, T \sim H$ the effective coupling constants a and b are of the same order as g^2 . Moreover, as will be clear from the discussion in the next Section, perturbation theory in de Sitter space diverges at $\frac{\lambda R^2}{m^2} \sim \frac{\lambda H^4}{m^4} \gg 1$ (Linde 1982a). Therefore eq. (5.10) for $V(\phi)$ becomes unreliable for $\lambda\phi^4 \ll R^2$, and just as in the case $R = 0$ we cannot prove rigorously that the Coleman-Weinberg theory with $m^2(0) = 0$ in de Sitter space does exist. However such as possibility seems plasuible, since exactly the same arguments as those we used above in favour of the existence of the Coleman-Weinberg theory in Minkowski space (see Section 2), can be applied in the case $R \neq 0$ as well. Therefore the theory with dynamical symmetry breaking and with the normalization condition (5.14) may actually exist. We shall call this theory the improved Coleman-Weinberg theory. In our opinion such a theory would be a most natural generalization of the usual Coleman-Weinberg theory with account taken of gravitational effects. However, in the next Section we will not restrict ourselves to a discussion of this particular theory, but rather consider the phase transition in the theories with different values of $m^2(0)$.

6 SCALAR FIELD FLUCTUATIONS AND THE PHASE TRANSITION IN THE EXPONENTIALLY EXPANDING FRIEDMANN UNIVERSE

In the previous Section we neglected the contribution of scalar particles to $V(\phi)$. Such an approximation is quite good for $g\phi \gg H$.

However for $g\phi << H$ the scalar field fluctuations give the leading contribution to the effective potential (Starobinsky 1982; Linde 1982c). Therefore, it is necessary to investigate the scalar field fluctuations in the exponentially expanding universe, and their role in the inflationary universe scenario.

For simplicity we shall consider here the scalar field theory with an effective potential which at small ϕ looks as follows

$$V(\phi) = \frac{m^2\phi^2}{2} + \frac{\lambda}{4}\phi^4 + V(0) . \qquad (6.1)$$

This effective potential at small $|m^2|$ and at $\lambda < 0$ imitates the effective potential in the Coleman-Weinberg theory in which $\lambda(\phi)$ is negative at small ϕ and depends only logarithmically on ϕ. The mass squared m^2 in (6.1) may contain an addition ξR to the bare scalar field mass squared m_o^2 in Minkowski space, as well as the quantum corrections $\Delta m^2(R)$ to m_o^2 :

$$m^2 = m_o^2 + \xi R + \Delta m^2(R) , \qquad (6.2)$$

where to the lowest order in λ, $\Delta m^2(R) = 3\lambda\langle\phi^2\rangle$, which yields

$$m^2 = M^2 + 3\lambda\langle\phi^2\rangle . \qquad (6.3)$$

Here $M^2 = m_o^2 + \xi R$ is the bare mass squared of the field ϕ in de Sitter space, which does not take into account quantum corrections, and m^2 is a total mass squared. We will study here the case $|m^2| << H^2$, or, to be more precise, $|m^2(\phi)| = |\frac{d^2V}{d\phi^2}| << H^2$, which is necessary for the realization of the new inflationary universe scenario. In the lowest approximation this yields $|M^2| << H^2$. In this case from the results of Bunch & Davies (1978) it follows that

$$\langle\phi^2\rangle = \frac{3H^4}{8\pi^2M^2} \qquad (6.4)$$

in de Sitter space (5.1) for $0 < M^2 << H^2$. However in the Friedmann universe, which becomes exponentially expanding only after some cooling down, an expression for $\langle\phi^2\rangle$ is much more complicated. To calculate $\langle\phi^2\rangle$ in such a case one should take into account temperature corrections to the mass squared of the scalar field, which are large at $T > H$. To imitate the situation in the Coleman-Weinberg theory let us suppose that the scalar field ϕ interacts with some gauge field with the coupling constant g, $g^2 >> \lambda$, so that the temperature correction to M^2 at $T \geq H$ is given by $\Delta m^2(T) = O(g^2)T^2$, see Section 4. A general normalized solution for the scalar field ϕ with account taken of the temperature corrections looks as follows

$$\phi = \frac{H}{4\sqrt{2}\pi} e^{i\vec{k}\vec{x}} \eta^{3/2} (A_k H_\nu^{(1)} (\sqrt{c^2+\vec{k}^2}.\eta) + B_k H_\nu^{(2)} (\sqrt{c^2+\vec{k}^2}.\eta)) . \quad (6.5)$$

Here $\eta = H^{-1}e^{-Ht}$, A_k, B_k are some functions of $|\vec{k}|$, such that $|B_k^2| - |A_k^2| = 1$, $A_k \to 0$ at $\vec{k}^2 >> c^2$; $c^2 = \frac{O(g^2)T_o^2}{4H^2}$, where T_o is the temperature at which the universe becomes exponentially expanding; $\nu = \frac{3}{2}\sqrt{1 + \frac{4M^2}{9H^2}} \approx \frac{3}{2} + \frac{M^2}{3H^2}$. Calculation of $<\phi^2>$ in this case is rather involved, but the leading contribution to $<\phi^2>$ at $|M^2| << H^2$ can be easily obtained and does not depend on any particular choice of A_k and B_k (Ford & Vilenkin 1982; Starobinsky 1982; Linde 1982c):

$$<\phi^2> = \begin{cases} \frac{3H^4}{8\pi^2 M^2} (1 - \exp[- \frac{2M^2}{3H} (t-t_o)]) & \text{for } M^2 > 0 , \\ \frac{H^3}{4\pi^2} (t-t_o) & \text{for } M^2 = 0 , \\ \frac{3H^4}{8\pi^2 |M^2|} c^{-\frac{2|M^2|}{3H^2}} (\exp[\frac{2|M^2|}{3H}(t-t_o)]-1) & \text{for } M^2 < 0 . \end{cases} \quad (6.6)$$

Here t_o is the time at which the value of $M^2(T) = M^2 + \Delta m^2(T)$ becomes smaller than $2H^2$. At $M^2 > 0$, $\Delta t = t-t_o >> \frac{3H}{2M^2}$ this result is in agreement with that of Bunch & Davies (6.4). However $\Delta t \sim \frac{3H}{2M^2}$ is a large time, $\Delta t >> H^{-1}$, which means that it is not easy for the exponentially expanding universe to forget its hot past.

Before investigating the consequences of our results let us discuss the reasons for the anomalous growth of $<\phi^2>$ at small m^2 (6.4), (6.6). The formal mathematical reason for such a behaviour of $<\phi^2>$ is connected with the contribution of zero modes of the scalar field ϕ in the Euclidean formulation of the quantum theory in de Sitter space. This contribution leads to the appearance of factors $\lambda \frac{H^4}{m^4}$ in each new order of perturbation theory in λ. Divergences, which appear in this theory at $m^2 \to 0$, are completely analogous to (but are stronger than) the infrared divergences, which appear due to the contribution of zero modes in quantum statistics of bose fields at $T \neq 0$ (Linde 1979), and which have led to many troubles in the theory of the phase transitions at finite temperature (Wilson & Kogut 1974). The physical reason for the singular behaviour of $<\phi^2>$ at $m^2 \to 0$ is connected with the anomalously large density of long-wave fluctuations of the field ϕ in de Sitter space. Even if initially the density of such fluctuations was suppressed by the high-temperature effects, then these fluctuations appear again merely due to the fact, that the short-wave fluctuations rapidly become long-wave ones in the exponentially expanding universe. This is just the reason for the time dependence of $<\phi^2>$ in eq. (6.6).

These results are valid only until the value of $<\phi^2>$ grows large, so that the quantum corrections become important. In order to improve these results one can use a self-consistent approximation, in which instead of M^2 one should substitute into (6.6) the time dependent mass squared of the field ϕ, $m^2(t) = M^2 + 3\lambda<\phi^2>$ (Starobinsky 1982, Linde 1982c). Analysis of the improved equations (6.6) is somewhat simplified if one takes into account that $c^{-2|M^2|/3H^2} \approx 1$ at $|M^2| << H^2$, and therefore, independently of the sign and value of M^2, at the initial stage of the process $<\phi^2> = \frac{H^3}{4\pi^2}(t-t_o)$ just as in the case $M^2 = 0$ (6.6). At this stage

$$m^2(t) = M^2 + 3\lambda<\phi^2> = M^2 + \frac{3\lambda H^3}{4\pi^2}(t-t_o) . \tag{6.7}$$

Depending on relations between M^2, H^2 and λ, this stage finishes either after the time

$$\Delta t_1 = t - t_o \approx \frac{3H}{2|M^2|} , \tag{6.8}$$

at which the growth of $<\phi^2>$ becomes nonlinear in $t-t_o$, or at the moment at which the value of $<\phi^2>$ becomes so large that the higher orders of perturbation theory become greater than the lowest ones and equations (6.6) become unreliable even after the substitution $M^2 \to M^2 + 3\lambda<\phi^2>$. Let us consider the behaviour of $<\phi^2>$ in the theory (6.1) for different relations between M^2, H^2 and λ (at $\lambda < 0$).

A. $M^2 < 0$, $M^4 \geq M_o^4$, where $M_o^4 = \frac{9|\lambda|H^4}{8\pi^2}$. In this case the typical time necessary for the phase transition to occur is

$$\Delta t_A = t - t_o \sim \Delta t_1 = \frac{3H}{2|M^2|} \leq \frac{\sqrt{2}\,\pi}{\sqrt{\lambda}\,H} . \tag{6.9}$$

During this time the back-reaction of the growing fluctuations $\langle\phi^2\rangle$ on the value of $m^2(t)$ is small, $m^2(t) \approx M^2$. Later on the fluctuations become large, $m^2(t)$ becomes large and negative, and $\langle\phi^2\rangle$ rapidly grows up to some value of the order of ϕ_o^2, where ϕ_o corresponds to the minimum of the effective potential. Long-wave fluctuations of the field ϕ in this scenario play the same role as the bubbles of the new phase in the first version of the new inflationary universe scenario (Linde 1982). Since the wavelength of the fluctuations grows in time as e^{Ht}, and a typical wavelength should grow at least e^{60} times during the exponential expansion period, one obtains a constraint on M^2 similar to that obtained in the previous Section, eq. (5.5): $|M^2| \le \frac{H^2}{40}$. This, together with the inequality $M^4 \ge M_o^4$, gives a constraint on λ:

$$|\lambda| \le \frac{8\pi^2 M^4}{9H^4} \sim \frac{\pi^2}{1800} \sim 5\times10^{-3} \; . \qquad (6.10)$$

Scalar field fluctuations generated in this scenario may produce fluctuations of metric and lead to the inhomogeneities which are necessary for the galaxy formation (Hawking 1982; Starobinsky 1982; Guth & Pi 1982; Bardeen et al. 1981).

B. $M^2 < 0$, $M^4 \le M_o^4$. In this case the back-reaction of $\langle\phi^2\rangle$ on the value of $m^2(t)$ becomes important after the period of time

$$\Delta t_B \sim \frac{4\pi^2|M^2|}{3|\lambda|H^3} \le \frac{\sqrt{2}}{\sqrt{\lambda}}\frac{\pi}{H} \; , \qquad (6.11)$$

after which $m^2(t)$ rapidly grows in magnitude, so that $m^2(t) < 0$, $|m^2(t)| >> M^2$, and a further growth of $\langle\phi^2\rangle$ proceeds almost instantaneously. In this sense the study of the back-reaction of $\langle\phi^2\rangle$ on $m^2(t)$ is not very important for us, since the duration of the exponential expansion stage is determined by the value of Δt_B (6.11). From the condition $\Delta t_B \ge 60\, H^{-1}$ and inequality $|\lambda| \ge \frac{8\pi^2 M^4}{9H^4}$ one obtains the same constraint on λ (6.10) as in case A.

C. $M^2 > 0$, $M^4 << M_o^4$. In such a case the value of $\langle\phi^2\rangle$ grows as $\frac{H^3}{4\pi^2}(t-t_o)$ up to

$$\langle\phi^2\rangle \sim \langle\phi^2\rangle_C = \frac{H^2}{2\sqrt{2}\,|\lambda|\pi} \qquad (6.12)$$

during the time of the same order as in the previous case:

$$\Delta t_C \sim \frac{\sqrt{2}\,\pi}{\sqrt{\lambda}\,H}\,, \tag{6.13}$$

after which the field ϕ acquires a negative mass squared growing in magnitude. At $t-t_o > \Delta t_C$ the rate of the process increases, so that the total duration of the exponential expansion period is given by Δt_C (6.13), which leads to the same constraint on λ (6.10) as before.

D. $M^2 > 0$, $M^4 >> M_o^4$. In this case the term $3\lambda<\phi^2>$ is always small as compared with M^2 , so that $m^2(t) \approx M^2$. Symmetry breaking proceeds due to formation of bubbles of the field ϕ . If the curvature of the effective potential $\frac{d^2V}{d\phi^2} = m^2(\phi)$ inside such a bubble is sufficiently small $(m\ (\phi) \leq \frac{H^2}{40})$, the new inflationary universe scenario in its form suggested by Linde (1982, 1982a) can be realized. Unfortunately, the theory of bubble formation in the Friedmann universe at the stage of exponential expansion is not elaborated so far. This problem has been investigated by many authors. However, about half of the papers discussing this question contain errors, whereas the correct investigation of bubble formation in curved space by Coleman & De Luccia (1980); Hawking & Moss (1982); Guth & Weinberg (1982); Parke (1982) and by Mottola & Lapedes (1982) is directly applicable only to the case of bubble formation in the eternally existing de Sitter space. The difference between the theory of bubble formation in a hot Friedmann universe with a de Sitter stage of expansion and in the eternally existing de Sitter space is very important. In the eternally existing de Sitter space one can choose a coordinate system in which the whole universe at some moment has a finite and rather small spatial volume $V \sim O(H^{-3})$. Therefore in the eternally existing de Sitter space the phase transition can occur simultaneously in the whole universe (Hawking & Moss 1982). However in the inflationary universe scenario the phase transition can start only when the temperature becomes less than H . If one takes into account that the exponential expansion in the SU(5) theory starts at $T \sim M_X$, one can easily show that even if the universe is closed, its spatial volume at the moment of the phase transition exceeds $(\frac{M_X}{H})^3 . O(H^{-3}) \sim 10^{12}\ O(H^{-3})$. Simultaneous phase transition in such a large space is practically impossible.

Our understanding of bubble formation in the Friedmann universe is still rather incomplete (Linde 1982 a,c; Goncharov & Linde 1982). Here we will simply take as an estimate of the curvature of the effective potential inside the bubble the value $m^2(\phi) \sim -M^2$ which seems reasonable and is in agreement with the results of the corresponding analysis of bubble formation in the eternally existing de Sitter space (Hawking & Moss 1982). This estimate leads again to the constraint (6.10) on the value of λ .

E. $M^2 > 0$, $M^4 \sim M_o^4 = \frac{9|\lambda|H}{8\pi^2}$. This case is most interesting and complicated. Indeed, as we have seen, $m^2(t) \approx M^2 > 0$ for $M^4 >> M_o^4$, whereas in the case $0 < M^2 << M_o^2$ the value of $m^2(t)$ becomes negative. It is clear that the boundary between these two regimes lies somewhere near $M^2 \lesssim M_o^2$. To analyse this question let us first write eq. (6.6) for $M^2 > 0$ in the limit $t \to \infty$ after the change $M^2 \to m^2 = M^2 + 3\lambda\langle\phi^2\rangle$:

$$\frac{3H^4}{8\pi^2 m^2} = \langle\phi^2\rangle = \frac{m^2-M^2}{3\lambda} . \tag{6.14}$$

From this equation one can conclude that the value $m^2 = 0$ is impossible. The minimal positive value of m^2 is achieved at $M^2 = 2M_o^2$ and is given by

$$m_C^2 = M_o^2 = \frac{3H^2}{4\pi}\sqrt{2\lambda} . \tag{6.15}$$

The value of $\langle\phi^2\rangle$ corresponding to this limiting case is equal to $\langle\phi^2\rangle_C$ (6.12). In the case $M^2 < 2M_o^2$ the value of $\langle\phi^2\rangle$ grows infinitely at $t \to \infty$. It may seem somewhat surprising that the boundary of the instability region lies not at $M^2 = 0$ or $m^2 = 0$, but at $M^2 = 2M_o^2$, $m^2 = M_o^2$. This is a purely quantum effect, connected with the instability with respect to the growth of the field $\langle\phi^2\rangle$. Analogous effects are known to appear in the $\lambda\phi^4$ theory and other theories studied in the self-consistent approximation similar to that used above (6.14), see e.g. Abbott et al. (1976), Linde (1977). However it is very difficult to justify validity of such an approximation at $m^2 \leq M_o^2$ (6.15), $\langle\phi^2\rangle \sim \langle\phi^2\rangle_C$ (6.12). A possible way to do it is to consider the O(N) theory $\frac{\lambda}{N}\phi^4$ with the scalar field ϕ in the vector representation. The leading approximation in $\frac{1}{N}$ in this theory coincides with our self-consistent approximation, which may justify our result (6.15) in the limit $N \to \infty$. However, one cannot use the same method in the theories with the scalar field ϕ in the adjoint representation (which are of the main interest for us), since the $\frac{1}{N}$ expansion in such theories is not elaborated. Some other approximations are possible, which effectively take into account leading infrared singular (at $m \to 0$) terms (Starobinsky 1982; Moss 1982). However since the infrared singularities at $m \to 0$ in the $\lambda\phi^4$ theory in de Sitter space are very strong, the nonleading terms may be also very important. Therefore the actual behaviour of $m(t)$ and $<\phi^2>$ in the case $m \lesssim M_o$ is far from being clear. In particular, it is quite possible that the boundary of the instability region actually lies not at $m_C^2 = M_o^2$ (6.15), but at $m^2 = 0$, which would correspond to the improved Coleman-Weinberg model advocated in the previous Section. A simplified analysis of the phase transition in the case $m \sim M_o$ gives the same estimate of the duration of the de Sitter stage and the same constraint on λ as in the cases A-D (Linde 1982c). However these estimates actually are very uncertain. For example, according to Moss (1982), infrared effects in de Sitter space make the effective potential very flat. This may decrease the effective coupling constant λ and, consequently, increase the duration of the exponential expansion stage (6.9). Therefore we believe that the case E is the most interesting one and deserves further investigation.

We see that there exist several different ways to realize the new inflationary universe scenario in the theories in which the Higgs boson mass m and the effective coupling constant at $|\lambda|\phi^4 < H^4$ are sufficiently small. Some natural possibilities of having sufficiently small m have been discussed in the previous Section. As for the coupling constant λ, at the first glance the constraint (6.10) does not seem very restrictive, since in the Coleman-Weinberg theory one just expects $|\lambda| \sim \alpha^2 \sim 10^{-2}$-$10^{-3}$. However, the effective coupling constant λ in the Coleman-Weinberg theory depends on ϕ, and at $\phi \le H$ usually λ is much greater than α^2, see Section 4. Moreover, in GUTs there are many types of Higgs bosons, and the role of the scalar field fluctuations increases, which leads to more stringent constraints on the scalar coupling constants. For example, in the SU(5) theory the constraint analogous to (6.10) is

$$\frac{|130a + 94b|}{15} \le \frac{\pi^2}{1800} \sim 5\times10^{-3} . \qquad (6.16)$$

If e.g. $a = b$, from (6.16) it follows that a should be less than 3×10^{-4}, whereas in Section 4 it is shown that at $\phi \le H$, $|a|\sim|b| \ge 3\times10^{-1}$. Therefore (if one forgets about the uncertainty of our estimates connected with the infrared problem in de Sitter space) it does not seem possible to implement the new inflationary universe scenario in the context of the minimal SU(5) Coleman-Weinberg theory. Some ways of overcoming this difficulty will be discussed later, after the discussion of another difficulty of the new inflationary universe scenario, connected with the homogeneity problem.

7 THE HOMOGENEITY PROBLEM AND THE GALAXY FORMATION

The main difficulty of the old inflationary universe scenario (Guth 1981) was the unacceptably large inhomogeneity of the universe produced by the colliding bubble walls. In the new inflationary universe scenario such a difficulty is absent, since all the observable part of the universe is created from the interior of a single bubble (Linde 1982). However, as we have mentioned in Section 3, even in this case the universe does not become absolutely homogeneous after the phase transition. The reason is that quantum fluctuations of metric (Mukhanov & Chibisov 1981, 1982) and of the scalar field ϕ (Hawking 1982; Starobinsky 1982; Guth & Pi 1982; Bardeen *et al.* 1982) grow during the phase transition, which gives rise to the density perturbations after the phase transition. The theory of this effect is rather complicated and will be considered in the papers of Hawking, Guth and Turner in this volume. Therefore here we will only briefly discuss their results.

The main result of all these authors is that the density perturbations, which appear after the phase transition, are almost scale-independent. This is a very good result from the point of view of the theory of galaxy formation (Zeldovich 1972). What is not quite so good is the magnitude of these perturbations. The estimates performed in the context of the SU(5) Coleman-Weinberg theory give the magnitude of the perturbations $\delta\rho/\rho$ about 10^5 times greater than the desirable value $\delta\rho/\rho \sim 10^{-4}$ (Hawking 1982; Starobinsky 1982; Guth & Pi 1982; Bardeen *et al.* 1982). The value of $\delta\rho/\rho$ decreases with a decrease of λ. To have $\delta\rho/\rho \sim 10^{-4}$ in this theory one should have $\lambda \le 10^{-10}$ (Starobinsky 1982), which is an even stronger constraint on λ than that obtained in the previous Section. As before, it is unclear whether these results are sufficient to exclude the SU(5) Coleman-Weinberg theory. Indeed, the expressions for $V(\phi)$ and for the correlation functions $\langle\phi(x)\phi(y)\rangle$ used in the above-mentioned papers have been obtained in a simplest approximation, which may give a wrong result in the most interesting case considered in the previous Section (case E). Moreover, the actual value of the effective coupling constant λ at $\phi \le H$ is unknown and it might be such that the scalar field fluctuations make the effective potential $V(\phi)$ so flat at $\phi \le H$ that $\lambda(\phi)$ vanishes at $\phi = 0$. This may reduce considerably the magnitude of the density perturbations.

Anyway, the results discussed above are alarming. Therefore one should try to suggest some theory, which has all the advantages of the SU(5) Coleman-Weinberg theory, but does not lead to large density perturbations. This question will be analysed in the next Section.

8 OTHER VERSIONS OF THE SCENARIO

1. Detailization of the new inflationary universe scenario

The new inflationary universe scenario discussed so far was somewhat simplified. For example, the phase transition in the SU(5) theory may proceed first to the phase $SU(4) \times U(1)$, and only later to the phase $SU(3) \times SU(2) \times U(1)$. By a proper choice of the coupling constants a and b and of the bare Higgs mass μ one can arrange things in such a way that the inflation of the universe will occur only during the first phase transition, or during the second one, or during each of the phase transitions. In our discussion we have neglected the phase transition with the symmetry breaking in the light Higgs boson sector. Usually it is quite possible, since this phase transition occurs at a much smaller temperature than the phase transition $SU(5) \to SU(3) \times SU(2) \times U(1)$. However in the Coleman-Weinberg theory the situation may be reversed, and the first phase transition may be just the phase transition in the light Higgs boson sector (Linde 1982a).

It does not seem likely that such a detailization of the simplified scenario discussed in the previous Sections will considerably improve the situation with the density perturbations. However one should not abandon this possibility without a detailed investigation.

2. Starobinsky model

Three years ago Starobinsky suggested a scenario of the early stages of the universe evolution, which is very similar to the inflationary universe scenario (Starobinsky 1979, 1980). He has pointed out that de Sitter space with a curvature $R \sim G^{-1}$ is a self-consistent solution of the Einstein equations with quantum corrections. Such a solution is unstable, and de Sitter space eventually evolves into the hot Friedmann universe. This transition proceeds rather slowly, just as the process of symmetry breaking in the new inflationary universe scenario. Consequently, in the context of the Starobinsky model one can solve all problems which can be solved in the context of the new inflationary universe scenario (except for the primordial monopole and domain wall problems).

The main difficulty of this scenario is the following. De Sitter space is nonsingular and has an infinite lifetime, $-\infty < t < \infty$, whereas the de Sitter stage in the Starobinsky model, just like in the inflationary universe scenario, can exist only during some finite time interval. This means that before the de Sitter stage the universe should be in some other state. In the inflationary universe scenario the de Sitter stage appears as a result of supercooling of the phase $\phi = 0$. However in the Starobinsky scenario the universe is assumed to be in the vacuum state before the transition to the hot Friedmann universe. Therefore it is much more difficult to understand what was before the de Sitter stage in this scenario. Some attempts to answer this question (as well as some other possibilities to realize the new inflationary universe scenario) will be discussed in Appendix A.

3. Superinflation

The value of the effective coupling constant $\lambda(\phi)$ in the Coleman-Weinberg theory is formed due to the heavy vector boson contribution, and is diminished when the contribution of heavy fermions is taken into account. In theories with broken super-symmetry these contributions almost exactly cancel each other, which makes λ very small and the effective potential $V(\phi)$ very flat. This leads to a greater inflation of the universe (Linde 1982c; Ellis et al. 1982; Albrecht et al. 1982) and also to smaller perturbations of density after the phase transition (Bardeen et al. 1982).

A very interesting example of such a super-symmetric generalization of the Coleman-Weinberg theory is suggested by Witten (1981) and by Dimopoulos & Raby (1982). Their model is discussed in the paper of Steinhardt in this volume. The effective potential in this model in the one-loop approximation looks as follows

$$V(\phi) = c_1 M_I^4 (1 - c_2 \ln \frac{\phi}{M_P}) , \qquad (8.1)$$

where $M_I \sim 10^{12}$ GeV is a scale of super-symmetry breaking, c_1 and c_2 are some small numerical constants, $M_P \sim 10^{19}$ GeV. The precise shape of the effective potential at $\phi \ll M_I$ and at $\phi \geq M_P$ is unknown, but it is known that $V(\phi) \geq 0$ (if gravitational effects are ignored). This effective potential is very flat, which leads to extremely large inflation (Albrecht et al. 1982). Density perturbations in this model are very small

(Bardeen et al. 1982), and presumably are even much smaller than is necessary for galaxy formation. However, this question is not quite clear, since the magnitude of the density perturbations depends on the details of the theory of reheating of the universe after the symmetry breaking, which in its turn depends on the unknown properties of $V(\phi)$ at $\phi \geq M_P$. The main difficulty of this scenario, which is very attractive in many other respects, is connected with the reheating of the universe and the baryon production. The field ϕ in the theory with the effective potential (8.1) grows so slowly that its growth does not lead to the creation of superheavy Higgs bosons with mass $M_S \geq 10^{10}$ GeV which are necessary for the baryon production (Ellis et al. 1982). This difficulty could be circumvented if there were a minimum of $V(\phi)$ at $\phi \geq M_P$ with curvature $\frac{d^2V}{d\phi^2} \geq M_S^2$. However the typical variation of the field ϕ at which the effective potential (8.1) changes considerably is $\Delta\phi \sim M_P$, whereas the typical value of $V(\phi)$ is $O(M_I^4)$. Therefore the typical curvature of $V(\phi)$ is of the order of $M_I^4/M_P^2 << M_S^2$, and the existence of a minimum of $V(\phi)$ with $\frac{d^2V}{d\phi^2} \geq M_S^2$ seems very unlikely.

4. Other possibilities

Let us try to compare the versions of the new inflationary universe scenario based on the Coleman-Weinberg theory and on the Witten-Dimopoulos-Raby model. In the Coleman-Weinberg theory the effective potential at $\phi \leq H$ is not sufficiently flat. Consequently, inflation of the universe is too small and the density perturbations are too large. The curvature of $V(\phi)$ at $\phi \sim \phi_0$ in this scenario is many orders greater than the minimal mass squared M_S^2 of the fractionally charged Higgs bosons, responsible for the baryon asymmetry generation ($M_S \geq 10^{10}$ GeV). Therefore these bosons are copiously produced during the phase transition, and the baryon asymmetry of the universe generated after the phase transition may be even greater than in the standard baryosynthesis scenario. On the other hand, the effective potential in the Witten-Dimopoulos-Raby model is extremely flat. Therefore the inflation of the universe in this scenario is very large. However, the density perturbations presumably are too small, and, due to the flatness of $V(\phi)$ at $\phi \sim \phi_0$, baryon asymmetry cannot be generated after the phase transition. What is necessary for a complete success of the new inflationary universe scenario is therefore a theory with the effective potential which is very flat at $\lambda\phi^4 \leq H^4$, and which is sufficiently curved at $\phi \sim \phi_0$, $\frac{d^2V}{d\phi^2} \geq M_S^2 \geq (10^{10}\text{ GeV})^2$. (The last condition in general can be relaxed if there is a possibility of baryon asymmetry generation at a mass scale much smaller than 10^{10} GeV .) Hopefully, these conditions can be satisfied in some improved version of the Coleman-Weinberg theory or of the Witten-Dimopoulos-Raby model with account taken of the gravitational corrections to $V(\phi)$.

Let us consider for example the effective potential $V(\phi)$ (2.8), (5.7) in the Coleman-Weinberg theory, scalar fluctuations being neglected. From eq. (5.7) it can be seen that by a proper choice of the parameter μ_3 ($\mu_3 = R$) one can obtain a theory in which the effective coupling constant λ at $e\phi \leq H$ vanishes (which justifies neglecting the scalar field contribution to $V(\phi)$ at small ϕ). In wuch a theory the effective potential $V(\phi)$ is very flat at $e\phi \leq H$, and it may be sufficiently curved at $\phi \approx \phi_o$, since $V(\phi_o) = 0$, all gravitational corrections to $V(\phi)$ at $\phi \approx \phi_o$ vanish, and $V(\phi)$ at $\phi \approx \phi_o$ is given by eq. (2.8). It can be shown, however, that in the SU(5) Coleman-Weinberg theory the choice of the parameter $\mu_3 = R$ would imply that $M_X \geq 10^{18}$ GeV , whereas actually in this theory $M_X \sim 10^{15}$ GeV . Nevertheless such a large scale of unification in general is quite possible, and is even desirable from the point of view of the theory of generation of density perturbations due to quantum gravity effects (Mukhanov & Chibisov 1981, 1982).

Potentially interesting possibilities are connected with symmetry breaking phase transitions in supergravity. However it would be premature to discuss these phase transitions now. Here we would like to note only that the Higgs scalar effective potential in the SU(5) theory coupled to $N = 1$ supergravity proves to be very flat (Ibañez 1982), which is an indication of a possible relevance of supergravity to the new inflationary universe scenario. One more possibility, connected with quantum gravity effects, will be considered in Appendix A.

9 CONCLUSIONS

Elementary particle physics is now in the very beginning of the long way through the desert (?) from energies $E \sim 10^2$ GeV to $E \sim 10^{14}$ GeV . Unfortunately, however, it is very unlikely that in the nearest future we will have any accelerators which will provide us with particles with $E \geq 10^7$ GeV (in the centre-of-mass system), see e.g. Okun' (1981). The same boundary exists also for the energies accessible in cosmic ray experiments. The only laboratory in which particles with energies $E \geq 10^{14}$ GeV have ever interacted with each other is our universe at the very early stages of its evolution. The main results of this cosmological experiment are the flatness, homogeneity, isotropy and baryon asymmetry of the universe, the absence of primordial monopoles and domain walls and the existence of galaxies. All these experimental data in principle can be explained in the context of the new inflationary universe scenario, and no other possibility to explain simultaneously all these data is known to us at present. Therefore it would be very desirable to find a natural realization of this scenario.

Investigation of the simplest versions of the new inflationary universe scenario indicates, however, that in some of these versions one obtains too small inflation of the universe and too large perturbations of density after the phase transition, whereas some other versions yield too small perturbations of density and too small baryon asymmetry of the universe. These results could seem somewhat disappointing. However one should remember that a year ago the difficulties of the inflationary universe scenario seemed absolutely unsurmountable (Guth 1981; Hawking et al. 1981; Guth & Weinberg 1982). At present the main problem of the new inflationary universe scenario is to find a theory in which the effective potential is sufficiently flat at small ϕ (ϕ may be the elementary Higgs scalar field or a condensate of the type of $\langle\bar{\psi}\psi\rangle$, $\langle G^a_{\mu\nu} G^a_{\mu\nu}\rangle$, $\langle\phi^2\rangle$, or the curvature scalar R , etc.), and has a large curvature at large ϕ . This problem still is not very easy, but it seems to be a solvable technical problem rather than a problem of principle. We hope very much that with a further development of grand unified theories and their supersymmetric extensions this problem will be solved and we will obtain an answer to many different questions which for a long time have been regarded as being almost metaphysical.

It is a pleasure to express my deep gratitude to G.V. Chibisov, P.C.W. Davies, V.P. Frolov, L.P. Grishchuk, A.H. Guth, J. Ellis, S.W. Hawking, R.E. Kallosh, D.A. Kirzhnits, M.A. Markov, V.F. Mukhanov, D.V. Nanopoulos, V.A. Rubakov, A.A. Starobinsky, A.V. Veryaskin, A. Vilenkin and Ya.B. Zeldovich for many valuable discussions. I am also thankful to S.W. Hawking and to Cambridge University for their hospitality during the Nuffield Workshop in Cambridge.

Appendix A. INFLATIONARY UNIVERSE SCENARIO AND THE SINGULARITY PROBLEM

One of the most important and intriguing cosmological problems is the problem of the general cosmological singularity. Physical intuition can hardly accept the idea of an instantaneous creation of all the universe from nothing at some moment of time, before which there was no time at all. Since the necessary ingredient of the inflationary universe scenario is the de Sitter stage, and de Sitter space is known to be nonsingular, it is very tempting to speculate about a possible connection between this scenario and the singularity problem. There are several different possibilities of such a connection to be discussed here.

1. All the observable part of the universe appears as a result of the exponential expansion of a single bubble (mini-universe), which is created due to quantum tunnelling inside the bare Minkowski space (Brout et al. 1978), or inside de Sitter space (Gott 1982; Linde 1982d). The initial singularity is avoided since the bare space is nonsingular.

Such a possibility at present seems rather unlikely (see, however, subsection 4 of this Appendix). Indeed, it can be easily understood that if the probability of bubble formation per unit time per unit volume in Minkowski space is finite, then the lifetime of the Minkowski space as a whole is finite also, and therefore this space could not exist at $t \to -\infty$.

In de Sitter space the situation is less obvious, since the total physical volume of the exponentially expanding space may grow very rapidly, so that the bubbles of the new phase will never cover all the space (Guth & Weinberg 1982; Linde 1982d). However, this also does not save de Sitter space from a complete decay. The geodesically complete extension of the exponentially expanding flat de Sitter space is a closed de Sitter space, in which the scale factor $a(t) \sim \mathrm{ch}\, Ht$. At $-\infty < t < 0$ the universe is contracting rather than expanding, and the lifetime of the contracting de Sitter space, which is unstable with respect to the bubble formation, proves to be finite. Therefore such a space could not survive during all the time interval $-\infty < t < 0$.

2. The universe initially has the Friedmann metric with a constant scale factor $a(t) = a_0$ (Atkatz & Pagels 1982). Then as a result of a quantum tunnelling which occurs at some moment t_0 , the universe as a whole transforms into de Sitter space with $a(t) \sim e^{H(t-t_0)}$. This scenario has the same difficulty as the previous one: Since the initial metric

has a finite lifetime, it could not exist at $t \to -\infty$, and therefore some other metric should exist before it.

3. De Sitter space has been created from nothing (Tryon 1973; Zeldovich 1981; Grishchuk & Zeldovich 1982; Vilenkin 1982a). This possibility may be used in particular to realize the Starobinsky scenario (see Section 8). However, it is not quite clear how one can describe creation of the universe from nothing. One possibility is that the universe initially has metric $g_{oo} = 1$, $g_{oi} = g_{io} = g_{ij} = 0$, and at some time t_o the universe as a whole transforms into de Sitter universe. Such a transformation can be regarded as a creation of the universe from "nothing", since initially the universe has vanishing physical volume. Unfortunately, this scenario seems to suffer from the same difficulty as the scenario discussed in the previous subsection (this scenario is a particular version of the previous one with $a_o = 0$). Another possibility is to interpret the familiar S^4 de Sitter instanton as a tunnelling trajectory from "nothing" to de Sitter space (Vilenkin 1982a). This interpretation, however, does not seem well motivated, since the analytic continuation of S^4 instanton is de Sitter space but not "nothing".

Despite all these difficulties, the possibility of creating the universe from nothing is very interesting and deserves further investigation. For a most detailed discussion of this possibility see a recent paper by Grishchuk & Zeldovich (1982a).

4. One of the most disturbing questions related to the singularity problem is the question what was before the universe was created. This question may seem absolutely metaphysical, but our experience with metaphysical questions teaches us that such questions sometimes have physical answers.

One possible way of answering this question is that the inflationary universe at the de Sitter stage of its evolution is created as a "child universe" spatially disconnected from the "mother universe" from which it originated. Such a universe creation may be a two-step process. Firstly, the mini-universe is created, which is connected with the "mother universe" by a wormhole (this part of the process is similar to that discussed in subsection 1). Then the wormhole evaporates (like a black hole) and the mini-universe becomes a child universe spatially disconnected from the mother universe (Sato et al. 1982). Unlike the process discussed in subsection 1, this process does not necessarily lead to a decay of the mother universe as a whole.

Another possibility is that the child universe is created without any wormhole formation and evaporation, but just as a quantum tunnelling event. In that case the child universe will be completely disconnected from the mother universe. For the inhabitants of the child universe this process will look as a process of creation of the universe from nothing. Therefore the theory of creation of topologically disconnected universes (if such a theory will be elaborated) may be used, at least formally, to make sense out of the concept of creation of the universe from nothing discussed in the previous subsection.

5. The last possibility to be discussed here is connected with the scenario of the universe evolution suggested by Markov (1981, 1982). He has supposed that the universe is closed, and at the moment of maximal contraction the energy-momentum tensor of matter becomes $T_{\mu\nu} = g_{\mu\nu}\Lambda$, so that the universe enters a de Sitter stage. Closed de Sitter space is nonsingular, $a(t) = H^{-1}\,\mathrm{ch}\,Ht$, and after the moment of maximal contraction $t = 0$ it becomes expanding. After some period of exponential expansion de Sitter universe transforms into the hot Friedmann universe according to the inflationary universe scenario (or to the Starobinsky scenario). Then the closed Friedmann universe contracts again, and again transforms into the de Sitter one, etc.

The possibility of the eternally oscillating nonsingular inflationary universe seems very attractive. The main problem of this scenario is that the hot Friedmann universe cannot evolve into de Sitter space without violation of the second law of thermodynamics. Indeed the pressure p in the hot universe is given by

$$p = -F = Ts - \varepsilon\ , \tag{A.1}$$

where ε is the energy density, s is the entropy density, F is the free energy density. Therefore in order to have $p = -\varepsilon$ ($T_{\mu\nu} = g_{\mu\nu}\Lambda$) one should have vanishing entropy s at the end of the universe contraction. This means, that at the end of the universe contraction the entropy of the universe should be decreasing (a similar possibility has been suggested earlier by Sakharov in his version of the oscillating universe scenario (Sakharov 1967, 1970, 1979, 1980)). At the first glance such a possibility seems absolutely forbidden. However eq. (A.1) is true only in the equilibrium quantum statistics, whereas in the nonequilibrium state near the would-be singularity eq. (A.1) does not hold and the equation of state

$p = -E$ becomes possible (Berezin 1982). Moreover, when temperature T exceeds $M_p \sim 10^{19}$ GeV, the time remaining till the moment of maximal contraction is of the order of $\frac{M_p}{T^2} < T^{-1}$. At that time the standard thermodynamic quantities become ill defined at the classical level, since the energies of particles $\sim T$ cannot be measured during the time interval which is smaller than T^{-1}.

Here we would like to suggest one more possibility which may help to realize the nonsingular oscillating inflationary universe scenario. Namely, it is known that at $R \geq G^{-1}$ perturbative results in quantum gravity become unreliable (here R is the curvature tensor, all indices are omitted). The only other non-Abelian theory which is well-known to us is quantum chromodynamics. Perturbative results in QCD are unreliable at small momenta, which is an indication of the colour confinement: The only objects which can exist as free particles (in- and out-states) are colourless, i.e. do not interact directly with the Yang-Mills field. Quarks are assumed to interact with each other with the force which (asymptotically) does not depend on the distance between them. Therefore any isolated quark would have an infinite energy, which just means that quarks cannot exist as free particles. Phenomenologically the existence of the distance-independent forces between two quarks can be described as the interaction between two quark currents j_1 and j_2 of the type of $\frac{j_1(k_1-k)j_2(k_2+k)}{k^4}$ in momentum space (though such an interpretation of the quark interactions in QCD is not quite precise). Recently a similar term of the type of $\frac{R^2(k_1-k)R^2(k_2+k)}{k^4}$ has been obtained by Green et al. (1982) and by Grisaru & Siegel (1982) as the one-loop correction to the effective action in the O(8) supergravity. We will not speculate here about a possible phenomenological significance of this result, but just assume that at sufficiently large curvature $R \geq G^{-1}$ in quantum gravity there occurs a phase transition to the gravitational confinement phase. We are unable to prove or disprove this conjecture at present, since it would require a much more deep understanding of quantum gravity than that we have now. Some estimates indicate that the regime of gravitational confinement may occur even at $R \ll G^{-1}$, but the confinement length in thise case is much greater than the size of the observable part of the universe. We hope to discuss this question somewhere else, and here we shall try only to understand to which consequences our assumption may lead.

First of all let us try to understand what the hypothesis of gravitational confinement could mean. As we have already mentioned, any isolated coloured particle in QCD would have an infinite energy, and therefore only colourless particles can exist as free particles in QCD. However, all particles are coloured with respect to gravitational interaction. Therefore, in the gravitational confinement phase (if such a phase can exist) any particle would have infinite energy. In other words, gravitational confinement phase is a vacuum state in a curved space, in which no free particles and, more generally, no inhomogeneities of the curvature tensor (which are also gravitationally "coloured") can exist. Therefore after the phase transition to the gravitational confinement phase the universe should transform into de Sitter space.

In order to investigate possible cosmological consequences of our hypothesis let us consider a hot closed universe contracting to a would-be singularity. The temperature of the universe increases, and at $T \geq 10^{14}$ GeV the phase transitions with symmetry restoration in GUTs occur. When the density of matter becomes greater than $\rho_p = G^{-2}$, the curvature tensor becomes of the order of G^{-1} and the phase transition to the gravitational confinement phase occurs. In this state no real particle excitations can exist (though virtual particles in general can exist), and the entropy of the universe in this state vanishes. This does not contradict the second law of thermodynamics, since any thermodynamical description of the gravitational confinement phase is impossible because of the absence of particles and of any possibility of disorder of matter. Since the universe is closed, after the transition to the gravitational confinement phase it evolves into the closed de Sitter universe, which contracts to some minimal value of the scale factor a_o and then expands as a_o ch Ht. Such a universe is stable with respect to the deconfining phase transition at the contraction stage, but is unstable at the expansion stage. Indeed, any bubble of the nonconfining phase (with $\rho \lesssim \rho_p$), which may appear inside the confining phase, at the contraction stage automatically transforms into the confining phase due to the increase of density during the contraction. On the other hand, at the exponential expansion stage sufficiently large bubbles with $\rho \lesssim \rho_p$ grows, the density inside them decreases, and all the universe eventually transforms into the hot Friedmann universe filled with the ordinary (nonconfining) phase with $\rho < \rho_p$. If this phase transition occurs slowly enough, the inflationary universe scenario can be realized at this stage of the universe expansion. The inflationary universe scenario can be also realized later during the phase transitions in GUTs. After these phase transitions the closed

Friedmann universe expands up to some maximal radius, contracts and again transforms into de Sitter universe in the confinement phase.

We have suggested therefore a possible realization of the Markov nonsingular oscillating inflationary universe scenario. Since the duration of de Sitter stage fluctuates from cycle to cycle, there will be large fluctuations in the duration of each cycle, but the average duration will be time-independent. This is one of the most important differences between the scenario discussed above and the standard oscillating universe model, in which a duration of each cycle grows infinitely with the number of the cycle due to the growth of the total entropy of the universe in each cycle (see e.g. Zeldovich & Novikov 1975). In our case all the entropy and all inhomogeneities of the contracting universe disappear in the purgatory of de Sitter stage, and then are generated anew in each new cycle.

Our discussion of the gravitational confinement hypothesis and of its cosmological consequences is, of course, far from being rigorous. The only purpose of this discussion was to reveal a rather unexpected possibility to realize the nonsingular oscillating inflationary universe scenario with the help of quantum gravity effects. For a discussion of quantum gravity and the singularity problem see also the paper by Hartle in this volume.

Appendix B. INFLATIONARY UNIVERSE SCENARIO AND THE ANTHROPIC PRINCIPLE

As we have mentioned in the Introduction, there exists a possible solution of the flatness, homogeneity, isotropy and baryon asymmetry problems based on the Anthropic Principle: It can be argued that in a curved, inhomogeneous and anisotropic universe without any baryon asymmetry no life would exist, and therefore nobody would ask any questions about homogeneity, isotropy, etc. This argument is very witty but is not quite convincing, since it cannot explain why the spectrum of inhomogeneities is almost scale-independent, why the universe is almost exactly isotropic and why the numerical value of $(n_B-n_{\bar{B}})/n_\gamma$ is $O(10^{-8})$.

In some cases, however, the results obtained by the use of the Anthropic Principle may be very informative. For example, one can obtain an "explanation" why our space is four-dimensional. Indeed, in the theories of electromagnetic and gravitational interactions in d-dimensional space-time at $d > 4$ any bounded systems such as atoms or planetary systems would be impossible, whereas at $d < 4$ free particles cannot exist. This indicates that the best conditions for the existence of life (or, at least, of our kind of life, based essentially on electromagnetic and gravitational interactions) can be realized just as in the four-dimensional space-time (Ehrenfest 1917). Similar considerations based on the study of conditions necessary for the existence of atoms, stars, galaxies, etc. lead to stringent constraints on the value of $\alpha = \frac{e^2}{4\pi}$, on the value of the gravitational constant G and on the masses of elementary particles (Carr & Rees 1979; Rosental 1980). The Anthropic Principle may help us to understand why the cosmological constant at present is zero (Hawking 1982a), why the elementary particle theory has the unbroken symmetry $SU(3) \times U(1)$ etc.

It could be argued, however, that our universe is unique, and it is meaningless to ask whether life can exist in space with $d \neq 4$ and with $\alpha \neq \frac{1}{137}$. One possible answer to this objection is that there may exist many disconnected universes (see Appendix A), and we live in just one of them, which is sufficiently suitable for the existence of intelligent life. Another possibility is related to the oscillating universe scenario discussed in Appendix A, since in this scenario the universe is created anew in each new cycle. However the simplest way to justify the Anthropic Principle (at least partially) can be found in the context of the new inflationary universe scenario.

Indeed, the phase transition in the new inflationary universe scenario occurs independently in each bubble, and after the phase transition the distance between the centres of the bubbles exceeds the size of the horizon $\sim ct$, where t is the age of the universe. Therefore the physical processes inside a bubble remain unaffected by the physical processes inside any other bubble even after the phase transition. In this sense any bubble behaves as a mini-universe almost completely isolated from all other mini-universes. Let us assume e.g. that the universe is open. After the phase transition this universe is divided into an infinite number of mini-universes, in each of which the phase transition occurs independently. This means that there will be infinitely many mini-universes with all possible symmetry breaking patterns. In particular, in the $SU(5)$ Coleman-Weinberg theory the phase transition may proceed with a comparable probability either to the phase $SU(3) \times SU(2) \times U(1)$ or to the phase $SU(4) \times U(1)$. Therefore after the phase transition the universe becomes divided into infinitely many mini-universes in the phase $SU(3) \times SU(2) \times U(1)$

and infinitely many mini-universes in the phase SU(4) × U(1) . This considerably simplifies the answer to the question why we are now in the phase SU(3) × U(1) but not in some other stable or metastable phase, corresponding to a local minimum of the effective potential. Indeed, the phase SU(3) × SU(2) × U(1) eventually evolves into the phase SU(3) × U(1) . Therefore this phase (if it is sufficiently stable) should exist inside infinitely many mini-universes, and all other "desert islands" (in which other kinds of life may exist) are of no importance for us.

There is some interest now in the theories in which our space originally has more than four dimensions, but extra dimensions are spontaneously compactified (see e.g. Cremmer & Sherk 1976; Witten 1981a). However it is not very simple to explain why just four dimensions remained uncompactified (Schwarz 1982). The new inflationary universe scenario makes it more easy to answer this question. In the context of this scenario it would be sufficient that the compactification to the space $d = 4$ is possible, but there is no need for the four-dimensional space to be the only possible space after the compactification. Indeed, if the compactification to the space $d = 4$ is possible, there will be infinitely many mini-universes with $d = 4$, in which intelligent life can exist.

Thus we see that the new inflationary universe scenario may serve as a basis for a kind of a Weak Anthropic Principle: All possible phases which may appear after the phase transition and which are sufficiently stable should exist in some of the mini-universes, and then (hopefully) we are free to choose the best mini-universe to live in. As was claimed by Guth (1982), the inflationary universe is the only example of a free lunch (all matter in this scenario is created from the unstable vacuum). Now we can add that the inflationary universe is the only lunch at which all possible dishes are available.

REFERENCES

Abbott, L.F., Kang, J.S. and Schnitzer, H.J. (1976). Bound states, tachyons, and restoration of symmetry in the 1/N expansion. Phys. Rev. D13, 2212-2226.

Abbott, L.F., Farhi, E. and Wise, M.B. (1982). Particle production in the new inflationary cosmology. Phys.Lett 117B, 29-34.

Albrecht, A., Dimopoulos, S., Fischler, W., Kolb, E.W., Raby, S. and Steinhardt, P.J. (1982) (in preparation).

Albrecht, A. and Steinhardt, P.J. (1982). Cosmology for grand unified theories with radiatively induced symmetry breaking. Phys.Rev. Lett. 48, 1220.

Albrecht, A., Steinhardt, P.J., Turner, M.S. and Wilczek, F. (1982). Reheating an inflationary universe. Phys.Rev.Lett. 48, 1437.

Allen, B. (1982). Phase transitions in de Sitter space. Nucl.Phys. (in press).

Atkatz, D. and Pagels, H. (1982). Origin of the Universe as a quantum tunelling event. Phys.Rev. D25, 2065-2073.

Bardeen, J., Steinhardt, P.J. and Turner, M.S. (1982). Spontaneous creation of (almost) scale-free density perturbations in an inflationary universe. Enrico Fermi Institute preprint.

Barrow, J.D. and Turner, M.S. (1981). Inflation in the universe. Nature 292, 35.

Berezin, V.A. (1982). Can the de Sitter space-time be a final state of the contracting universe? In Quantum Gravity, Plenum, ed. by M.A. Markov.

Billoire, A. and Tamvakis, K. (1982). The Coleman-Weinberg mechanism in early cosmology. Nucl. Phys. B200 [FS4], 329-344.

Birrell, N.D. and Davies, P.C.W. (1982). Quantum Fields in Curved Space. Cambridge University Press.

Brout, R., Englert, F. and Gunzig, E. (1978). The creation of the universe as a quantum phenomenon. Ann.Phys. (N.Y.) 115, 78-106.

Bunch, T.S. and Davies, P.C.W. (1982). Quantum field theory in de Sitter space: renormalization by point-splitting. Proc.Roy.Soc. A360, 117-134.

Carr, B.J. and Rees, M.J. (1979). The anthropic principle and the structure of the physical world. Nature 278, 605-612.

Coleman, S. and De Luccia, F. (1980). Gravitational effects on and of vacuum decay. Phys.Rev. D21, 3305-3315.

Coleman, S. and Weinberg, S. (1973). Radiative corrections as the origin of spontaneous symmetry breaking. Phys.Rev. D7, 1888-1910.

Collins, C.B. and Hawking, S.W. (1973). Why is the universe isotropic? Astrophys.J. 180, 317-334.

Cremmer, E. and Scherk, J. (1976). Spontaneous compactification of space in an Einstein-Yang-Mills-Hills model. Nucl.Phys. B108, 409-416.

Dicke, R.H. (1961). Nature 192, 440.

Dolan, L. and Jackiw, R. (1974). Symmetry behaviour at finite temperature. Phys.Rev. D9, 3320-3341.

Dolgov, A.D. and Linde, A.D. (1982). Baryon asymmetry in inflationary universe. Phys.Lett. 116B, 239-334.

Ehrenfest, P. (1917). Proc. Amsterdam Acad. 20, 200.

Ellis, J., Linde, A.D. and Nanopoulos, D.V. (1982). Inflation can save the gravitino. CERN preprint TH.3356.

Ellis, J., Nanopoulos, D.V., Olive, K.A. and Tamvakis, K. (1982). Cosmological inflation cries out for supersymmetry. Phys.Lett. 118B, 335-339.

Ford, L.H. and Vilenkin, A. (1982). Gravitational effects upon cosmological phase transitions. Phys.Rev. D26, 1231.

Gibbons, G.W. and Hawking, S.W. (1977). Cosmological event horizons, thermodynamics and particle creation. Phys.Rev. D15, 2738.

Goncharov, A.S. and Linde, A.D. (1982). Bubble formation in the inflationary universe (in preparation).

Gott, J.R. (1982). Creation of open universes from de Sitter space. Nature 295, 304-307.

Green, M.B., Schwarz, J.H. and Brink, L. (1982). N=8 Yang-Mills and N=8 supergravity as limits of string theories. Nucl.Phys. (to be published).

Grisaru, M.T. and Siegel, W. (1982). The one-loop, four-particle S-matrix in extended supergravity. Phys.Lett. 110B, 49-54.

Grishchuk, L.P. and Zeldovich, Ya.B. (1982). Complete cosmological theories. In Quantum structure of space-time, eds. M. Duff and C. Isham. Cambridge University Press.

Grishchuk, L.P. and Zeldovich, Ya.B. (1982a). Complete cosmological theories (extended version). Space Res.Inst. preprint Pr-726.

Guth, A.H. (1981). The inflationary universe: a possible solution to the horizon and flatness problems. Phys.Rev. D32, 347.

Guth, A.H. (1982). 10^{-35} seconds after the Big Bang. MIT preprint CTP # 991.

Guth, A.H. and Pi, S.-Y. (1982). Fluctuations in the new inflationary universe. Phys.Rev.Lett. 49, 1110.

Guth, A.H. and Tye, S.-H.H. (1980). Phase transitions and magnetic monopole production in the very early universe. Phys.Rev.Lett. 44, 631.

Guth, A.H. and Weinberg, E. (1982). Could the universe have recovered from a slow first-order phase transition? Nucl.Phys. 212, 321.

Hawking, S.W. (1982). The development of irregularities in a single bubble inflationary universe. Phys.Lett. 115B, 295-297.

Hawking, S.W. (1982a). The cosmological constant and the Weak Anthropic Principle. In Quantum structure of space-time, eds. M. Duff and C. Isham. Cambridge University Press.

Hawking, S.W. and Moss, I.G. (1982). Supercooled phase transition in the very early universe. Phys.Lett. 110B, 35-38.

Hawking, S.W., Moss, I.G. and Stewart, J.M. (1981). Bubble collisions in the very early universe. Phys.Rev. D26, 268.

Hu, B.L. and Parker, L. (1978). Anisotropy damping through quantum effects in the early universe. Phys.Rev. D17, 933-945.

Ibanez, L. (1982). Locally supersymmetric SU(5) grand unification. CERN preprint TH.3374.

Kibble, Y.W.B. (1976). Topology of cosmic domains and strings. J.Phys. A9, 1387.

Kirzhnits, D.A. (1972). Weinberg model and the hot universe. JETP Lett. 15, 529-531.

Kirzhnits, D.A. and Linde, A.D. (1972). Macroscopic consequences of the Weinberg model. Phys.Lett. 42B, 471-474.

Kirzhnits, D.A. and Linde, A.D. (1974). A relativistic phase transition. ZhETF 67, 1263-1274 (JETP 40 (1975), 628-634).

Kirzhnits, D.A. and Linde, A.D. (1976). Symmetry behaviour in gauge theories. Ann. Phys. (N.Y.) 101, 195-238.

Lazarides, G., Shafi, Q. and Walsh, T.F. (1982). Cosmic strings and domains in unified theories. Nucl.Phys. B195, 157-172.

Linde, A.D. (1976). Dynamic reconstruction of symmetry and limitations on the masses and coupling constants in the Higgs model. JETP Lett. 23, 64-67.

Linde, A.D. (1977). 1/N-expansion, vacuum stability and quark confinement. Nucl.Phys. B125, 369-380.

Linde, A.D. (1979). Phase transitions in gauge theories and cosmology. Rep.Progr.Phys. 42, 389-437.

Linde, A.D. (1981). Decay of the false vacuum at finite temperature. Lebedev Phys.Inst. preprint No.265 (submitted to Nucl.Phys.).

Linde, A.D. (1982). A new inflationary universe scenario: a possible solution of the horizon, flatness, homogeneity, isotropy and primordial monopole problems. Phys.Lett. 108B, 389-393.

Linde, A.D. (1982a). Coleman-Weinberg theory and the new inflationary universe scenario. Phys.Lett. 114B, 431-435.

Linde, A.D. (1982b). Temperature dependence of coupling constants and the phase transition in the Coleman-Weinberg theory. Phys.Lett. 116B, 340-342.
Linde, A.D. (1982c). Scalar field fluctuations in expanding universe and the new inflationary universe scenario. Phys.Lett. 116B, 335-339.
Linde, A.D. (1982d). Nonsingular regenerating inflationary universe. DAMTP preprint.
Markov, M.A. (1981). Is perpetually oscillating Friedmann universe possible that passes through the region near the classical singularity in a state of metastable de Sitter universe? Inst.Nucl. Research preprint P-0227.
Markov, M.A. (1982). Limiting energy density as a universal law of nature. Pisma ZhETF 36, 214-216.
Moss, I.G. (1982). Infrared problems in de Sitter space, a talk at the Nuffield Workshop on the Very Early Universe.
Mottola, E. and Lapedes, A. (1982). The inflationary universe with gravity. Santa Barbara preprint NSF-ITP-82-101.
Mukhanov, V.F. and Chibisov, G.V. (1981). Quantum fluctuations and the nonsingular universe. JETP Lett. 33, 532-535.
Mukhanov, V.F. and Chibisov, G.V. (1982). Vacuum energy and the large-scale structure of the universe. ZhETF 82, 475.
Okun', L.B. (1981). Present status and perspectives of the high energy physics. Usp.Fiz.Nauk 134, 3-44.
Parke, S. (1983). Gravity, the decay of the false vacuum and the new inflationary universe scenario. Phys.Lett. 121B, 313-315.
Parke, S. and Pi, S.-Y. (1981). The phase structure of the early universe in the minimal SU(5) grand unified theory. Phys.Lett. 107B, 54-58.
Preskill, J.P. (1979). Cosmological production of superheavy magnetic monopoles. Phys.Rev.Lett. 43, 1365-1368.
Rozental, I.L. (1980). Physical laws and numerical values of fundamental constants. Usp.Fiz.Nauk 131, 239-256.
Rubakov, V.A., Sazhin, M.V. and Veryaskin, A.V. (1982). Graviton creation in the inflationary universe and the grand unification scale. Phys.Lett. 115B, 189-192.
Sakharov, A.D. (1967). Breaking of CP-invariance, C-symmetry and the baryon asymmetry of the universe. JETP Lett. 5, 27.
Sakharov, A.D. (1970). A multisheet cosmological model. Inst.Appl.Mathem. preprint.
Sakharov, A.D. (1979). Baryon asymmetry of the universe. ZhETF 76, 1172-1181.
Sakharov, A.D. (1980). Cosmological models with a reverse of the time arrow. ZhETF 79, 689.
Sato, K., Kodama, H., Sasaki, M. and Maeda, K. (1982). Multi-production of universes by first-order phase transition of a vacuum. Phys.Lett. 108B, 103-107.
Schwarz, J.H. (1982). Superstring theory. Caltech preprint CALT-68-911.
Sher, M. (1982). Importance of the coupling-constant temperature dependence in supercooled phase transitions. Phys.Rev. D24, 1699-1701.
Shore, G.M. (1980). Radiatively induced spontaneous symmetry breaking and phase transitions in curved spacetime. Ann.Phys. 128, 376-424.
Starobinsky, A.A. (1979). Spectrum of primordial gravitational radiation and the initial state of the universe. Pisma ZhETF 30, 719-723.
Starobinsky, A.A. (1980). A new type of isotropic cosmological models without singularity. Phys.Lett. 91B, 99-102.

Starobinsky, A.A. (1982). Dynamics of phase transitions in the new inflationary universe scenario and generation of perturbations. Phys.Lett. 117B, 175-178.

Tamvakis, K. and Vayonakis, C.E. (1982). A possible solution to the problem of cosmic domain walls. Phys.Lett. 109B, 15-18.

Tryon, E.P. (1973). Is the universe a vacuum fluctuation? Nature 246, 396-397.

Vilenkin, A. (1982). Gravitational effects in Guth cosmology. Phys.Lett. 115B, 91-94.

Vilenkin, A. (1982a) Creation of universes from nothing. Phys.Lett. 117B, 25-28.

Weinberg, S. (1974). Gauge and global symmetries at high temperature. Phys.Rev. D9, 3357-3378.

Weinberg, S. (1979). Cosmological production of baryons. Phys.Rev.Lett. 42, 850-853.

Wilson, K.G. and Kogut, J. (1974). The renormalization group and the ε-expansion. Phys.Rep. 12C, 75-199.

Witten, E. (1981). Mass hierarchies in supersymmetric theories. Phys.Lett. 105B, 267-271.

Witten, E. (1981a). Search for a realistic Kaluza-Klein theory. Nucl.Phys. B186, 412-428.

Yoshimura, M. (1978). Unified gauge theories and the baryon number of the universe. Phys.Rev.Lett. 41, 281-284.

Zeldovich, Ya.B. (1970). Particle creation in cosmology. JETP Lett. 12, 307.

Zeldovich, Ya.B. (1972). Mon.Not.Roy.Astr.Soc. 160, 1.

Zeldovich, Ya.B. and Khlopov, M.Yu. (1978). On the concentration of relic magnetic monopoles in the universe. Phys.Lett. 79B, 239-241.

Zeldovich, Ya.B., Kobzarev, M.Yu. and Okun', L.B. (1974). Cosmological consequences of spontaneous breaking of a discrete symmetry. ZhETF 67, 3-11.

Zeldovich, Ya.B. and Novikov, I.D. (1975). Structure and evolution of the universe. Moskow, Nauka.

Zeldovich, Ya.B. and Starobinsky, A.A. (1972). Particle production and vacuum polarization in an anisotropic gravitational field. JETP 34, 1159.

NATURAL INFLATION

Paul Joseph Steinhardt
Department of Physics, University of Pennsylvania, Philadelphia, Pennsylvania 19104

Abstract. It is shown how the inflationary Universe scenario can be implemented in a special class of supersymmetric models without any unnatural fine-tuning of parameters. The special class of models utilizes O'Raifeartaigh-type supersymmetric breaking and employs the Witten "reverse hierarchy" scheme. A prime test example, the Dimopoulos-Raby "geometric hierarchy" model, has been studied in collaboration with A. Albrecht, W. Fischler, S. Dimopoulos, E. Kolb & S. Raby and some results of that study are presented. Implications of this new kind of inflationary model for our understanding of the fluctuations necessary for galaxy formation, the initial cosmological singularity and the very large scale structure of the Universe are also briefly discussed.

Recently, it has been shown by Linde (1982a), Albrecht and Steinhardt (1982a,b) that the Guth (1981) inflationary scenario can be successfully implemented in grand unified theories (GUTs) in which the GUT symmetry is broken through radiative corrections to the tree effective potential — so-called "Coleman-Weinberg (C-W) models" (Coleman and Weinberg, 1973). Unlike the case for ordinary GUTS, (Guth and Weinberg, 1982), the C-W models result in enormous exponential expansion of the Universe during the GUT phase transition without preventing the phase transition from being completed and from homogenizing the Universe. The key conceptual difference is the following: For ordinary GUT inflationary scenarios, the exponential expansion of the Universe occurs while the Universe is supercooling in a metastable (SU(5)) phase; the metastable regions expand so rapidly that bubble nucleation (tunneling) cannot occur rapidly enough for the phase transition to the stable phase to be completed. In C-W GUTs, the major inflation occurs after fluctuations have driven different regions of the Universe a small distance away from the SU(5) symmetric state towards different spontaneous symmetry breaking (SSB) minima; the interiors of the fluctuation regions have a small Higgs expectation value which evolves inevitably towards a stable minimum with large expectation value (due to the potential), but because of the special nature of the C-W potential the evolution occurs very slowly (sometimes referred to as the "slow roll-over" of the scalar field) (Steinhardt, 1982). During the slow roll-over the energy density in the fluctuation region is nearly that of the symmetric phase (because the Higgs expectation value is small) and each fluctuation region begins to expand to a size many orders of magnitude times the size of the observable Universe today. Because the expansion occurs in the evolving stable phase, rather than the metastable phase, the problem of completion of the phase transition that occurred in Guth's original inflationary scenario is obviated. Therefore, the C-W models provide a possible explanation of the cosmological homogeneity, isotropy, flatness and monopole problems (Linde, 1982a; Albrecht & Steinhardt, 1982a, Guth, 1981) and, it has been more recently shown, a natural solution to the problem of baryon asymmetry (Albrecht, et al., 1982a; Abbott, et al., 1982; Dolgov and Linde, 1982) and the problem of providing the fluctuations neccessary for galaxy formation (Chibisov and Mukhanov, 1981; Hawking, 1982; Bardeen, et al., 1982; Starobinski, 1982; Guth and Pi, 1982).

The one drawback to the C-W inflationary scenario is that it requires the fine-tuning of the effective mass parameter of the theory in the de Sitter phase to a value close to zero (see, for example, Albrecht and Steinhardt, 1982b). It can be argued that such a model is special and that the setting of a parameter to zero is not so unnatural. However, since there is no unbroken symmetry that can result in such a finely tuned condition, such arguments are suspect. The C-W models provide a possible but unlikely support for the inflationary scenario.

In this paper I plan to outline a first attempt at finding a theory which leads to inflation without fine-tuning of parameters. The models I will discuss utilize the lesson learned from the C-W models -- inflation must occur in evolving stable phase -- but the inflation will occur for a wide and natural choice of all parameters.

Models which possess supersymmetry appear to be obvious candidates for at least two reasons. Firstly, supersymmetry is supposed to solve the unnatural hierarchy problem; shouldn't the same notion solve the unnatural inflation problem? Secondly, supersymmetric models are known to lead to radiative symmetry breaking without fine-tuning of parameters. (I use the term "radiative symmetry breaking" rather than "Coleman-Weinberg" because the latter term has come to be associated with potentials that are very flat for small values of the scalar field expectation value. Both terms refer to the breaking of symmetry through one-loop radiative corrections to the effective potential, but in the supersymmetry case this does not naturally lead to a potential that is flat near the region of small scalar field.)

The natural radiative symmetry breaking occurs especially in supersymmetry models with O'Raifeartaigh (O'R) symmetry breaking (O'Raifeartaigh, 1975). For example, consider a model with three left-handed chiral fields, A, X, Y with a superpotential:

$$W = \lambda_1 XA^2 + \lambda_2 Y\,(A^2 - M^2) \quad . \tag{1}$$

The scalar potential in tree approximation is given by:

$$\begin{aligned} V &\equiv \sum_i \left|\frac{\partial W}{\partial \phi_i}\right|^2 \equiv \sum_i \left|F_i\right|^2 \\ &= \lambda_1^2\,|A|^4 + \lambda_2^2\,|A^2 - M^2|^2 + |2\lambda_1 AX + 2\lambda_2 AY|^2 \quad , \end{aligned} \tag{2}$$

where i runs over the various fields and the same symbol has been used for the superfields and their scalar components. There is no state with $V = 0$ so supersymmetry is (O'R) spontaneously broken with a global minimum at $\langle A\rangle = \lambda_2 M/(\lambda_1^2 + \lambda_2^2)^{1/2}$. The value of $\langle X\rangle$ and $\langle Y\rangle$ remain undetermined at tree level, only constrained by

$$|X| = -\frac{\lambda_2}{\lambda_1}\,\langle Y\rangle \ . \tag{3}$$

Thus, the mass of the A field, given by $\sim 2\,\lambda_1\langle X\rangle$ (for large X), can be arbitrarily large. This kind of degeneracy is natural to O'R breaking since to break supersymmetry and have $V \neq 0$ at the minimum of the potential, there must be an algebraic inconsistency in setting all the F_i's to zero; to obtain inconsistency for one field expectation value (in this case the A field) at least one other field expectation value usually remains undetermined.

The degeneracy is broken by one-loop corrections to the effective potential - just the radiative symmetry breaking discovered by Coleman and Weinberg (1973):

$$V_{1\text{-loop}} = \sum_i \frac{(-1)^F}{64\pi^2} M_i^{\,4}(\phi)\, \ln\, \frac{M_i^{\,2}}{\mu^2} \tag{4}$$

where μ is a renormalization mass and $(-1)^F$ is +1 (-1) for bosons (fermions). For $X \gtrsim M$, the only X-dependent masses come from A and its supersymmetric partners. Their masses are such that the one-loop corrections are given by (Witten, 1981; Huq, 1976):

$$V_{1\text{-loop}} = \alpha M^4 \ln \frac{|X|}{\mu} \tag{5}$$

where α is a function of the scalar coupling constants. This form is very special because, unlike C-W GUT models, the only X-dependence appears inside the logarithm; this crucial behavior occurs because of special cancellations that occur in Eq. (4), as a remnant of the broken supersymmetry. The net result for the pure scalar theory is that the state with $X = 0$ is stabilized. For a gauge theory, the analog of Eq. (5) is (Witten, 1981)

$$V_{1\text{-loop}} = (\alpha' - \beta g^2)M^4 \ln \frac{|X|}{\mu} \tag{6}$$

where α' and β are functions of the scalar couplings and g is the gauge coupling constant. If $(\alpha' - \beta g^2) < 0$, the state with $\langle X \rangle = 0$ is destabilized compared to the state with large $\langle X \rangle$. (NOTE: The potential has only been determined for $X \gtrsim M$ for which it is independent of parameters; the shape of the potential for $X \lesssim M$ depends upon the choice of parameters and the $\langle X \rangle = 0$ state may be metastable or unstable). At large X ($\sim Me^{1/\alpha}$), the potential appears to become negative, which is impossible for a globally supersymmetric theory. In fact, asymptotic freedom forces g to decrease with large X and the sign of the logarithm term changes sign; thus V develops a stable minimum at this large value of X. Beginning with a single mass scale, M, we have generated a much larger mass scale. Witten (1981) hoped to be able to use this idea to generate the GUT scale from a theory in which the only weak scale has been introduced as a fundamental mass. Because usual hierarchy solutions attempt to generate the weak scale from a fundamental GUT scale, this sort of model has come to be known as a "reverse hierarchy" model.

Dimopoulos and Raby (1982) attempted to extend this idea to produce both the weak and GUT scales through radiative corrections to a theory with only a single fundamental intermediate scale. In their model particles appear with masses of order $M_{GUT} = 10^{19}$ GeV $= M_I\, e^{1/\alpha}$; $M_W = M_I\, e^{-1/\alpha} = 10^5$ GeV; and $M_I = 10^{12}$ GeV. (The model is referred to as the "geometric hierarchy" model because $M_I^2 \approx M_{GUT} M_W$). The theory has many interesting properties that are discussed by Dimopoulos and Raby (1982). For the purpose of this paper, the most important feature is that the effective zero temperature scalar potential for the X field in the range $M_I \lesssim X \lesssim M_{GUT}$ can be written (in analogy with the toy model discussed above):

$$V_{eff} = c_1 M_I^4 - c_2 M_I^4 \ln \frac{|X|}{M_{GUT}} \tag{7}$$

where $c_1 = (\lambda_1^2 \lambda_2^2/(\lambda_1^2 + 30 \lambda_2^2)$, $c_2 = c_1\lambda_2^2(29 \lambda_1^2 - 50 g^2)/(80\pi^2(\lambda_2^2 + \lambda_1^2/30))$, and the λ_i are the scalar coupling constants. (For the purpose of illustration, we set $c_1 M_I^4$, the constant term in the effective potential, at the value determined by global supersymmetry. The appropriate adjustment of the constant term, which sets the value of the cosmological constant, is determined by setting V evaluated at the global minimum to zero. This point is examined more closely in the Addendum.) Once again it must be emphasized that there has been no special choice of parameters to obtain this form for the potential. (In a recent preprint, Hall & Hinchcliffe (1982) have shown that, without tuning the values of the couplings, the coefficient of the logarithm term in Eq. (7) according to the renormalization group equations changes sign when $X \sim 10^3 M_I$ or less for the simplest models, so that M_C/M_I is much less than supposed by Dimopoulos & Raby (1982). This tuning may be avoided by adding additional terms to the superpotential to change the renormalization group equations. In general, the effect they note can be made insignificant, if α is somewhat smaller (α~1/30) than they assume.) The shape of the potential is shown in Fig. 1, where nothing has been assumed for the shape of the potential for $X \gtrsim M$. We will assume that a transition takes place in at least some region (of size $\lesssim M_I^{-1}$) of the Universe which takes us from a state with $X = 0$ to a state with $X \lesssim M_I$; eventually, due to the potential, the state evolves to a value of $\langle X \rangle$ of order M_I and a value of $\langle X_I \rangle$ of order M_I^2; these will form the initial conditions for our analysis. The insensitivity of our results to the prehistory of the region is, of course, what makes this scenario natural and attractive.

I will now outline two means by which tremendous inflation can be produced through the subsequent evolution of the X field. These methods have been studied

Fig. 1. The zero-temperature effective potential for the geometric hierarchy model.

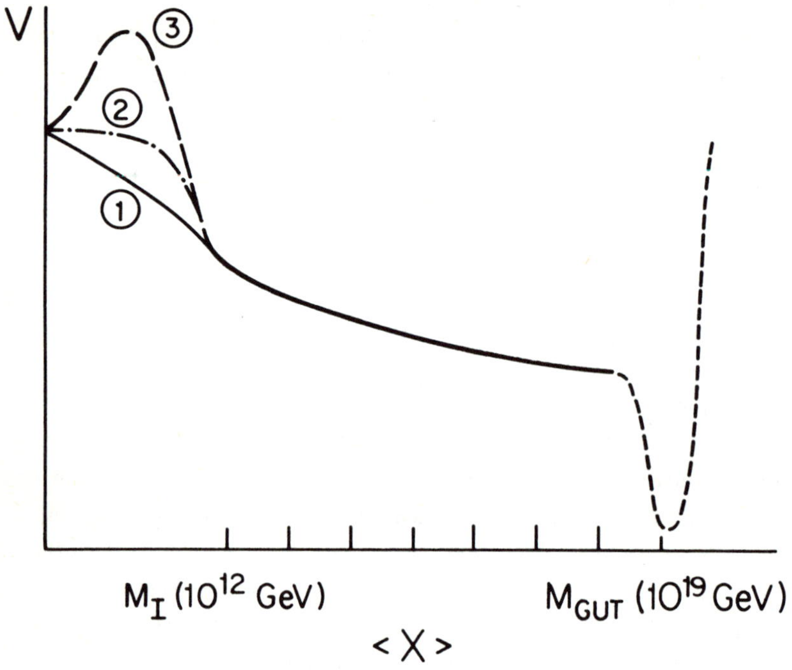

in collaboration with A. Albrecht, S. Dimopoulos, E Kolb, W. Fischler and S. Raby. A paper discussing our results associated with both methods will appear in a forthcoming publication (Albrecht, et al., 1982b).

Method I: The equation for the evolution of the X-field in the expanding Universe is given by:

$$\ddot{X} + 3H\dot{X} + V'(X) = 0 \tag{8}$$

where $V'(X) = -c_2 M_I{}^4/X$; H is the Hubble constant:

$$H^2 \simeq \frac{8\pi}{3}\frac{V(X)}{M_p{}^2}, \; M_p = \text{Planck mass} \simeq 10^{19} \text{ GeV}, \tag{9}$$

which gives rise to an effective damping term in Eq. (8). If one writes $V'(X) = -m^2(X)X$ where

$$m^2(X) = \frac{c_2 M_I^4}{X^2}, \tag{10}$$

the damping term dominates Eq. (8) provided that

$$4m^2(X) << 9H^2 \quad \text{or} \quad X^2 >> \frac{c_2}{6\pi c_1} M_p^2 \; . \tag{11}$$

Suppose that the domination of the damping term occurs for some large value, $X = X_o$ which is $\approx .1\, M_{GUT} = 10^{18}$ GeV. According to Eqs. (9, 10, 11), this requires $c_1 \gtrsim 50\, c_2$. If we assume that $\lambda_1 \approx \lambda_2 \approx \lambda = .1$, we find from Eq. (6) that $c_1 > 27000\, c_2$ (for g small) so that Eq. (11) can be satisfied without fine-tuning of parameters (see Addendum). If the damping term dominates, Eq. (8) is solved approximately by:

$$\dot{X} = c_2 M_I^4 / 3HX \tag{12}$$

or

$$\Delta t = \frac{3}{2} \frac{H}{c_2 M_I^4} \Delta X^2 \; . \tag{13}$$

Suppose that the evolution of X results in

$$\Delta X^2 \approx \frac{1}{e^2} M_{GUT}^2 - X_o^2 \; ,$$

where we assume Eq. (7) becomes unreliable for $M \sim M_{GUT}/e$ (the factor of 1/e is chosen just for the purpose of illustration; general choice of parameters is discussed by Albrecht, et al., 1982b). Then, the roll-over time, t_r, is given approximately by

$$\Delta t_r = \frac{3}{2} \frac{M_{GUT}^2}{e^2\, c_2\, M_I^4} H \; . \tag{14}$$

The scale parameter expands by a factor (Eqs. 9, 14):

$$\frac{R(t_r)}{R(t_o)} = \exp\,(H \Delta t_r) = \exp\, \frac{4\pi}{e^2} \frac{c_1}{c_2} \tag{15}$$

so that for $c_1 > 50\ c_2$, a fluctuation region of order M_I^{-1} $((10^{12}\mathrm{GeV})^{-1} = 2 \times 10^{-26}\mathrm{cm})$ expands to a size of order 10^{11} cm, much greater than the size of our observed Universe ($\sim 10^3$ cm) at the time of the completion of the transition. The inflation occurs near the very end of the logarithmic portion of the potential where $X \sim M_{GUT}$ and the degree of inflation is controlled by the ratio c_1/c_2. The fact that M_{GUT} is of order the Planck mass (Dimopoulos and Raby, 1982) in the geometric hierarchy model is unfortunate, but it is not required for the inflation to occur. Provided that c_1/c_2 is sufficiently large, the value of M_{GUT} can be reduced to $10^{-2}\ M_{Planck}$ where quantum gravitational effects are not bothersome. Thus, as a generic potential, Eq. (7) results in inflation without any special tuning of the parameters of the theory and without any detailed dependence on the shape of the potential for $X \lesssim M_I$.

Method II: For the second method, the inflation depends on radiation damping due to the particle production resulting from the time variation of the X field. In Albrecht, et al. (1982a), the time variation of the Higgs field was studied in relation to the problem of reheating at the end of the inflationary scenario. Suppose, as in that case, one postulates that the time varying field radiates an energy density per unit time given by

$$\delta \approx a(\dot{X})^d\ X^{5-2d} \tag{16}$$

where d is arbitrarily chosen. As a first test, suppose that the X field in our case has couplings such that $\delta = a\ X\ \dot{X}^2$; the reasonableness of this guess will be discussed at the end of the analysis. The equation of motion for the X field is given by

$$\ddot{X} + 3H\dot{X} + aX\dot{X} + V'(X) = 0 \tag{17}$$

where the third term is the effect due to δ. For $X \gtrsim M_I$, the δ-term dominates the time evolution equation. The approximate solution to Eq. (17) is given by

$$\dot{X} = c_2\ M_I^4/aX^2 \tag{18}$$

or

$$\Delta t = \frac{a}{3c_2\ M_I^4}\ \Delta X^3 \quad . \tag{19}$$

If X rolls from $X \sim M_I$ to $X \sim M_{GUT}$, the rollover time is given by

$$\Delta t_r = a\, M_{GUT}^3 / c_2\, M_I^4 \quad . \tag{20}$$

If the Hubble constant, H, is given by

$$H^2 \simeq \frac{8\pi}{3} \frac{c_1 M_I^4}{M_p^2} \tag{21}$$

we find that the scale factor after the inflation is given by

$$\frac{R(t_r)}{R(t_o)} = \exp(H\Delta t_r) = \exp\left[\frac{8\pi\, c_1}{3} \frac{a\, M_{GUT}^3}{c_2\, M_p\, M_I^2}\right] > e^{10^{12}} \quad ! \tag{22}$$

The resulting expansion is enormous, more than enough to solve the cosmological puzzles if we begin with a region with an initial size $\sim M_I^{-1}$.

Two major warnings should be given concerning Method II, however. Firstly, the results depend crucially on d in Eq. (16) having a value $1 < d < 2.5$; this was not the case for the studies of Albrecht et al., (1982a). The assumption is only reasonable if X couples in a special way to light (nearly massless) particles. Secondly, it is unlikely that the X particle couples strongly to any light particles. Dimopoulos and Raby have especially designed the model to ensure that X decouples from light fields as it evolves in order to maintain their weak scale masses. Even in a more general theory, we expect the particles that couple to X to have masses of order <X>, which becomes very large. Only in a very special model could Method II be possible, but the huge inflation effect that results makes it worth searching for such a model. Perhaps supersymmetry can permit the X to couple to fields which remain light due to cancellations coming from coupling to other fields with growing expectation values?

After the inflation must come the reheating, and here may lie the Achilles heel of this scenario. As X gets large, asymptotic freedom leads to the decrease of the gauge coupling constant until the sign of the logarithmic potential in Eq. (7) changes, producing a stable minimum for X. If the curvature of the effective potential (and the mass of the X, M_X) remains small ($\sim \frac{M_I^2}{M_P}$), X reaches the minimum and stops (due to the damping term); $\dot{X}$ remains small, so no reheating occurs. However, when X becomes large, two other effects become important: (a) perturbation theory breaks down and (b) gravitational corrections become important. The breakdown of perturbation is not likely to change the curvature near the minimum. The minimum occurs where the coefficient of the logarithm changes sign at which point the values of g and λ are still small. The renormalization group, which should still be reliable at the minimum, suggests that the curvature at the minimum is only corrected by logarithmic factors, and so remains too small for significant reheating. The gravitational effects can be estimated assuming one wishes to imbed this model in a locally supersymmetric model (Weinberg, 1982). (I will assume that the Planck mass, M_P, is treated as a fundamental parameter rather than a dynamically generated parameter or else the gravitational corrections would be scaled by $\langle X \rangle$ and become important for all X. In order to ensure that gravitational effects become important only for large X, M_P must be fixed by hand.) It appears from crude computations a model might be adjusted so that there is sufficient inflation due to the logarithmic tree potential, but also sufficient reheating due to gravitational effects to a temperature $\gtrsim 10^{-3}$ M_I (see Addendum for discussion). (Here the constant term, $c_1 M_I^4$, that appears in Eq. (7) is generated dynamically as the difference in energy density between the potential at $X \sim M_I$ and the minimum $X \sim M_p$ (where our Universe now exists with a cosmological constant of zero).) The details of our study will be presented in Albrecht, et al. (1982b).

Reheating is crucial because (a) the thermal energy must be recovered so that the entropy of the Universe rises to a large value to correspond with observation, and (b) baryon asymmetry must be generated <u>after</u> the inflation. Since there exist color triplet Higgs mesons in the theory which have masses $\sim M_I$, if the Universe can recover to a temperature even as small as $\sim 10^{-3}$ M, it might be sufficient to ensure that the usual sort of baryon asymmetry scenarios (in which light mesons begin from near-equilibrium and then fall out of equilibrium (Dimopoulos and Raby, 1982; Albrecht, et al., 1982b) succeeds.

Much work is required to check whether the reheating can be accommodated with the rest of the features of the geometric hierarchy model, but if it is possible <u>all the successes of C-W inflation are obtained without any fine-tuning of parameters.</u> In fact, it appears that the problem of producing the inhomogeneities that grow into galaxies is more easily solved in models for which the inflation occurs when the scalar expectation value is large compared to H (as occurs naturally in this model, but not in the C-W models). The perturbation $\delta\rho/\rho$, evaluated when a scale enters its horizon has been shown (e.g. Bardeen et al., 1982) to be proportional to the value of $\Delta\phi/\bar{\phi}$ during the inflation epoch, where $\Delta\phi$ is the fluctuation about the mean value, $\bar{\phi}$. In C-W inflation, $\Delta\phi \sim H$ and $\bar{\phi}<H$, so $\Delta\phi/\bar{\phi}$ is large; as a result $\delta\rho/\rho$ is found to be too large compared to the value required by astrophysics ($\sim 10^{-4}$) to explain galaxy formation without producing a large background microwave anisotropy. In the supersymmetry model, $\Delta X \sim H << X$ during inflation and one can adjust parameters so that $\delta\rho/\rho \sim 10^{-4}$. Furthermore, the de Sitter scalar field fluctuations ($\propto H^2$) which can cause problems with C-W inflation scenarios (Vilenkin and Ford, 1982; Linde, 1982b) can be shown to be insignificant in this case, again because X is large compared to H in the inflationary phase (Albrecht et al. 1982b; Vilenkin, private communication).

As an amusement, it is interesting to speculate how the behavior of the effective potential for $X \leq M_I$ can affect the inflationary cosmology. The three possibilities are shown in Fig. 1, corresponding to (1) negative, (2) zero (analog of C-W) or (3) positive curvature of the potential near X = 0. As has been emphasized, which case is obtained depends upon the choice of parameters and inflation occurs in any case; therefore, which case one obtains has no effect on the predictions for scales of the size of our observable Universe. However, there can be different predictions for scales much larger (by a factor of 10^{30} or more) than the size of our observed Universe. Since such predictions are in principle unable to be tested by any human (protonic) observer, I call this branch of cosmology "metaphysical cosmology" or meta-cosmology for short. Each of the three cases shown in Fig. 1 can lead to a distinctive meta-cosmology:

(1) Spinodal Metacosmology - The potential has negative curvature at X = 0. At some finite temperature, the X = 0 phase becomes unstable and fluctuations drive different regions towards different SSB minima; the phenomenon of dividing the whole Universe (no region remains in the metastable phase) into different developing phase regions - fluctuation regions - is analogous to spinodal decomposition in condensed matter physics (Steinhardt, 1982). The different SSB minima are related by continuous or discrete symmetries. When different regions come together for which the phases contained therein differ by a discrete (continuous) symmetry, one expects domain walls (monopoles) to be formed. Of course the inflation pushes these topological defects to distances beyond which we can ever hope

to observe and before we could even reach them, the domain walls collapse gravitationally (Hawking, private communication). We conclude for the Spinodal Universe, that even though the Universe near us is very nearly homogeneous and isotropic, the Universe far away from us (near the edges of fluctuation regions) might be quite different. This is a startling possibility from a philosophical point of view.

(2) Single Universe Metacosmology: With special fine-tuning of parameters, the effective potential for the supersymmetry case can be made to be flat (in the C-W GUT sense, curvature $\lesssim H^2$) near the origin. In this case, the Hawking-Moss (1982) analysis suggests that the universe rolls all at once into the same SSB phase. There are no monopoles or domain walls formed as topological defects in the transition because there is only one "fluctuation region". The total Universe is uniform, homogeneous and isotropic. This metacosmology yields the simplest picture of our Universe.

(3) Regenerative Meta-Cosmology: If parameters are chosen so that there remains a barrier with a curvature $\gtrsim H^2$ at $X \approx 0$, the $X = 0$ phase remains metastable even as T approaches zero. Because $Z = 0$ is metastable, rather than stable, quantum fluctuations of the type studied by Coleman and Deluccia (1980) - analogs of flat space bubbles - are produced rarely in the metastable Universe. Their interiors contain evolving stable phase. These bubbles occupy only a small fraction of the total Universe, most of which remains in the metastable phase and continues to expand forever. The expectation value of X begins $\lesssim M_I$ and evolves along the potential. The resulting bubble inflates to a size much larger than our observed Universe. The interior is perfectly Robertson-Walker. Observers inside the bubble can never observe the bubble wall because the bubble expands at nearly the speed of light. Bubbles are produced so rarely that they almost never collide. When a new bubble is formed, it regenerates a new Universe which can never contact our own. New Universes are regenerated forever because the Universe is only percolated (fractally) as time approaches infinity. (See Albrecht, et al., 1982b for more elaborate discussions.) The regenerative scenario is probably the most surprising and radical cosmological possibility.

Linde (1982c) has noted (in response to this talk) that a regenerative metacosmology can avoid the usual problem of the initial cosmological singularity! In almost any other cosmology which has an early epoch in which the Universe was very hot, extrapolating backwards in time leads to an unavoidable space-time singularity; this naturally makes physicists uncomfortable. In the regenerative metacosmology the Universe could begin (t = 0) in the metastable phase at T = 0 (actually $T_{Hawking}$); The metastable region never has to be hot so there need never be an initial singularity. The Universe even today could be mainly in this phase, with only a few rare bubbles in which true thermal matter and ourselves can exist.

Why should the Universe have begun in such a metastable phase? Perhaps because the X = 0 phase is a state of highest symmetry. It should also be noted that there remains a problem if time is continued for $t < 0$. A full de Sitter spacetime has a contraction phase for $t < 0$ and an expansion phase for $t > 0$. If the Universe began in a metastable symmetric phase with T = 0 and $t = -\infty$, it would complete the transition before t = 0 and there would be no inflation phase $t > 0$. If one insists upon only beginning at t = 0, the singularity problem is only solved somewhat artificially (Linde, private communication).

What is not emphasized in Linde's paper, however, is that the CW GUTs inflation scenario leads, more likely, to a spinodal metacosmology because the barrier has a height $\ll H^4$ as $T \to 0$. On the other hand, in a model like the one considered in this paper, where the inflation does not depend on the properties of the potential for small values of the order parameter, a high barrier and, thus, the regenerative metacosmology clearly occur for a wide range of parameters. Thus, it appears possible that one more fundamental mystery of early cosmology might, in principle, be explained with an appropriate inflationary Universe.

This sort of amusing speculation only serves to emphasize the importance of no fine-tuning in the inflationary scenario. The "geometric hierarchy" model under consideration may not be a correct model for particle physics, but it illustrates how inflation can be achieved without unnatural assumptions. The "tricks" used for this case, Methods I and II, may well be useful for a more realistic theory not yet discovered. By increasing our catalog of techniques for accomplishing inflation, we can hope to raise the inflationary Universe hypothesis from the level of Possible to the level of Likely.

I thank the Nuffield Foundation for the gracious support during the Workshop on the Early Universe. I especially thank S. Hawking, G. Gibbons and S. Siklos and J. Fella for their kind hospitality during my stay in Cambridge and for arranging for such a stimulating conference. I also wish to thank A. Albrecht, S. Dimopoulos, W. Fischler, E. Kolb and

S. Raby for their collaborative assistance and A. Linde and S. Hawking for useful discussions.

This work was supported in part by the U.S. Department of Energy under Contract No. EY-76-C-02-3071, and by a DOE Outstanding Junior Investigator Grant.

REFERENCES

Abbott, L., Farhi, E. & Wise, M.B., (1982). Particle Production in the New Inflationary Universe. Phys.Lett. 117B, 29.

Albrecht, A. & Steinhardt, P., (1982a). Phys. Rev. Lett. 48, 1220.

Albrecht, A. & Steinhardt, P., (1982b). Coleman-Weinberg Models and the Inflationary Universe. Penn preprint, to be published.

Albrecht, A., Steinhardt, P.J., Turner, M.S. & Wilczek, F., (1982a) Phys. Rev. Lett. 48, 1437.

Albrecht, A., Dimopoulos, S., Fischler, W., Kolb, E., Raby, S. & Steinhardt, P.J., (1982b). Proceedings of the Third Marcel Grossman Conference, to be published.

Bardeen, J., Steinhardt, P.J. & Turner, M.S., (1982). Spontaneous Creation of Almost Scale-Free Density Perturbations in an Inflationary Universe. Penn preprint UPR-2020T.

Coleman, S. & DeLuccia, R., (1980). Phys. Rev. D21, 3305.

Coleman, S. & Weinberg, E., (1973). Phys. Rev. D7, 788.

Dimopoulos, S. & Raby, S. (1982). "Geometric Hierarchy", Los Alamos preprint, LA-UR-82-1282, to be published.

Dolgov, A. & Linde, A., (1982). Baryon Asymmetry in Inflationary Universe. Phys. Lett. 116B, 329.

Guth, A., (1981). Phys. Rev. D23, 347.

Guth, A. & Pi, S.-Y., (1982). Phys. Rev. Lett. 49, 1110.

Guth, A. & Weinberg, E., (1982). Could the Universe have Recovered from a Slow First-Order Phase Transition, to be published in Nucl. Phys. B212, 321.

Hall, L. & Hinchliffe, I., (1982). Phys. Lett. 119B, 128.

Hawking, S., (1982). The Development of Irregularities in a Single Bubble Inflationary Universe. Phys. Lett. 115B, 295.

Hawking, S. & Moss, I., (1982). Phys. Lett. 110B, 35.

Huq, M., (1976). Phys. Rev. D14, 3548.

Linde, A., (1982a). Phys. Lett. 108B, 389.

Linde, A., (1982b). Scalar Field Fluctuations in Expanding Universe and the New Inflationary Universe. Phys. Lett. 116B, 335.

Linde, A., (1982c). Non-singular Regenerating Inflationary Universe, unpublished, Lebedev preprint.

O'Raifeartaigh, L., (1975). Nucl. Phys. B96, 331.

Starobinski, A., (1982). Talk given at the Nuffield Workshop on the Very Early Universe.

Steinhardt, P., (1982). A New Cosmology. Proceedings of the Second Moriond Astrophysics Workshop, ed. P. Crane, T. Gaisser, D. Hegyi and J. Van Tranh.

Vilenkin, A. & Ford, L., (1982). Gravitational Effects upon Cosmological Phase Transitions. Phys. Rev. D26, 1231.

Weinberg, S., (1982). to be published.

Witten, E., (1981). Phys. Lett. 105B, 267.

ADDENDUM

Eq. (7) only represents an initial guess for the form of the effective potential based solely on an analysis of global supersymmetry. The constant term that is computed ensures that the potential, V, is greater than zero at the minimum, as required by supersymmetry. The analysis began by using this constant as a rough estimate for setting the cosmological constant in the equations of motion for the evolution of the Universe. Without some concrete calculation to determine the effective potential near $\langle X\rangle = M_{GUT}$, the constant cannot be adjusted so as to make the cosmological constant at the minimum of the potential equal to zero.

As an improvement on this guess we (Albrecht, et al., 1982b) have analyzed the lowest order gravitational corrections to the effective potential. Near $\langle X\rangle = M_{GUT} \sim M_p$, higher order gravitational corrections certainly become important, but we have considered the lowest order corrections only to obtain a qualitative feeling of what to expect. We have also assumed that the Planck scale is introduced as a fundamental scale and not generated dynamically. If it were generated dynamically by the expectation value of X, gravitational corrections would be incalculable and important for all X. By introducing the Planck scale as a fundamental scale we must accept the fact that we have, to some degree, violated the spirit of the reverse hierarchy scheme, but at least we can begin to calculate.

Instead of Eq. (2), V with first order gravitational corrections takes the form (Weinberg, 1982):

$$V = \exp(8\pi GD)\left[\sum_{i,j} h_{ij} F_i F_j^* - 24\pi G|W|^2\right]$$

$$F_i = \frac{\partial W}{\partial \phi_i} + 8\pi GW\left(\frac{\partial D}{\partial \phi_i}\right) \quad , \tag{23}$$

where D is the supersymmetry term derived from the gauge interactions:

$$D = -g([A^*,A] + [X^*,X] + \text{irrelevant terms}) \tag{24}$$

Inserting Eqs. (1,24) into Eq. (23), we find that the corrections are given by:

$$\Delta V_g \sim c_3 \frac{M_I^4}{M_p^4}(X^4 - c_4 M_p^2 X^2) + \Lambda \ , \tag{25}$$

where c_3 and c_4 are constants that we cannot accurately compute.

Consider the effects of adding V_g to the effective potential, Eq. (6). First, note that V_g becomes important compared to the term logarithmic in X when $X \sim M_p$; for large X the correction leads to a minimum near $X \sim M_p$ with a curvature that is generally greater than that obtained from the logarithmic term alone. The constant, Λ, is chosen so as to set the cosmological constant at the minimum of the potential (corresponding to our present Universe) equal to zero. We have computed numerically the roll-over of the X field for the combined potential and we have discovered that for a wide (and natural) range of parameters sufficient inflation can be achieved due simply to the effects of the 3HX term in the equation of motion, just the same effect as Method I used for Eq. (7). To analyze the reheating problem (which is aided by the potential which is steeper near the minimum) we have considered various forms for δ:

$$\delta = aX^5\left(\frac{\dot{X}}{X^2}\right)^a \left(\frac{M_I}{X}\right)^b \tag{26}$$

We find that for a = 2, $b \leq 2$ reheating can be achieved for a wide range of parameters; for a = 2 and $b \geq 2$, almost no reheating occurs. This is an improvement on the situation in the absence of gravitational effects where none of these choices resulted in reheating. We have also been rather conservative in our allowed choices of the parameters, and a more precise calculation of gravitational effects could help matters. As it stands, we are presently examining which is the appropriate choice for δ for the supersymmetric model. For the Dimopoulos-Raby model it appears that because of the decoupling of X from light fields in that model, $b \geq 6$. However, it is not yet clear whether a variation on this class of models might not yield a case where $b \leq 2$ is appropriate. More details will be provided in Albrecht, et al. (1982b).

PERTURBATIONS OF A DE SITTER UNIVERSE

John D. Barrow
Astronomy Centre, Sussex University, Brighton BN1 9QH, U.K.

Abstract: We have calculated the scalar, vector and tensor modes of small metric perturbations to a matter-filled de Sitter universe and show that it is stable but not asymptotically stable against these perturbations of the metric.

INTRODUCTION

There has recently been a resurgence of interest in the properties of the de Sitter Universe (de Sitter 1917) stimulated by the unusual phase portraits of grand unified gauge theories at very high energy. As discussed by many other authors in this volume, if the phase transition associated with the spontaneous symmetry breakdown in a grand unified theory (GUT) is first order, then the Universe can find itself in a metastable symmetric phase in which the vacuum energy is ~ M_u^4 where M_u is the energy scale of symmetry breaking. When the Universe cools below M_u the vacuum energy will dominate over the radiation density and, because of its Lorentz invariant form, force the cosmic scale factor, a(t), to grow exponentially with proper time, $a(t) = \exp(t/t_u)$ where $t_u \sim m_p M_u^{-2}$, (here $m_p = G^{-1/2}$ is the Planck mass and $c=\hbar=k_B=1$). If the Universe remains in this symmetric phase for a sufficient period ($\gtrsim 70 t_u$) then an explanation may be provided for the proximity of the present Universe to spatial flatness, the absence of magnetic monopoles and the correlation of cosmic properties over scales exceeding $\sim 10^{15} M_\odot$, (Guth 1981, Barrow & Turner 1981). The problem of the return to a standard radiation-dominated Friedman Universe may have been overcome by the recent suggestion by Linde (1982) and Albrecht and Steinhardt (1982) that the spontaneous symmetry breakdown is of the Coleman-E. Weinberg form, (Coleman & Weinberg, 1973).

The aim of this note is to study the response of the de Sitter phase to small gravitational perturbations. The universe cannot be precisely isotropic and homogeneous at t_u; there must inevitably exist some small, and perhaps large, fluctuations of the metric. These would correspond to density, velocity and gravitational wave distortions typically. If any of these modes were found to grow as a(t) increases during the vacuum-dominated de Sitter phase then the subsequent details of the phase transition and the observables emerging from it would be strongly affected. To demonstrate that this will not happen we need to prove what Hawking and Moss call a cosmic 'no-hair' theorem, (Hawking & Moss, 1982) and show that all expanding universes with $\Lambda>0$ approach de Sitter space. We demonstrate a more modest result, that small perturbations of a de Sitter Universe do not grow as $a(t) \to \infty$. We show that perturbations of the metric tensor associated with gravitational wave and density perturbations tend to a constant value asymptotically. Universes that are close to de Sitter when inflation begins at t_u stay close during the subsequent inflationary phase.

FORMALISM

The stability of expanding homogeneous and isotropic universes with $\Lambda=0$ has been extensively studied (Lifshitz & Khalatnikov 1963, Bardeen 1980). We shall generalize the analysis of Lifshitz & Khalatnikov to the case of $\Lambda\neq 0$. The Einstein equations are (notation is as in Landau & Lifshitz 1975),

$$R_i^k - \tfrac{1}{2} R\, \delta_i^k - \Lambda\, \delta_i^k = 8\pi G\, T_i^k \tag{1}$$

The positive curvature de Sitter metric is

$$ds^2 = a^2(\eta)\left[\, d\eta^2 - d\chi^2 - \sin^2\chi(\sin^2\theta d\phi^2 + d\theta^2)\,\right] \tag{2}$$

where η is conformal time and (χ,θ,ϕ) are spherical coordinates. The negative curvature model is obtained by $(a,\eta,\chi) \to i(a,\eta,\chi)$. If we assume, for simplicity, the matter content is dust (the results are unaffected by this assumption) then the Friedmann equation for $a(\eta)$ is

$$\left(\frac{da}{d\eta}\right)^2 = Aa - a^2 + \frac{\Lambda a^4}{3} \equiv F(a) \tag{3}$$

where $A = 4GM/3\pi$ and M is the total mass of dust. Clearly, the equations for the perturbations of (1) have no explicit dependence on the cosmological constant Λ; that dependence only enters implicitly through (3). We have

$$\delta R^k_i - \tfrac{1}{2}\,\delta R\,\delta^k_i = 8\pi G\,\delta T^k_i \tag{4}$$

Defining $h_{ab} \equiv \delta g_{ab}$ and picking a gauge with $h_{oo} = h_{o\alpha} = 0$ we follow Lifshitz and Khalatnikov and decompose into spherical harmonics

$$h^\beta_\alpha = \lambda(a)\,P^\beta_\alpha + \mu(a)\,Q^\beta_\alpha \tag{5}$$

where the tensors Q^β_α and P^β_α are formed from the nth order scalar harmonics Q

$$Q^\beta_\alpha = 1/3\,\delta^\beta_\alpha\,Q\;;\; P^\beta_\alpha = \frac{Q_{;\alpha}^{\;\;;\beta}}{n^2-1} + Q^\beta_\alpha \tag{6}$$

We may now consider the tensor, vector and scalar modes in turn.

GRAVITATIONAL WAVE PERTURBATIONS

These are associated with the set of tensor spherical harmonics, ψ^α_β, with which there are no corresponding scalars and vectors so $\psi^\alpha_\alpha = 0$, $0 = \psi^\beta_{\alpha;\beta}$. Physically, this corresponds to a gravitational potential perturbation during which the matter distribution remains smooth and at rest. Setting

$$h^\beta_\alpha = y(a)\,\psi^\beta_\alpha \tag{7}$$

the equations (4) become ($' \equiv d/da$)

$$y'' + y'\left[\frac{F'}{2F} + \frac{2}{a}\right] + \frac{(n^2-1)y}{F} = 0 \tag{8}$$

As $a \to \infty$, $F \to \Lambda a^4/3$ and $y \to y_\infty$ where, (A,B constants);-

$$y_\infty = a^{-1}\left[A\left(\frac{a\sin(\alpha/a)}{\alpha} - \cos(\alpha/a)\right) + B\left(\frac{a\cos(\alpha/a)}{\alpha} + \sin(\alpha/a)\right)\right] \tag{9}$$

and where $\alpha \equiv \left[3(n^2-1)\Lambda^{-1}\right]^{\frac{1}{2}}$. Asymptotically we see that $y_\infty \to$ constant. This solution corresponds to a real change in the metric which cannot be removed by a coordinate transformation. Both long and short wavelength gravitational wave mode perturbations of de Sitter freeze-in with constant amplitude as $a \to \infty$

ROTATIONAL PERTURBATIONS

These correspond to changes in the velocity but not the density of matter when the metric potential is perturbed. Formally,

$$h_\alpha^\beta = \sigma(a) \; S_\alpha^\beta \tag{10}$$

where $S_\alpha^\beta = S_\alpha^{;\beta} + S^\beta_{;\alpha}$ and $S^\alpha_{;\alpha} = 0$. We have, from (4),

$$\sigma'' + \sigma'\left[\frac{F'}{2F} + \frac{2}{a}\right] = 0 \tag{11}$$

The solution σ= constant can be removed by a suitable coordinate transformation and the physically relevant solution is

$$\sigma \propto \int \frac{da}{a^2 F^{\frac{1}{2}}} \tag{12}$$

Therefore as a $\rightarrow \infty$ we have $\sigma \propto a^{-3}$ and the associated velocity perturbation falls off as $\delta v^\alpha \propto a\sigma' F^{\frac{1}{2}} \propto a^{-1}$ in accord with angular momentum conservation. All metric perturbations to de Sitter corresponding to vortical perturbations of the matter motion die out as $a \rightarrow \infty$

DENSITY PERTURBATIONS

Perturbations of the matter density are associated with metric perturbations of the form (5). The density perturbation is,

$$\frac{\delta\rho}{\rho} = \frac{a^2}{9(F-a^2)}\left[(n^2-4)(\lambda+\mu) + \frac{3F^{3/2}}{a}\mu'\right] Q \tag{13}$$

and in the de Sitter background we have

$$\lambda'' + \lambda'\left[\frac{2}{a} + \frac{F'}{2F}\right] - \frac{(n^2-1)}{3F}(\lambda+\mu) = 0 \tag{14}$$

$$\mu'' + \mu'\left[\frac{2}{a} + \frac{F'}{2F}\right] + \frac{(n^2-4)}{3F}(\lambda+\mu) = 0 \tag{15}$$

As $a \rightarrow \infty$ we find $\lambda \sim a^{-2}$, $\mu \sim$ a constant and $\frac{\delta\rho}{\rho} \rightarrow$ constant to leading order.

CONCLUSIONS

The results presented above illustrate the stability of the de Sitter inflationary phase against small perturbations. It is probable that a similar stability would hold for larger perturbations than the first order analysis suggested here. However, such considerations are unlikely to be relevant for the inflationary Universe scenario because, as shown by Barrow and Turner (1981), if the anisotropy dominates the expansion dynamics at t_u then the de Sitter phase does not occur in the early Universe and there is no inflationary phase. Thus, large gravitational wave mode perturbations exclude the inflationary phase and for this reason the inflationary model cannot explain the isotropy of the Universe. For some reason, presumably associated with quantum phenomena, the Universe must be close to the isotropic solution at t_u, (Barrow, 1982).

ACKNOWLEDGEMENTS

I would like to thank S.T.C. Siklos and G. Gibbons for discussions.

REFERENCES

Albrecht, A. & Steinhardt, P.J. (1982) Phys. Rev. Lett., 48, 1220.
Bardeen, J.M. (1980) Phys. Rev., D22, 1882.
Barrow, J.D. & Turner, M.S. (1981) Nature, 292, 35.
Barrow,J.D. (1982) Quart. J. Roy. Astron. Soc., 23, 344.
Coleman, S. & Weinberg, E. (1973) Phys. Rev., D7, 788.
de Sitter, W. (1917) Mon. Not. Roy. Astron. Soc. 78, 3.
Guth, A. (1981) Phys. Rev., D23, 347.
Hawking, S.W. & Moss, I.G. (1982) Phys. Lett., 110B, 35.
Landau, L.D. & Lifshitz, E.M. (1975) The Classical Theory of Fields. Oxford: Pergamon Press. 4th revised edition.
Lifshitz, E.M. & Khalatnikov, I.M. (1963) Adv. in Phys., 12, 185.
Linde, A.D. (1982) Phys. Lett., 108B, 389.

COSMIC BALDNESS

W. Boucher
G.W. Gibbons
D.A.M.T.P.,
Cambridge University

INTRODUCTION

Some years ago (Gibbons & Hawking 1977) it was suggested that solutions of Einstein's equations with a positive cosmological term should eventually settle down to a state which is stationary inside the cosmological event horizon of any future inextendible timelike curve. This stationary state, it was suggested, would, if no black holes were present, be de Sitter space. The recent work on the inflationary scenario in the early universe has reawakened interest in this topic and the purpose of this article is to review the mechanism whereby a universe dominated by vacuum energy, so that it satisfies the equation

$$R_{\alpha\beta} = \Lambda g_{\alpha\beta} , \quad \Lambda > 0 , \tag{1}$$

relaxes to an asymptotically de Sitter state inside the event horizon of any observer. There is some overlap with the article by Barrow but the emphasis here is rather different.

GEOMETRY OF DE SITTER SPACE

De Sitter space may be thought of as the hyperboloid in 5-dimensional Minkowski space given by

$$-(X^0)^2 + (X^1)^2 + (X^2)^2 + (X^3)^2 + (X^5)^2 = 3/\Lambda \tag{2}$$

where Λ is the cosmological constant which is related to Hubble's constant, H , or the surface gravity of the horizon, κ , by

$$H = \kappa = (\frac{\Lambda}{3})^{\frac{1}{2}} \tag{3}$$

It is useful to coordinatize the spacetime in two different ways, depending upon whether we wish to think of it as an expanding F.R.W. universe or a static universe with an event horizon. We use (s,χ,θ,ϕ) in the first case and (t,r,θ,ϕ) in the second. They are given by (Hawking & Ellis 1973; Hawking & Gibbons 1977):

$$r\sin\theta\cos\phi = X^1 = H^{-1}\cosh Hs\sin\chi\sin\theta\cos\phi \tag{4}$$

$$r\sin\theta\sin\phi = X^2 = H^{-1}\cosh Hs\sin\chi\sin\theta\sin\phi \tag{5}$$

$$r\cos\theta = X^3 = H^{-1}\cosh Hs\sin\chi\cos\theta \tag{6}$$

$$(H^{-2}-r^2)^{\frac{1}{2}}\cosh Ht = X^5 = H^{-1}\cosh Hs\cos\chi \tag{7}$$

$$(H^{-2}-r^2)^{\frac{1}{2}}\sinh Ht = X^0 = H^{-1}\sinh Hs \tag{8}$$

The metric thus becomes:

$$d\mathcal{s}^2 = -(1 - H^2r^2)\,dt^2 + (1 - H^2r^2)^{-1}\,dr^2 + r^2d\Omega^2 \tag{9}$$

or

$$d\mathcal{s}^2 = -ds^2 + H^{-2}\cosh^2 Hs\,(d\chi^2 + \sin^2\chi\,d\Omega^2) \tag{10}$$

The event horizon of an observer situated at $r = 0$ is given by $r = H^{-1}$ and so since (4) - (6) imply that

$$\sin\chi = Hr\,(\cosh Hs)^{-1}$$

$$= Hr\,(1 - H^2r^2\tanh^2 Ht)^{-\frac{1}{2}}\,(\cosh Ht)^{-1} \tag{11}$$

we see that as $s,t \to \infty$ an exponentially smaller portion of the 3-spheres $s = \text{const.}$ is included within the event horizon of this observer. This is the key to understanding the decay of perturbations. From the point of view of the static frame we expect them to be radiated through the cosmological event horizon just as in the familiar black hole case (see Price 1972). From the point of view of the expanding universe frame we shall see — as pointed out by Starobinsky (1977) that gravitational wave perturbations do not decay but rather they are frozen in as $t \to \infty$. However this is not in contradiction with the cosmic No Hair Theorem since as time proceeds the observer sees these perturbations on an exponentially smaller scale and so the region inside his/her event horizon appears to become more and more accurately de Sitter.

BEHAVIOUR OF LINEARIZED SCALAR AND GRAVITATIONAL WAVE PERTURBATIONS

Following Lifshitz & Khalatnikov (1963) we consider perturbations of the metric form (10) given by

$$d\mathcal{s}^2 = -ds^2 + H^{-2} \cosh^2 Hs \, (d\chi^2 + \sinh^2\chi \, d\Omega^2 + \sum_n \nu_n(s) \, G^{(n)}_{ij}(\chi,\theta,\phi) \, dx^i dx^j) \tag{12}$$

where the $G^{(n)}_{ij}$ are tensor harmonics on S^3 and the $\nu_n(s)$ are the amplitude of the gravitational wave perturbations. It turns out that if one considers solutions of the massless, minimally coupled scalar wave equation in the de Sitter background of the form

$$\phi = \nu_n(s) \, Q^{(n)}(\chi,\theta,\phi) \tag{13}$$

where the $Q^{(n)}$ are scalar harmonics on S^3, the coefficients $\nu_n(s)$ in (13) satisfy the same equation as those in (12) and are given up to a constant factor by

$$\nu_n(s) = (\text{in sech } Hs + \tanh Hs) \exp in(\tan^{-1} \sinh Hs) \tag{14}$$

In neither the scalar nor the gravitational case do the perturbations die away, rather they tend to constants at late times. Thus the scalar field ϕ can have any functional form, $\Phi(\chi,\theta,\phi)$, at late times. However using the relation given by (11) between χ and the coordinates r and t we see that as $t \to \infty$ for all r inside the event horizon

$$\phi(t,r,\theta,\phi) \to \Phi(o) \tag{15}$$

Thus ϕ tends to a constant inside the event horizon exponentially fast which is in accordance with the fact that there are no static solutions of the wave equation which are regular inside and on the event horizon other than the constant one.

In the gravitational case similar results hold; locally the constant gravitational wave modes are pure gauge, even though they are not pure gauge globally over the entire S^3. We shall see this in more detail when we consider the fully non-linear asymptotic form of the metric. Note that in contrast to the black hole case the perturbations die away exponentially fast; there is no power law fall-off of the sort discussed by Price (1972).

FULLY NON-LINEAR ASYMPTOTIC ANALYSIS

Starobinsky has pointed out to us that a general asymptotic solution of the equation $R_{\alpha\beta} = \Lambda g_{\alpha\beta}$ takes the form

$$d\mathscr{s}^2 = -ds^2 + \exp 2Hs\ a_{ij}(\underline{x})\,dx^i dx^j + O(1) \tag{16}$$

where a_{ij} is an arbitrary 3-metric. This clearly indicates that the gravitational waves do not decay globally over the entire 3-surface s = constant. The geometry of this surface never settles down to that of a smooth 3-sphere. The curve $x^i = 0$ is a geodesic. By means of a linear coordinate transformation of the x^i's we may, with no loss of generality, set

$$a_{ij}(0) = \delta_{ij} \tag{17}$$

Now introduce coordinates y^i and t by

$$y^i = e^{Hs} x^i \tag{18}$$

$$e^{Ht} = (1 - H^2y^2)^{-\frac{1}{2}} e^{Hs} \tag{19}$$

It is now an easy exercise to show that in the coordinates (t, y^i), which are only valid within the event horizon of the timelike observer at $y^i = 0$, the general metric (16) approaches the exact metric of de Sitter space (equation (9)) exponentially fast. Thus as far as every freely falling observer is concerned the observable universe becomes quite bald.

GENERALIZED ISRAEL'S THEOREM

If we exclude the possibility of a central black hole it would seem likely from the above analysis that de Sitter space is the only exactly static solution of the equations $R_{\alpha\beta} = \Lambda g_{\alpha\beta}$, $\Lambda > 0$, surrounded by a regular event horizon and with a regular centre — i.e. such that the interior of the event horizon is diffeomorphic to the product of an open ball in R^3 with the real line. This last proviso is necessary to exclude the Nariai (1951) metric. We have tried unsuccessfully to generalize the standard Israel theorems for black holes (Israel 1967; Muller zum Hagen, et al. 1973; Robinson 1977) to this case. If the metric is static the boundary conditions on the horizon are just those required to justify analytically continuing the metric to the Euclidean regime by putting $t = i\tau$. The resulting metric is positive definite, defined on a manifold diffeomorphic to S^4 , and satisfies $R_{\alpha\beta} = \Lambda g_{\alpha\beta}$. The only known such metric is of course the standard Einstein metric on S^4 which is just the analytic continuation of de Sitter space obtained by setting $t = i\tau$ in (9) and identifying τ modulo $2\pi(3/\Lambda)^{\frac{1}{2}}$. The natural conjecture to make — which is stronger than the generalised Israel theorem suggested above — is that the only such metric is the standard one. Mathematicians we have asked do not know whether this is true but the corresponding statement is false for S^{4n+3} , $n \geq 1$ (Jensen 1973). If there is another such metric it cannot be continuously deformed into the standard one as one can readily check by perturbing the Einstein equations on S^4 as described in (Gibbons and Perry 1978).

ACKNOWLEDGEMENTS

We would like to thank S.W. Hawking, A. Starobinsky, S. Siklos and J. Barrow for useful discussions and suggestions on the material presented above.

REFERENCES

Gibbons, G.W., and Hawking, S.W. (1977). Cosmological event horizons, thermodynamics, and particle creation. Phys. Rev. D15, 2738.
Gibbons, G.W. and Perry, M.J. (1978). Quantizing gravitational instantons. Nucl. Phys. B146, 90.
Hawking, S.W., and Ellis, G.F.R. (1973). The large scale structure of space-time. Cambridge: Cambridge University Press.
Israel, W. (1967). Event horizons in static vacuum space-times. Phys. Rev. 164, 1776.
Jensen, G.R. (1973). Einstein metrics on principal fibre bundles. J. Diff. Geom. 8, 599.

Lifshitz, E.M. and Khalatnikov, I.M. (1963). Investigations in Relativistic Cosmology. Adv. Phys. 12, 185.

Müller zum Hagen, H., Robinson, D.C., and Seifert, H.J. (1973). Black holes in static vacuum space-times. Gen. Rel. Grav. 4, 53.

Nariai, H. (1951). On a new cosmological solution of Einstein's field equations of gravitation. Sci. Rep. Tôhoku Univ. 35, 62.

Price, R.H. (1972). Nonspherical perturbations of relativistic gravitational collapse, I: Scalar and gravitational perturbations. Phys. Rev. D5, 2419. Nonspherical perturbations of relativistic gravitational collapse, II: Integer-spin, zero-rest-mass fields. Phys. Rev. D5, 2439.

Robinson, D.C. A simple proof of the Generalization of Israel's Theorem. Gen. Rel. Grav. 8, 695.

Starobinsky, A.A. Spectrum of relict gravitational radiation and the early state of the universe. JETP Lett. 30, 682.

BUBBLE COLLISIONS IN GENERAL RELATIVITY

S. T. C. Siklos

Newnham College, Cambridge, England.

Z. C. Wu

Department of Applied Mathematics and Theoretical Physics,
Silver Street, Cambridge and
University of Science and Technology of China, Hofei,
Anhwei, China.

Abstract. The collision of two bubbles of true vacuum in a background of false vacuum is considered in the context of General Relativity. It is found that in the thin wall approximation, the problem can be solved exactly. The region to the future of the collision is described by the pseudo-Schwarzschild de Sitter metric. The parameters in this metric are found by solving the junction conditons at each collision.

1. Introduction

An essential feature of the various inflationary models described in this volume is the existence of a first order phase transition in the early universe. This transition can arise in Grand Unified Theories if the potential $U(\varphi)$ is of the form shown in Figure 1.

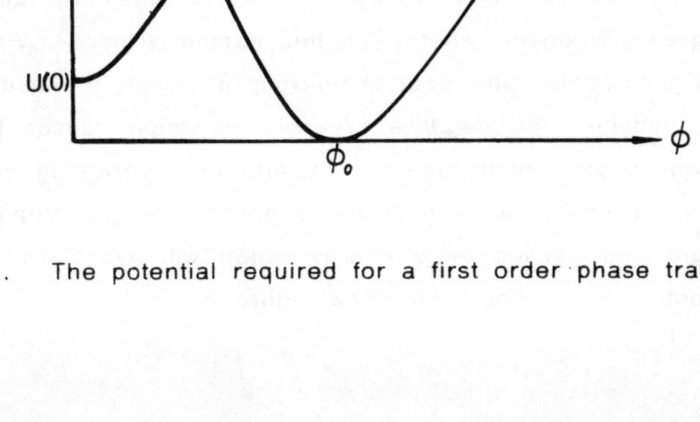

Figure 1. The potential required for a first order phase transition.

In the classical theory corresponding to such a potential there are two stable equilibrium states, but in the quantum version, the state with $\varphi = 0$ is a false vacuum, because it can decay by barrier penetration into the true vacuum $\varphi=\varphi_0$. In the false vacuum, the non-zero potential U(0) gives rise to a cosmological term Λ, of order $8\pi\kappa$ U(0), where κ is the gravitational coupling constant Gc^{-4}. The energy momentum tensor is Λg_{ab} which becomes diag $\Lambda(1,-1,-1,-1)$ when measured in an orthonormal frame. If the universe is initially in the false vacuum state, quantum fluctuations will cause bubbles of true vacuum to form in the same way as thermal fluctuations cause bubbles of liquid to form in a supercooled vapour. Relative to the surrounding false vacuum, the vanishing energy momentum tensor inside the bubble represents a negative energy, but a positive pressure. There is also a surface energy in the wall of the bubble which must compensate exactly for the lost energy of the false vacuum, since energy is conserved in the tunneling process. The total surface energy is proportional to the surface area of the bubble and the energy released is proportional to its volume so there will be a critical bubble size below which only a virtual bubble can exist, and above which the bubble will be created with energy in the form of kinetic energy of outward moving walls. In flat space this critical size would be given by $4\pi R^2\sigma = 4/3\pi R^3$ U(0) where σ is the surface energy density, but in curved space, non-Euclidean effects must be taken into account. In either case the probability that a bubble will form decreases exponentially with its size, so most real bubbles will form with close to the critical size and with walls instaneously at rest.

Once formed, a bubble develops according to the classical field equations. The positive relative pressure inside causes the wall to accelerate rapidly outwards eventually colliding with the walls of other bubbles. The energy of the false vacuum which is eaten up during the expansion is stored as kinetic energy in the bubble walls. When they collide the walls bounce off one another leaving a region of false vacuum between the two bubbles, the negative pressure of which causes the walls to accelerate towards each other again, resulting in a series of collisions. At each collision, a phase wave may be given off, which propogates at the speed of light and carries away energy which will eventually be thermalised by interaction with other fields (see Figure 2).

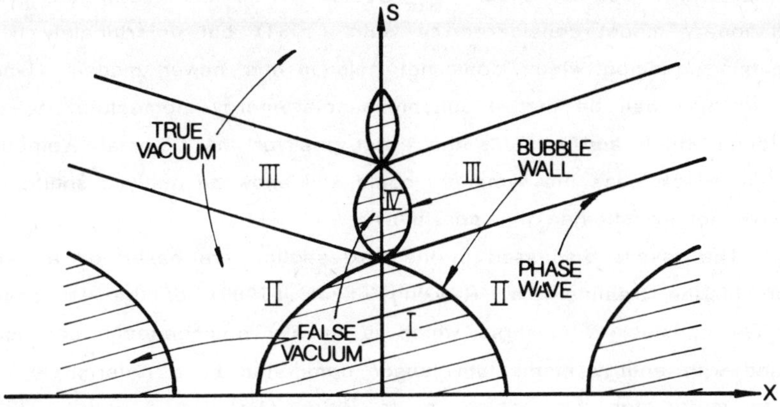

Figure 2. Collision of two bubbles.

In order to make progress in an analytical investigation of this situation, one has to make two assumptions. The first is that the bubbles are initially formed with spherical symmetry. This has been proved in the absence of gravity by Coleman et al., (1978) and it seems highly unlikely that a less symmetric situation could have a lower Euclidean action in the more general case. But if this were the case the problem would be almost certainly intractable in General Relativity where a high degree of symmetry is essential even in computer calculations. The symmetry of the initial data for a two-bubble collision persists at later times because the field equations are also invariant, and this allows one to consider only a small class of spacetimes called the pseudo-Schwarzschild de Sitter spaces which are discussed below.

The second key assumption is that the "thin wall" approximation holds, i.e. that the region in which φ changes from its value at the centre of the bubble ($\varphi=\varphi_0$) to its value at ∞ ($\varphi = 0$) is small compared with the radius (of curvature) of the bubble. This allows the wall to be regarded as a discontinuity in the gravitational field on the one side of which the Λ-term is present and on the other it is not. It has been shown by Coleman (1977) that the thin wall approximation holds if the potential difference between the real and false vacua is much smaller than the potential barrier between them $U_{max} \gg U(0)$). This was true in the original inflationary model considered by Guth (1981) but unfortunately it is precisely this condition which does not hold in the newer models (Linde 1983). It may well be that a suitably simple energy momentum tensor can be found which approximates the exact one for the Coleman-Weinberg potential, in which case the symmetry could still allow an analytic solution, but we have not investigated this possibility.

The results discussed in the next section are based on a generalisation of the treatment by Hawking et al.,(1983) of the flat space case. We consider a universe which is initially homogeneous and isotrpoic, and with energy momentum tensor dominated by a Λ-term. The appropriate metric for this region is de Sitter (Hawking & Ellis 1973). The Λ-term arises from a complex Higgs scalar field with effective Lagrangian

$$L = -(D\varphi)^{+}.(D\varphi) - U(\varphi)$$

where D is the gauge covariant derivative and $U(\varphi)$ has the form shown in Figure 1.

Then two bubbles of true vacuum form and if (as we shall assume) they are spacelike separated, the hypersurface through them can be chosen to be one of constant time, $t = 0$. The minimum at φ_0 in Figure 1 really represents a whole family of degenerate minima, so in general the values of φ will be different in the two bubbles. These values can always be connected by a U(1) subgroup, so a representation can be chosen such that the difference is just a phase factor α.

In the case of a single bubble being formed the spherical symmetry means that most of the O(4,1) symmetry of de Sitter space remains; it is reduced to O(3,1), the Lorentz group, which acts on given 3-hyperboloids which pass through points at constant distance from the centre of the bubble at formation. When two bubbles form there is a preferred axis of symmetry, say the x-axis, which passes through the centres of the bubbles. The symmetry on the surfaces is reduced to O(2,1) acting on two dimensional spacelike surfaces of constant negative curvature. The problem is then to find solutions of the Einstein equations with this symmetry, and match them across the bubble walls and phase waves.

2. The Metrics

The most general metric form with O(2,1) (Bianchi VIII) symmetry acting on spacelike 2-surfaces is given by

$$ds^2 = -e^{\gamma(s,x)}ds^2+e^{\alpha(s,x)}dx^2+s^2(d\theta^2+\sinh^2\theta d\varphi^2) \qquad (1)$$

By writing out the field equations, one can easily prove a theorem analogous to the generalisation of Birkhoff's theorem given in Kramer et al., (1980). The result is that in vacuum the solution to Einstein's field equations with a cosmological term has

$$e^{\alpha(s,x)} = e^{-\gamma(s,x)} = (1-2m/s+1/3\Lambda s^2). \qquad (2)$$

or a limiting form of this. When m and Λ are positive, as is appropriate here, the global structure of the s-x sections of these spaces is similar to the r-t sections of the Schwarzschild metric rotated by $\pi/2$, except that each point represents a Lobatchevski plane, and $s = \infty$ is not an asymptotically flat region, even in the $\Lambda = 0$ case (see Figure 3). The metric is called pseudo-Schwarzschild de Sitter.

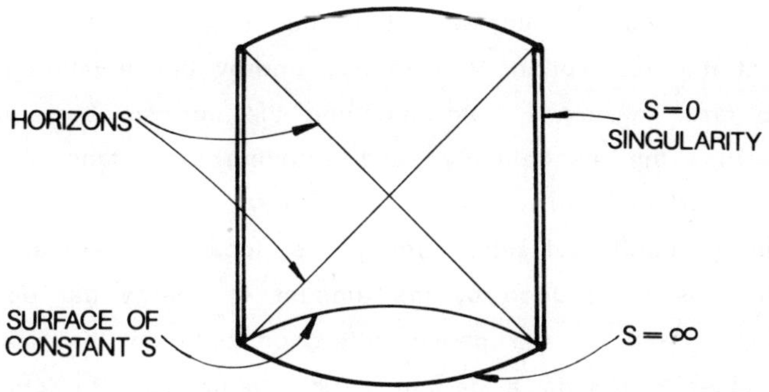

Figure 3. A representation of the pseudo-Schwarzschild de Sitter spaces with $\Lambda > 0$ and $m > 0$.

As can be seen from the diagram, the singularity in these spaces is naked, that is, it can be seen from infinity instead of being hidden behind a horizon as in Schwarzschild.

The principles used to determine the parameters m and Λ in the various regions of the space-time have been described by Wu (1983) and can be summarised as follows. Before the collision, the region inside the bubbles is flat, while the exterior region is de Sitter, just as in the case of a single bubble. The bubbles collide at $s = s_1$, $x = 0$ in the co-ordinates of Equation (1). The formula for s_1 is the same as that given by Hawking et al., (1983)

$$s_1 = (b^2 - R^2)^{1/2}$$

where 2b is a constant representing the initial separation of the two bubbles and R is the initial radius. In this case R is slightly smaller than in flat space:

$$R = \frac{3\sigma}{6\pi\kappa\sigma^2 + U(0)}$$

where $\sigma \equiv 1/3(U_{max})^{1/2}\varphi_0$ in the thin wall approximation is the surface energy density.

This result was derived by Coleman & De Luccia (1980) using the principle of minimum Euclidean action, but in General Relativity it is a consequence of the assumptions (symmetry, energy conservation in tunneling) made so far. It comes from matching the intrinsic curvature of the bubble wall across the discontinuity, and equating the change of extrinsic curvature to the surface energy density (Wu 1983).

During each collision, energy is locally conserved, so the energy in the walls is reduced by the amount of energy carried away by the phase waves. If α, the phase difference between the scalar fields in the two bubbles, exceeds a certain critical value α_c, the phase waves carry off all the energy, and the bubble ceases to exist. The thickness at $s = s_1$ of the phase waves and the new bubble walls is taken to be the same as that of the incoming bubble walls at s_1, except in the case $\alpha > \alpha_c$ when the phase waves must have extra thickness.

The region to the future of the phase wave, but within the new bubble walls (marked III in Figure 2) is described by the pseudo-

Schwarzschild metric (Equation (2) with $\Lambda=0$). If $\alpha=0$ so that the phase waves do not occur, this region would be flat ($m = 0$) but if $\alpha\neq0$ the parameter m must be positive. In the Schwarzschild solution, m could arise from matching a spherical dust shell of mass m to flat space, and similarly in the pseudo-Schwarzschild solution m could arise from a pseudo-spherical pressure discontinuity. The limiting case as this discontinuity becomes a null hypersurface gives a phase wave.

Just to the future of the collision, the phase of the scalar field takes the compromise value of $\alpha/2$. In the presence of an electromagnetic field, the phases of the scalar fields would rotate away from $\alpha/2$ as s increased ($s>s_1$) and phase waves would also be formed at the next collision. Although electromagnetic fields can be incorporated in this scheme relatively easily, we shall assume for simplicity that their effect is negligible so that phase waves only occur at s_1

Taking into account the Lorentz factors $b/R(1+s_1^2\Lambda/3)^{-1/2}\equiv\cosh\theta$ and $b/R\equiv\cosh\mu$ which relate the incoming wall with static observers in Regions I and II at $s = s_1$ respectively, one finds that the original thickness of the bubble walls will have decreased there and the energy/unit area in the phase wave when the phase jumps by $\alpha/2$ is $\alpha^2\sigma\cosh\theta$ and the corresponding value of m in Region III is

$$m = \tfrac{1}{2}s_1\alpha^2\left[1-e^{2\theta-2\mu}\right]<\tfrac{1}{2}s_1$$

The inequality follows because $\alpha<\alpha_c\approx1$ (Hawking et al., 1983), and its significance is that Region III of Figure 2 must correspond to the region of Figure 3 which lies to the future of the horizons at $s = 2m$. Fortunately, this region does not include the naked singularity at $s = 0$. In fact, one would not expect the collision of two bubbles to be associated with any form of singular behaviour, because the surfaces of symmetry are non-compact and so the matter never becomes sufficiently concentrated to cause gravitational collapse however severely the bubbles collide.

Finally, one must calculate the parameter m in the Region IV where Λ has its false vacuum value, and the equation of the trajectory of the new bubble walls. The considerations are similar to those discussed above. The surface energy density σ of the walls is given by the field theory and is therefore unchanged. The thickness of the walls immediately before and after the collision is taken to be the same, so that the effect of the energy lost in the phase wave is to reduce the velocity of the new bubble walls. Then matching the intrinsic curvature and the jump in the extrinsic curvature as before gives the solution. In general, this turns out to be too complicated to determine analytically, but in principle one can now investigate subsequent collisions by an iteration procedure. One interesting qualitative feature which emerges from this analysis, but which does not occur in a flat space approximation is that viewed from the real vacuum inside the bubbles, the walls always move outwards. Each collision appears as a kink in the increasing velocity.

Acknowledgement

We are most grateful to Prof. S.W. Hawking and Dr. I.G. Moss for many useful discussions and suggestions.

References

Coleman, S. (1977). Fate of the false vacuum: Semiclassical Theory, Phys. Rev. D15, 2929.

Coleman, S. & De Luccia, F. (1980). Gravitational effects on and of vacuum decay. Phys. Rev. D21, 3305.

Coleman, S., Glaser, V & Martin, A. (1978). Action minima among solutions to a class of Euclidean scalar field equations. Commun. Math. Phys. 58, 211.

Guth, A.H. (1981). Inflationary universe: A possible solution to the horizon and flatness problems. Phys. Rev. D23, 347.

Hawking, S.W., Moss, I.G. & Stewart, J.M. (1983). Bubble collisions in the very early universe. Phys. Rev. D26, 2681.

Hawking, S.W. & Ellis, G.F.R. (1973). The Large Scale Structure of Space-time. Cambridge: CUP.

Kramer, D., Stephani, H., MacCallum, M. & Herlt, E. (1980). Exact Solutions of Einstein's Equations. Berlin: VEB Deutscher Verlag der Wissenschaften.

Linde, A. (1983) Review Article in this Volume.

Wu, Z.C., (1983) Gravitational Effects in Bubble Collisions. Preprint submitted to Phys. Rev. D.

EUCLIDEAN APPROACH TO THE INFLATIONARY UNIVERSE

S.W. Hawking

Department of Applied Mathematics and Theoretical Physics, Silver Street, Cambridge, England.

The aim of this article is to show how one can use the Euclidean approach to study the inflationary universe. Although this formulation may appear counterintuitive in some respects, it has the advantage that it defines a definite quantum state and provides a framework for calculating quantities of interest such as correlation functions or tunnelling probabilities. By contrast, in the more usual approach in real Lorentzian spacetime, it is not so clear what the quantum state should be or how to evaluate such quantities.

One may suppose either that the universe was initially in an exponentially expanding stage or that it started off as a radiation dominated Friedman Universe at a temperature of the order of the Planck energy 10^{19} GeV and subsequently entered an exponentially expanding stage as a result of being trapped in a metastable state of unbroken symmetry. In either case, as explained in the article by Boucher and Gibbons, in this volume, the metric inside the event horizon of any observer would approach that of de Sitter space. This does not mean that the global structure of the space-time would approach that of de Sitter space, let alone any particular one of the three $k = 1, 0$ or -1 forms of the de Sitter metric in Friedman co-ordinates. However, this does not matter because the whole point of the inflationary scenario is that it explains the present observed homogeneity and isotropy of the universe by the hypothesis that the whole of the presently observed universe expanded from less than one horizon volume in the exponentially expanding stage. In particular, it does not matter whether the universe is spatially open or closed because this will make a difference only at times much later than the present.

There is a close analogy with the No Hair theorems for black holes. Figure 1 shows a Kruskal diagram of a star collapsing to form a black hole. Null geodesics are at $\pm 45^0$ to the vertical and each point except those on the axis on the left of the diagram corresponds to a two-sphere. The region occupied by the matter is shown shaded.

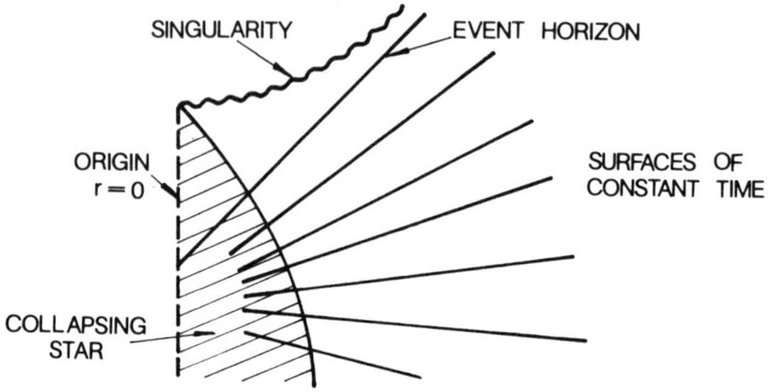

Figure 1 - A star collapsing to form a Black Hole

An observer who does not fall into the black hole can see only the region in his past i.e. within his event horizon which coincides with what is normally called the event horizon of the black hole. Surfaces of constant time as measured by the observer are represented in the diagram as straight lines at less than 45° to the horizontal. As time goes to infinity they approach the event horizon. The No Hair theorem states that within the observer's past i.e. within his event horizon, the metric rapidly approaches that of one of the Kerr-Newman solutions. It does not say anything about the metric beyond the event horizon and in general it will be very different. Because the Kerr-Newman metric is time independent, one can perform a time translation on Figure 1 to make a surface of large time horizontal. This squashes the surface of the star down on to the line at 45° that would have been the past event horizon of the complete Kerr-Newman solution (Figure 2). It is therefore a very good approximation to use the exact Kerr-Newman solution to calculate measurements made by the observer. In particular the quantum field theory for the observer can be formulated on the Euclidean i.e. positive definite, section of the complexified Kerr-Newman metric (Hartle & Hawking 1976; Gibbons & Hawking 1977a; and Hawking 1981). The facts that the past event horizon is not present and that the metric may be very different beyond the event horizon do not matter.

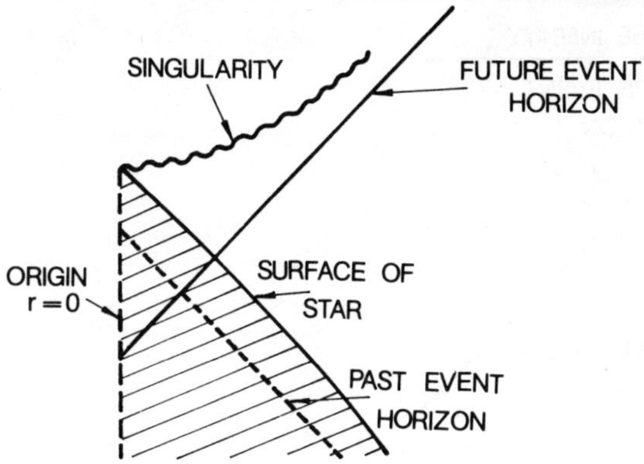

Figure 2 - A Black Hole at late times

The situation with the inflationary universe is similar. Figure 3 is a diagram of a universe that starts with a big bang and then enters an exponentially expanding stage. It may or may not be spatially closed.

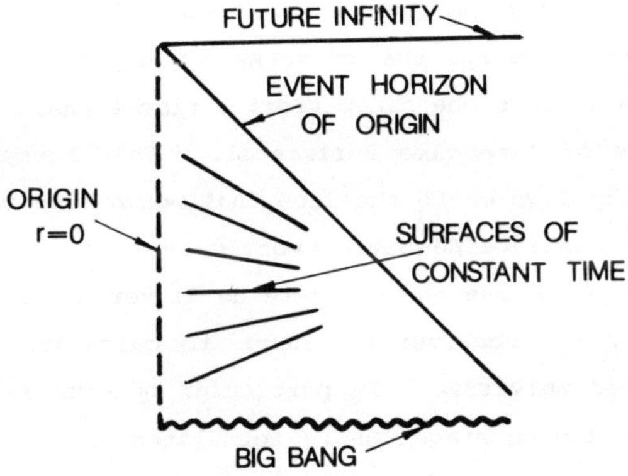

Figure 3- An exponentially expanding universe

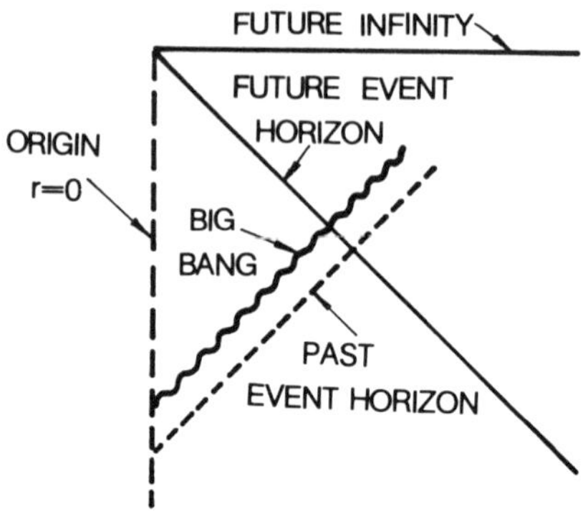

Figure 4 –An exponentially expanding universe at late times

The cosmic No Hair theorem states that the metric approaches that of de Sitter space within the event horizon of an observer who, in this diagram, is taken to be on the axis. Surfaces of constant time for the observer are represented by straight lines at less than 45^{0} to the horizontal. As time goes to infinity, they approach the observer's event horizon. The de Sitter metric is time independent and the volume within the event horizon is constant although the universe is exponentially expanding. As in the black hole case one can perform a time translation on Figure 3 to make a surface of late time horizontal. This squashes the initial big bang singularity down on to the line that would have been the past event horizon in the complete de Sitter metric (Figure 4). It is therefore a good approximation to use the complete de Sitter metric to calculate measurements made by the observer and hence to calculate the state of the presently observed universe. In particular quantum field theory in the exponentially expanding stage can be formulated on S^4, the Euclidean section of de Sitter space. If one is dealing with inflation caused by the vacuum energy of a Grand Unified Theory, the four-sphere will have a radius H^{-1} where $H \approx 10^{10}$ GeV is the rate of expansion. In this case one can ignore quantum gravitational effects and treat the matter fields as quantum fields on a fixed background geometry. However, one would have to take quantum gravity into account if one was dealing with Planck era inflation (Starobinsky 1980).

Most treatments of the inflationary universe assume that quantum field theory in an exponentially expanding stage behaves as it would in flat spacetime. However the fact that the theory is defined on S^4 rather than R^4 makes a number of important differences. In flat spacetime one normally adds to the action a term $\int J(x)\Phi(x)dv_x$ where $J(x)$ is a source coupled to the Higgs field $\Phi(x)$. One then defines $Z(J) = \exp{-W(J)}$ by performing a path integral in Euclidean space over all gauge and fermion fields that approach zero at infinity and over all Higgs fields that approach a constant value at infinity. One can then define the vacuum expectation value of the Higgs field by

$$\langle\Phi(x)\rangle = \frac{\delta W}{\delta J(x)} \qquad (1)$$

Because of the boundary conditions at infinity, the vacuum expectation value can remain non-zero when the source $J(x)$ tends to zero even if the action has a continuous or discrete global symmetry such as $\Phi \to -\Phi$. Another way of looking at it is that there is zero amplitude to tunnel from the state with $\langle\Phi\rangle$ to the state with $\langle-\Phi\rangle$ because the action of an interpolating field configuration would be infinite as the spatial volume of R^4 is infinite. In the case of quantum fields on S^4 however there is a non-zero amplitude to tunnel from Φ to $-\Phi$ because the volume is finite and so the vacuum expectation value of the Higgs field will always be zero in the absence of sources if the action is invariant under $\Phi \to -\Phi$ as is normally assumed. Thus the basic assumptions of most treatments of the inflationary universe, that there is an effective action Γ which is a function of $\langle\Phi\rangle$ and that the evolution of $\langle\Phi\rangle$ is determined by extremising Γ are false.

One can overcome these problems by adding to the action a term

$$\int \Phi(x)K(x,y)\Phi(y)\ dv_x dv_y \qquad (2)$$

Where $K(x,y)$ is a two point function coupled to the Higgs field (Hawking & Moss 1982b). One then defines $W(K)$ as before by integrating over all field configurations on S^4. By functionally differentiating W with respect K one obtains the two point correlation function for Φ

$$\langle\Phi(y)\Phi(x)\rangle = \frac{\delta W}{\delta K(x,y)} = \rho(y,x) \tag{3}$$

Taking x = y and regularising, one obtains $\langle\Phi^2(x)\rangle$. One can also calculate an effective action Γ as a function of $\rho(x,y)$. In the inflationary universe, this will have the approximate form

$$\Gamma = \tfrac{1}{2}tr[(-\Box+m^2)\rho-\log\rho] + \Gamma_0 + \int d\mu_x \frac{3e^4}{32\pi^2}\rho_1^2 \log[M^{-2}(e^2\rho_1+\tfrac{1}{2}H^2)] \tag{4}$$

where $\rho_1(x)=\rho(x,x)_{REN}=\langle\Phi^2(x)\rangle$, e is the gauge coupling constant, m is the effective mass of the Higgs field including any coupling to the scalar curvature, M is a renormalization constant and Γ_0 is the vacuum energy in the unbroken symmetry phase (Hawking & Moss 1982b). The derivative of the effective action with respect to ρ is the source term K

$$\frac{\delta\Gamma}{\delta\rho(x,y)} = -K(y,x) \tag{5}$$

Thus the equation of motion for ρ is

$$\frac{\delta\Gamma}{\delta\rho} = 0 \tag{6}$$

$$(-\Box+m^2)\rho = \delta - \frac{3e^4}{16\pi^2}\rho\rho_1\lambda \tag{7}$$

where λ is the logarithm in Equation (4) and the variation of the logarithm has been neglected.

In the hot early stages of the universe, the symmetry would have been unbroken and $\langle\Phi^2\rangle$ would have had the thermal value cT^2 where c is a numerical constant of order one. As the universe expanded, the temperature T would have dropped until the vacuum energy dominated over the thermal energy. The universe would then have entered an exponential expanding stage with

$$H^2 = \frac{m_p^2}{\pi\Gamma_0} \tag{8}$$

The temperature would have fallen rapidly to the de Sitter value $T_S = \frac{H}{2\pi}$ (Gibbons & Hawking 1977b). It is not possible for quantum fields in de

Sitter space to have a temperature lower than T_S but this temperature is automatically taken into account by formulating the field theory on S^4.

As the temperature fell to T_S, $\langle\Phi^2\rangle$ would have fallen to the order of H^2 at some time $t = t_0$. The subsequent evolution of $\langle\Phi^2\rangle$ would have been governed by Equation (7). If $m^2 \ll H^2$, as is necessary for the inflationary scenario, $\langle\Phi^2\rangle$ will initially increase linearly with time

$$\langle\Phi^2\rangle \approx \frac{H^3}{4\pi^2}(t-t_0) \tag{9}$$

This agrees with the results of Linde (1982 and this volume) and Vilenkin and Ford (1982).

If $m^2 > O(e^2H^2)$, the increase of $\langle\Phi^2\rangle$ according to Equation (9) will cease when

$$\langle\Phi^2\rangle = \frac{3H^4}{8\pi^2m^2} \tag{10}$$

The effective action Γ has a local minimum at this value of $\langle\Phi^2\rangle$. However the global minimum is at $\langle\Phi^2\rangle = e^{-2}M^2$. The system will therefore tunnel quantum mechanically through the potential barrier separating the two minima at the rate $A\exp{-B}$ where $A \approx H^4$ is a determinant and B is the difference between the action of the tunnelling solution of Equation (7) and the local minimum. If $m^2 < 2H^2$, the tunnelling solution will have $\langle\Phi^2\rangle$ constant over S^4 (Hawking & Moss 1982a,b). This result has been widely misunderstood. It does not mean that the tunnelling occurrs simultaneously over the whole universe, which may be infinite in spatial extent. What it means is that the tunnelling will happen in a spacially homogeneous manner over a horizon volume. There are quantum fluctations of Φ^2 about the local minimum (10). These are characterized by $\langle\Phi^4\rangle - \langle\Phi^2\rangle^2$. Eventually there will be a fluctation which is just big enough to get over the barrier in the effective action. While Φ^2 is near the top of the barrier, it will be changing only very slowly with time. This corresponds to the fact that the tunnelling solution for $\langle\Phi^2\rangle$ is constant over S^4. However the tunnelling solution is unstable so Φ^2 will start running down hill to the global minimum on a time scale τ of order $3Hm^{-2}$. The surfaces of constant Φ^2 in the physical space-time may be open or closed but they will have a typical radius of spatial curvature of order H^{-1}. In order to increase this radius to the present lower bound of 10^{10} light-years, one requires $H\tau > 65$ i.e. $m<0.2H$.

When $\langle\Phi^2\rangle$ becomes greater than $e^{-2}H^2$ at a time $t = t_1$, the non-linear terms in Equation (7) will become important. If $m^2 < e^2H^2$, this will happen without any tunnelling. The solution is then

$$\langle\Phi^2\rangle = \rho_1(t) \approx \frac{8\pi^2 H}{e^4\lambda(t_2-t)} \qquad (11)$$

where $t_2-t_1 \approx \lambda^{-0.5}e^{-2}H^{-1}$ (cf. Hawking 1982, and Guth & Pi 1982). Equation (11) holds until a time t_3 where $t_2 - t_2 = H^{-1}$ i.e. until about one expansion time before t_2. After this $\langle\Phi^2\rangle$ will increase to, and oscillate about, its symmetry breaking value of $e^{-2}M^2$ in a time short compared to H^{-1}. These oscillations can be regarded as a large number of Higgs particles in a coherent state. They will decay and reheat the universe to a temperature near the Grand Unified Temperature. The subsequent evolution of the universe will be the same as in the standard hot big bang model.

Although the expectation value of Φ^2 will be constant on surfaces of constant time t, there will be spatial fluctuations of Φ^2. These can be characterized by their two point correlation function $\langle\Phi^2(x)\Phi^2(y)\rangle$. For x and y on the same surface of constant $t < t_3$,

$$\langle\Phi^2(x)\Phi^2(y)\rangle = (\rho_1(t))^2 + 2\rho_1(t)\rho_0(x,y) \qquad (12)$$

(Hawking & Moss 1982b) where $\rho_0(x,y)$ is the analytic continuation of the Euclidean Green function on S^4 with the $\ell = 0$ mode projected out. If the separation of the points x and y is large compared to the horizon distance H^{-1},

$$\rho_0(x,y) \approx -\frac{H^2}{4\pi^2}\log(H|\underline{x}-\underline{y}|) \qquad (13)$$

where $|\underline{x}-\underline{y}|$ is the proper distance between x and y on the surface of constant t.

It is convenient to change to a new time co-ordinate, $\bar{t} = t + \delta t$, which is constant on surfaces of constant Φ^2. The energy momentum tensor will be nearly constant on these surfaces because it is dominated by the potential term in the effective action and this is a function of Φ^2. However the rate of expansion $H+\delta H$ will have spatial fluctuations with the two point function

$$\langle \delta H(x)\delta H(y)\rangle = \frac{\rho_1 \nabla_x^2 \nabla_y^2 \rho_0(x,y)}{9(\dot{\rho}_1)^2} \tag{14}$$

$$= \frac{\lambda}{9\pi^2}\left[\frac{e^2}{4\pi}\right]^2 H^{-2}|\underline{x}-\underline{y}|^{-4} \tag{15}$$

when $\tilde{t} = t_3$. Before t_3 one can neglect the gravitational effects of the fluctuations in the Higgs field and treat the spacetime as de Sitter space. After t_3 the universe will reheat and become radiation dominated in a time short compared to the expansion time H^{-1}. The details of the reheating do not matter apart from the fact that it occurs simultaneously in the $\tilde{t}$ co-ordinate. Local conservation of energy will then imply that the radiation dominated universe starts off with a uniform energy density but with fluctuations in the rate of expansion given by Equation (15). Comoving regions that are separated by more than the horizon distance will evolve like independent Friedman universes with the same initial density but different initial rate of expansion. Equation (15) will continue to hold until $|\underline{x}-\underline{y}| = H^{-1}$, i.e. until x and y are separated by less than the horizon distance. The amplitude $\frac{\delta H}{H}$ of fluctuations when they reenter the horizon will be

$$\left[\frac{\lambda}{9\pi^2}\right]^{1/2} \frac{e^2}{4\pi} \tag{16}$$

independent of the scale of the fluctuation. One therefore has an exactly scale-free spectrum of perturbations as suggested by Harrison (1970) and Zel'dovich (1972). The amplitude given by (16) is smaller than previous estimates by Hawking (1982), Guth & Pi (1982) and Bardeen, et al., (1982). For standard GUTs for which $\frac{e^2}{4\pi} \approx \frac{1}{40}$, it would still be inconsistent with the upper limit of about 10^{-4} given by observations of the isotropy of the microwave background. However, in the case of supersymmetric GUTs, the effective self-coupling of the Higgs field would be smaller because the fermion loops contribute with the opposite sign to the vector loops. The predicted flucutations might therefore be in agreement with the universe that we see around us.

REFERENCES

Bardeen, J.M., Steinhardt, P.J. & Turner, M.S.,(1982) Spontaneous Creation of Almost Scale Free Density Perturbations in an Inflationary Universe, (Preprint UPR-2020T)

Gibbons, G.W. & Hawking, S.W., (1977a) Action Integrals and Partition Functions in Quantum Gravity, Phys.Rev.D.15 2752-2756

Gibbons, G.W. & Hawking, S.W. (1977b) Cosmological Event Horizons, Thermodynamics and Particle Creation, Phys.Rev.D.15, 2738-2751.

Guth, A.H. & Pi, So-Young, (1982) Fluctuations in the New Inflationary Universe. Phys.Rev.Lett.49, 1110

Harrison, E.R. (1970) Fluctuations at the Threshold of Classical Cosmology, Phys.Rev.D.1, 2726

Hartle J.B. & Hawking, S.W., (1976) Path Integral Derivation of Black Hole Radiance, Phys.Rev.D.13, 2188

Hawking S.W. & Moss, I.G., (1982a) Supercooled Phase Transitions in the Very Early Universe, Phys.Lett. 110B, 35.

Hawking, S.W. & Moss, I.G.(1982b) Fluctuations in the Inflationary Universe, D.A.M.T.P. Preprint

Hawking, S.W. (1981) Interacting Quantum Fields around a Black Hole, Commun.Maths.Phys.80,421-442

Hawking, S.W. (1982) The Deveopment of Irregularities in a Single Bubble Inflationary Universe, Phys. Lett.B.115,295

Linde, A.D. (1982) Scalar Field Fluctuations in the Expanding Universe and the New Inflationary Universe Scenario. Phys.Lett.B.116

Starobinsky, A.A. (1980) A New Type of Isotropic Cosmological Models without Singularity. Phys.Lett. 91B, 99

Vilenkin, A. & Ford, L.H., (1982), Gravitational Effects upon Cosmological Phase Transitions. Phys.Rev.D.26, 1231

Zel'dovich, Ya.B. (1972) A Hypothesis, Unifying the Structure and the Entropy of the Universe. Mon.Not.R.astr.Soc.160, IP.

THE ORIGIN OF DENSITY FLUCTUATIONS IN THE 'NEW INFLATIONARY UNIVERSE'

Michael S. Turner
Astronomy and Astrophysics Center, The University of Chicago, Chicago, IL 60637, USA

Abstract. The 'new inflationary scenario' of Linde, and Albrecht and Steinhardt provides natural explanations for the homogeneity, isotropy, flatness, baryon asymmetry, and monopole problems. I report on work done by Bardeen, Steinhardt, and myself on the origin of the density perturbations necessary for the eventual formation of structure. We find that quantum fluctuations in the scalar field which is responsible for spontaneous symmetry breaking at the grand scale lead to a nearly scale-free spectrum of density inhomogeneities after inflation, whose magnitude when they enter the horizon is: $(\delta\rho/\rho) = 4H\Delta\phi/\dot{\phi}(t_o)$, where $\Delta\phi \simeq O(H/2\pi)$ is the fluctuation in the scalar field ϕ, H is the expansion rate, and $\dot{\phi}(t_o)$ is the value of $\partial\phi/\partial t$ when the scale in question leaves the horizon during the deSitter epoch. For the simplest model of 'new inflation', SU(5) with a Coleman-Weinberg scalar potential, $(\delta\rho/\rho) \simeq O(100)$ since $\dot{\phi} \ll H^2$ during the deSitter phase. [This agrees with the results of Guth and Pi; Hawking; and Starobinskii.] To remedy this disaster, a potential which is flat for $\phi \gg H$ is needed, so that when the relevant scales cross the horizon during the deSitter epoch $\phi \gg H$ and $\dot{\phi} \gg H^2$. We analyze a supersymmetric potential which has this property and which can lead to fluctuations of the desired magnitude.

INTRODUCTION

The standard hot big bang model (Friedmann-Robertson-Walker (FRW) cosmology) is a remarkably successful and economical model of the evolution of the Universe. It nicely accounts for the universal expansion, the 3K cosmic microwave background radiation, and through primordial nucleosynthesis the large abundances of ^{4}He (about 25% by mass) and of the weakly-bound deuteron (about 3×10^{-5} by mass). In short, the model provides a reliable framework for describing the evolution of the Universe from $t \simeq 0.01$ s until the present ($t \simeq 3 \times 10^{17}$ s). This in itself is a great achievement.

However, there are a handful of very fundamental and puzzling observations which to date it has failed to elucidate. My list includes: (1) the isotropy; (2) the large-scale ($d \gg 10$Mpc) homogeneity; (3) the small-scale ($d \lesssim 10$Mpc) inhomogeneity; (4) the oldness/flatness; and (5) the baryon asymmetry. The marriage of particle physics, unified gauge theories in particular, and cosmology while providing a framework for understanding the origin of the baryon asymmetry, results in two additional problems: (6) the monopole problem (which is discussed in detail in Preskill's contribution to this volume); and (7) the puzzle of why the cosmological constant is so small today.

In spontaneously-broken gauge theories the energy density of the vacuum depends upon which phase the Universe is in. For a given theory the energy of the vacuum can be computed as a function of an order parameter (e.g., the classical value of the Higgs field) - up to an arbitrary constant. A constant energy density acts just like a cosmological constant ($\Lambda = 8\pi G\rho_v$). Phenomenologically we can determine the constant since today ρ_v must be less about $2\rho_c \simeq 10^{-46}\text{GeV}^4$. This is very small compared to the difference in the vacuum energies of the broken and symmetric phases, which is typically $0(M^4)$, where M is the scale of spontaneous symmetry breaking (SSB). Since the grand scale $M_G \gtrsim 0(10^{14}\text{GeV})$, their ratio is $\lesssim 10^{-102}$! To date, theory has shed very little light on why this is so (see, e.g., Wilczek's contribution to this volume).

Guth (1981) pointed out that if one just accepts this 'experimental determination' of the arbitrary constant, then although the vacuum energy is not relevant today, it could have played an extremely important role in the very early history of the Universe, perhaps providing natural explanations for puzzles (1) - (6). Although (as Guth himself pointed out) his original inflationary Universe scenario had a fatal flaw: the 'graceful return' to the standard cosmology, the basic idea is extremely attractive (for further discussion see Guth's contribution to this volume). Recently, Linde (1982) and Albrecht and Steinhardt (1982) independently have proposed a variant of the original scenario, so-called 'new inflation', which circumvents the 'graceful return' problem. Hawking and Moss (1982) have discussed the Euclidean formulation of the new scenario. 'New inflation' is discussed in detail in Linde's and Steinhardt's contributions to this volume; I will briefly summarize the relevant aspects.

Their original model of 'new inflation' is based upon an SU(5) theory which has a scalar potential of the kind first discussed by Coleman and E. Weinberg (1973). The key features of this potential are:

(i) a temperature-dependent barrier at $\phi \simeq 0(T)$ between the false ($\phi=0$) and the true ($\phi = \sigma \simeq 4.5 \times 10^{14}$GeV) vacua for $T \lesssim 0(M_G) \simeq 0(10^{14}$GeV); (ii) beyond the barrier the potential is very flat, for $T \lesssim M_G$, $V(\phi) = B\phi^4 \ln(\phi^2/\sigma^2) + \frac{1}{2}B(\sigma^4 - \phi^4) \simeq M_G^4 - b\,\phi^4/4$ (for $\phi \ll \sigma$), where $B \simeq 90\alpha^2$ and α is the gauge coupling constant; (iii) near the true minimum ($\phi=\sigma$) V is very steep, $V''(\phi) \simeq 0(M_G^2)$.

For $T \gtrsim T_c \simeq 0(M_G)$, the critical temperature for SSB, the $\phi=0$ vacuum has the lowest free energy, and it is assumed that the Universe begins in this state. For $T \lesssim T_c$ the free energy of the broken phase ($\phi=\sigma$) is lower; however, the Universe remains 'stuck' in the false vacuum due to the barrier. The vacuum energy of the false vacuum, which is $0(M_G^4)$, dominates the total energy density since the thermal contribution is only $0(T^4)$, and the Universe commences a deSitter phase in which the scale factor grows exponentially, $S = S_0 \exp(t/t_e)$, where $t_e \simeq 0(m_{pl}/M_G^2) \simeq (10^{10}\text{ GeV})^{-1} \simeq 10^{-34}$ s ($m_{pl} \simeq 10^{19}$GeV). At a temperature $T_1 \simeq 0(10^8 - 10^9$ GeV) thermal fluctuations can drive ϕ over the barrier and the Universe undergoes 'spinodal decomposition' into fluctuation regions of size $\simeq 0(t_e) \simeq 10^{-23}$ cm - the size of the 'effective particle horizon'. The scalar field is assumed to be constant within a given fluctuation region. Once over the barrier ϕ begins its slow, but inevitable evolution to the true vacuum. Until $\phi \simeq 0(\sigma)$ the vacuum energy which is $0(M_G^4)$ dominates the dynamics, and the Universe expands exponentially. The evolution of ϕ within a given fluctuation region is governed by its classical equation of motion:

$$a\ddot{\phi} + 3aH\dot{\phi} + V'(\phi) + (\delta/\dot{\phi}) = 0, \tag{1}$$

where 'dot' indicates a derivative with respect to proper time t, 'prime' indicates a derivative with respect to ϕ, $H \equiv \dot{S}/S \simeq t_e^{-1}$, and a is a group-theoretic factor which, for an adjoint 24 of Higgs which breaks $SU(5) \rightarrow SU(3) \times SU(2) \times U(1)$, is 15/2. The δ term represents the energy drain caused by the radiation of particles which couple to the time-varying classical field ϕ (Albrecht etal. 1982; Abbott, Farhi, and Wise 1982). Because of the flatness of V near $\phi=0$, ϕ evolves very slowly at first, and to a good approximation the $\ddot{\phi}$ and δ terms can be neglected. For $\phi \ll \sigma$ $V' \simeq -b\phi^3$, and thus eqn. (1) becomes: $\dot{\phi} \simeq (b/3aH)\phi^3$, which is easily solved

$$\phi(t) \simeq H(3a/2b)^{\frac{1}{2}}(\beta - Ht)^{-\frac{1}{2}}. \tag{2}$$

Here $\beta = 3aH^2/(2b\phi_i^2)$, $b = 4B\ell n(\sigma^2/\phi_i^2) \simeq 350\alpha^2 \ell n(\sigma^2/\phi_i^2)$, and $\phi_i \simeq O(T_1)$ is the initial value of ϕ. It takes ϕ a time of $O(\beta t_e)$ to begin to increase significantly, during which the fluctuation region grows by $O(e^\beta)$. For $\phi_i \simeq 3 \times 10^8$GeV, $\beta \simeq 200$ and $e^\beta \simeq 10^{80}$.

When ϕ is $O(\sigma)$ V steepens and ϕ oscillates rapidly (timescale $\simeq \sigma^{-1} << t_e$) around $\phi=\sigma$. The rapid oscillation of ϕ causes the vacuum energy to be quickly (timescale $<<t_e$) and efficiently converted into radiation through the δ term, and the Universe is reheated to a temperature $O(M_G)$. A numerical calculation of the evolution of ϕ is shown in Fig. 1 (for details see Albrecht etal. 1982).

Fig. 1 - Evolution of ϕ and ρ_r for $\phi_i = 3 \times 10^8$GeV; $t_e = H^{-1} \simeq (7 \times 10^9 \text{GeV})^{-1} \simeq 10^{-34}$ s. Note the changes in the scales.

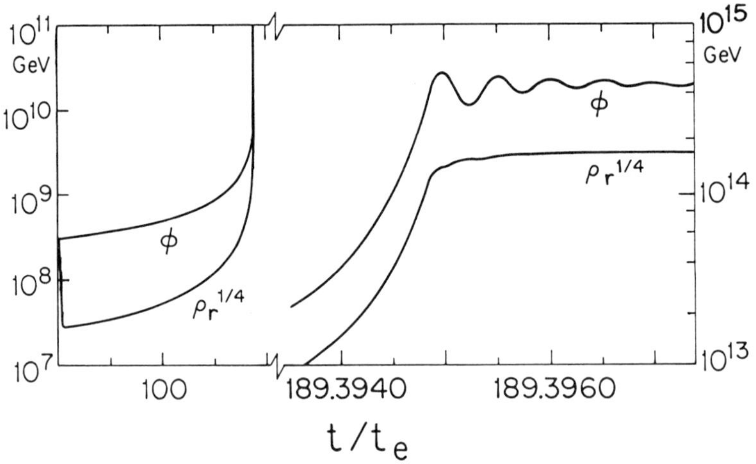

A single fluctuation region grows to a size $O(t_e e^\beta) >> 10$ cm - the size of the present observable Universe when the Universe was at a temperature of 10^{14}GeV. This growth and the reheating to $O(M_G)$ resolve all the puzzles except (3) - the small-scale inhomogeneity, and (7) - the cosmological constant. The assumption that ϕ is constant within a given fluctuation region results in complete uniformity after reheating, whereas density inhomogeneities on a mass scale $M_? \simeq 10^{12\pm5} M_\odot$ of magnitude $(\delta\rho/\rho)_{HOR} \simeq 10^{-3\pm1}$ when that scale enters the horizon seem to be required for the structure which is so conspicuous today to grow up. In addition, fluctuations on all scales less than $M_?$ should not be as large as O(1)

when they enter the horizon otherwise too many black holes will be formed, and fluctuations on scales up to the present horizon ($\simeq 10^{22}\ M_{\odot}$) should be smaller than $0(10^{-3})$ to avoid producing fluctuations in the 3K background which are larger than those observed. Although the 'new inflation scenario' fails to account for the necessary density inhomogeneities, it does provide for the first time a smooth background model upon which the production of inhomogeneities can be analyzed. In fact, it is well known that due to quantum flucutations ϕ should not be constant, even within a single fluctuation region. I will now discuss the role the quantum fluctuations in the scalar field ϕ play in the production of density inhomogeneities.

PRODUCTION AND EVOLUTION OF DENSITY PERTURBATIONS IN THE 'NEW INFLATIONARY UNIVERSE'

In discussing and calculating the evolution of density perturbations we have followed the treatment of Bardeen (1980). I will begin by introducing our notation and conventions. Before discussing the details of our calculations, I will present our results in a more qualitative and heuristic manner, stressing the physics, in order to prepare the reader for the more detailed discussion which follows. Finally, I will discuss the application of our results to the simplest model of 'new inflation', the SU(5) model with Coleman-Weinberg SSB, and then to a supersymmetric model. This entire section is a summary of work done by Bardeen, Steinhardt, and myself (1982); similar results have been obtained for the Coleman-Weinberg SU(5) model by Guth and Pi (1982); Hawking (1982); and Starobinskii (1982).

Notation and Conventions

All perturbations are measured with respect to an unperturbed, isotropic and homogeneous background model. The background model is described by the scale factor S(t), whose evolution is determined by

$$H(t)^2 \equiv (\dot{S}/S)^2 = 8\pi\rho_o/3m_{pl}^{\ 2}\ , \tag{3}$$

where ρ_o is the total energy density. The spatial curvature term has been neglected because in the inflationary scenario its value relative to ρ_o is small. The isotropic pressure p_o and energy density ρ_o are related by $w \equiv p_o/\rho_o$ and $c_s^{\ 2} \equiv dp_o/d\rho_o$.

During the inflationary phase we have

$$\rho_o = V(\phi) + a\dot{\phi}^2/2 + \rho_r , \tag{4a}$$

$$p_o = -V(\phi) + a\dot{\phi}^2/2 + \rho_r/3 , \tag{4b}$$

$$w = -1 + (\rho_o + p_o)/\rho_o , \tag{4c}$$

$$\alpha \equiv -3(1 + c_s^2) = H^{-1}\frac{d}{dt}\ell n(\rho_o + p_o) , \tag{4d}$$

$$\dot{\rho}_r + 4H\rho_r = \delta. \tag{4e}$$

As before the evolution of ϕ is given by eqn. (1). Until reheating $\rho_r \ll \rho_o$ and ρ_o remains essentially constant; to a good approximation $H(t) = H$ and $S = S_o \exp(Ht)$. [Unless the time dependence of H is explicitly displayed, H refers to its value during the quasi-deSitter epoch.] However, since ϕ is evolving the Universe is not exactly deSitter ($\rho_o + p_o \simeq a\dot{\phi}^2 \neq 0$), and because of this fact $c_s^2 \neq -1$. In fact $c_s^2 < -1$ and $\alpha > 0$ - this will have important consequences for the evolution of perturbations.

After reheating, which for simplicity we shall assume occurs very quickly at $t = t_*$ ($\Delta t \ll H^{-1}$), the Universe is once again radiation-dominated: $\rho_o = \rho_r$, $p_o = \rho_o/3$, $w = 1/3$, $\alpha = -3(1+c_s^2) = -4$, and

$$H(t) = H(S(t)/S_*)^{-2} , \tag{5a}$$

$$S(t) = S_*[2H(t-t_*)+1]^{\frac{1}{2}} , \tag{5b}$$

where $S_* = S(t_*)$. [The validity of ignoring the details of reheating is justified in Bardeen, Steinhardt, and Turner (1982)].

A perturbation is described by its comoving wavenumber and comoving wavelength as measured in the background model: k and $\lambda (\equiv k^{-1})$, which, of course, remain constant as the Universe expands. The physical (or proper) wavenumber and wavelength are related to k and λ by: $k_{ph} = k/S$ and $\lambda_{ph} = S\lambda$, and do change as the Universe expands.

Consider a density (scalar) perturbation whose magnitude, as measured in the comoving frame ($T^{oi} = 0$), is $\varepsilon \equiv \delta\rho/\rho_o$. The evolution of ε is completely determined by

$$\ddot{z} + (1-\alpha)H(t)\dot{z} - [\alpha + 3(1+w)]H^2(t)z + (k^2 c_s^2/S^2)\, z = -\eta w H^2(t), \tag{6}$$

where $z = (H(t)S/k)^2\varepsilon$, and η, the non-adiabatic stress perturbation, measures the difference between the actual fractional pressure perturbation and that expected from the background equation-of-state: $\eta \equiv (\delta p - \delta\rho\, dp_o/d\rho_o)/p_o$. Before considering the detailed evolution of z from the inflationary phase through the subsequent FRW phase, I will discuss the evolution of z qualitatively.

The Qualitative Picture

During the evolution of a perturbation the 'effective particle horizon' $\equiv H(t)^{-1}$ is an important length scale. In a time interval of $O(H(t)^{-1})$, $S(t)$ and all physical lengths e-fold. This means that microphysical processes can only operate coherently on length scales $\lesssim O(H(t)^{-1})$, and so $H(t)^{-1}$ sets the scale for microphysics. During the de Sitter phase H^{-1} is nearly constant, while during the subsequent FRW-phase $H(t)^{-1} \propto S^2 \propto t$.

Consider a density perturbation (on a scale λ) which is created during the deSitter epoch. Early on $\lambda_{ph} = S\lambda$ is less than H^{-1}, and microphysical processes (quantum fluctuations, etc.) can affect the amplitude of the perturbation - mathematically, through the η term in eqn. (6). Suppose $S(t_o)\lambda = H^{-1}$, then for $t > t_o, \lambda_{ph} > H^{-1}$ and the evolution of z is determined solely by gravitational effects. During most of the deSitter phase $z \simeq$ constant. Within a few expansion times of reheating, z grows by a factor $\gamma \simeq O(M_G^4/\dot{\phi}(t_o)^2)$, which up to logarithmic terms does not depend on t_o. This growth occurs because near $t = t_*$ $\alpha >> 1$ ($c_s^2 << -1$); if the Universe were exactly deSitter ($\alpha = 0$) this growth would not occur.

During the subsequent radiation-dominated phase $H(t)^{-1}$ is no longer constant, but increases $\propto t$. Until $t = t_H$ and $\lambda_{ph} = H(t)^{-1}$ once again, z remains constant. Once inside the 'effective particle horizon' pressure forces can operate, and the perturbation oscillates as a sound wave of constant amplitude $\varepsilon(t_H)$ until the Universe becomes matter-dominated.

The value of $z \equiv (HS/k)^2\varepsilon = \varepsilon$ at $t = t_o$ is set by microphysics. During the deSitter epoch the Universe is essentially time-translationally invariant - H is nearly constant and ϕ changes only by a factor of O(1) (see Fig. 1). The physical size of the scale λ at $t = t_o$ is H^{-1} and so the value of $\varepsilon(t_o)$ set by the microphysics should be independent of λ (again, this is true up to logarithmic terms). As each scale leaves the 'effective particle horizon' $z(t_o) = \varepsilon(t_o)$ is set to a value, say ε_o, which is <u>independent of the scale λ</u>.

At $t = t_H$ when the perturbation reenters the horizon $H(t)S/k = 1$ once again, and $\varepsilon(t_H) = z(t_H)$. Since $z(t_H) = \gamma z(t_o)$, $\varepsilon(t_H)$, the amplitude of the sound wave, is just equal to $\gamma\varepsilon_o$ - independent of λ. This is the Zel'dovich (1972) - Harrison (1970) scale-invariant spectrum.

It is clear that the time-translation invariance of the deSitter epoch plays a crucial role in creating this spectrum. To summarize the evolution of $\varepsilon = \delta\rho/\rho_o$: at $t = t_o$ $z(t) = \varepsilon(t) = \varepsilon_o$ is set by microphysics; then, until reheating $z \simeq$ constant and $\varepsilon(t) \propto S^{-2}$; around reheating z grows by a factor γ; until the perturbation reenters the horizon $z \simeq$ constant and $\varepsilon(t) \propto S^2 \propto t$; finally, at $t = t_H$ $\varepsilon(t) = z(t) = \gamma z(t_o) = \gamma\varepsilon_o$, and $\varepsilon(t)$ remains constant until the Universe becomes matter-dominated.

The Quantitative Picture

During the inflationary epoch both the η-term and the k^2-term in eqn. (6) are smaller than the other terms by $O(S^2/S(t_o)^2)$ since $\eta \simeq O(\varepsilon)$ and $H\,S(t_o)/k = 1$. Thus they can be neglected and eqn. (6) becomes,

$$\ddot{z} + (1-\alpha)H\dot{z} - \alpha H^2 z = 0\,. \tag{7}$$

An exact solution for the 'growing mode' which has $z(t_o) = z_o$ is: $z(t) = z_o\int_{t_o}^{t} H\exp(H(t' - t))E(t')/E(t_o)dt' + z_o\exp(-H(t - t_o))$, where $E(t) = \rho_o + p_o = a\dot{\phi}^2 + 4\rho_r/3$. If the inflationary epoch were exactly deSitter ($\rho_o + p_o \equiv 0$), then the 'growing mode' solution would instead be $z(t) = z_o$.

During the inflationary epoch the stress energy of the Universe is dominated by the coherent field energy of the scalar field. Since $T^{oi}(\phi) \propto \dot{\phi}\partial_i\phi$, in the comoving gauge ϕ is constant on constant time hypersurfaces. Thus the perturbation $\delta\rho = \Delta(a\dot{\phi}^2/2) = a\Delta\phi\ddot{\phi}$, and the initial value of z is $z_o = a\Delta\phi\ddot{\phi}_o/\rho_o$ where $\dot{\phi}_o = \dot{\phi}(t_o)$.

During almost the entirety of the deSitter epoch $\dot{\phi}$ is small ($<< M_G^2$); at the very end of this epoch ($t_* - t \sim O(M_G^{-1})$) $\dot{\phi}$ increases rapidly to $O(M_G^2)$ as ϕ oscillates around the true minimum ($\phi=\sigma$), and the coherent field energy is converted into radiation. The conversion is very efficient and occurs in a time $O(M_G^{-1})$. Using these facts it is simple to compute $z(t_*)$ and $\dot{z}(t_*)$: $z(t_*) = z_o(4\rho_o/3a\dot{\phi}_o^2)\,O(H/M_G)$ and $\dot{z}(t_*) = z_o(4H\rho_o/3a\dot{\phi}_o^2)\,[1 - O(H/M_G)]$, where $\rho_o = 3m_{pl}^2 H^2/8\pi$ is the almost constant energy density of the deSitter phase. In addition, we have assumed that at $t = t_*$, $\dot{\phi} = 0$ and $\rho_r = \rho_o$ (perfect conversion of vacuum energy into radiation). Because ϕ is changing very rapidly during the final expansion time $\dot{z}(t_*)$ is very large; in fact it is $\dot{z}(t_*)$ and not $z(t_*)$ which determines the asymptotic value of z during the subsequent FRW phase.

During the radiation-dominated epoch, which commences at $t=t_*$, eqn. (6) becomes

$$\ddot{z} + (5/2t)\dot{z} + (k^2/3S^2)z = 0, \tag{8}$$

since $\eta = 0$. The solution to eqn. (8) is: $z(t) = a_1(X^{-3}\sin X - X^{-2}\cos X) + a_2(X^{-3}\cos X + X^{-2}\sin X)$, where $X = k/(\sqrt{3}H(t)S) \propto t^{\frac{1}{2}}$. It is a simple matter to match the solution for $t < t_*$ to the solution above for $t > t_*$. For $X << 1$ ($t_* \lesssim t << t_H$), $z(t) \simeq a_1/3 + a_2X^{-3}$, so that the 'growing mode' bit of $z \simeq a_1/3$ remains constant while $\varepsilon \propto a_1 t$ grows. For $X >> 1 (t >> t_H)$ ε oscillates with amplitude $4\Delta\dot{\phi}/\dot{\phi}_o$, i.e., when a given scale reenters the horizon it oscillates as a sound (pressure) wave.

It is straightforward to relate $\Delta\dot{\phi}$, the fluctuation in $\dot{\phi}$ when a given scale crosses outside the 'effective particle horizon' ($S\lambda \simeq H^{-1}$) as measured in the comoving gauge, to the fluctuation in the scalar field as measured in the synchronous or uniform-Hubble flow gauge: $\Delta\dot{\phi} \simeq H\Delta\phi$. Up to factors of $O(1)$, the zero-point fluctuation in ϕ in deSitter space on a length scale $\lambda_{ph} \simeq H^{-1}$ is: $\Delta\phi \simeq a^{-\frac{1}{2}}(H/2\pi)$. Bringing everything together, we find that the quantum fluctuations in ϕ during the deSitter epoch result in density inhomogeneities during the subsequent radiation-dominated phase which cross the horizon with amplitude,

$$(\delta\rho/\rho)_{HOR} \simeq (2a^{-\frac{1}{2}}/\pi)H^2/\dot{\phi}(t_o), \tag{9}$$

on the comoving scale λ, where $S(t_o)\lambda = H^{-1}$.

RESULTS IN TWO SPECIFIC MODELS OF 'NEW INFLATION'

SU(5) With Coleman-Weinberg SSB

This is the simplest model of 'new inflation', the one first discussed by Linde (1982), and Albrecht and Steinhardt (1982). To apply the results of the previous section we need only compute $\dot{\phi}(t_o)$. From eqn. (2) it follows that $\dot{\phi}(t_o) = H^2(3a/8b)^{\frac{1}{2}}(\beta - Ht)^{-3/2}$. At $t=t_o$, $\lambda_{ph} = S(t_o)\lambda = H^{-1}$; from $t = t_o$ until $t = t_* \simeq \beta H^{-1}$, S grows by a factor of $\exp(\beta - Ht_o)$, so that $\lambda_{ph}(t_*) \simeq H^{-1}\exp(\beta - Ht_o)$. The physical size of a given perturbation at $t = t_*$ is related to its size today by: $\lambda_{ph}(t_*) \simeq \lambda_{ph}(\text{today}) \times (3K/\rho_o^{\frac{1}{4}})$, where $\rho_o^{\frac{1}{4}} \simeq T(t_*) \simeq M_G$. Using these relations it is straightforward to compute $(\beta - Ht_o)$ for a given mass scale M: $(\beta - Ht_o) \simeq 50 + \ell n(M/10^{12}M_\odot)/3 + \ell n(M_G/10^{14}\text{GeV})$. Thus we find that

$$(\delta\rho/\rho)_{HOR} \simeq 50b^{\frac{1}{2}}\left(1 + \frac{\ell n(M/10^{12}M_\odot)}{150} + \frac{\ell n(M_G/10^{14}\text{GeV})}{50}\right)^{3/2}. \tag{10}$$

The scale dependence is very slight - over the mass range $10^{-6}M_\odot$ to $10^{22}M_\odot$ $(\delta\rho/\rho)_{HOR}$ varies by only a factor of 2. Recall that b $\simeq$ $350\alpha^2 \ell n(\sigma^2/\phi_i^2)$; the calculations of Albrecht and Steinhardt (1982) suggest $\phi_i \simeq 10^8 - 10^9$GeV, and sufficient inflation requires $\phi_i \lesssim 7 \times 10^8$GeV. For $\phi_i \simeq 3 \times 10^8$GeV, the amplitude of the spectrum is $O(10^2)$ - a disaster for the simplest model of 'new inflation'. The root of the problem is clear: $(\delta\rho/\rho)_{HOR} \simeq O(H^2/\dot{\phi})$, and during inflation $\dot{\phi} \lesssim O(H^2)$.

One possible solution is obvious: a potential which is flat for $\phi >> H$ rather than (or as well as) for $\phi \lesssim H$, so that during inflation $\phi >> H$ and $\dot{\phi} >> H^2$.

Reverse Hierarchy Supersymmetric Models - 'Newer Inflation?'

Supersymmetric (SS) unified models which employ the Witten (1981) reverse hierarchy scheme with O'Raifeartaigh (1975) - type SSB have just the desired feature; namely, the potential is flat for large values of the scalar field. For definiteness, we have considered the geometric hierarchy model of Dimopoulos and Raby (1982). In this model there are 3 scales: $M_I \sim 10^{12}$GeV which is the scale of SS breaking, and two other scales, the grand scale $M_G \sim m_{pl}$ and the weak scale $M_W \sim M_I^2/M_G$, both of which are generated radiatively. The intermediate scale M_I determines the properties of the phase transition which leads to inflation. Let ϕ be the scalar field whose vacuum expectation value leads to the SSB of the unified gauge theory; for $m_{pl} >> \phi >> M_I$, the scalar potential can be written as

$$V(\phi) \simeq M_I^4 [c_1 - c_2 \ell n(\phi/M_G)] . \qquad (11)$$

The shape of the potential near $\phi = 0$ will not be of interest here, and the shape near the SSB minimum ($\phi \simeq M_G$) has not yet been completely determined.

The key feature of this potential is that it becomes very flat for large values of ϕ, $V' \simeq -c_2 M_I^4/\phi$. Albrecht etal. (1983) have shown that if $c_1/c_2 \gtrsim O(10^2)$, then the potential is sufficiently flat for $\phi \gtrsim (3c_2/8\pi c_1 a)^{1/2} m_{pl}$ so that ϕ evolves very slowly and inflation occurs. During this period the $\ddot{\phi}$ and δ terms in eqn. (1) can be neglected, and $3aH\dot{\phi} \simeq - V'$. The solution to this is:

$$\phi(t) \simeq (3c_2/8\pi c_1 a)^{1/2}(1 + \tfrac{2}{3}Ht)^{1/2} m_{pl}, \qquad (12)$$

where $H^2 \simeq 8\pi c_1 M_I^4/3m_{pl}^2$ and t = 0 marks the start of the 'slow growth epoch'. When $Ht_* = \beta \equiv 4\pi c_1 a/c_2$ $\phi \simeq O(m_{pl})$, and hopefully the potential steepens to ensure efficient reheating. [At present it appears that in this particular model the light fields do not couple with sufficient strength to ϕ to reheat the Universe; see Albrecht etal. (1983), or Steinhardt's contribution to this volume.]

As before, it is a simple matter to relate β-Ht_o to M and M_I: $Ht_o \simeq \beta - 55 - \ell n(M/10^{12}M_\odot)/3 - \ell n(M_I/10^{12}\text{GeV})$. As long as $\beta >> 55 + \ell n(M/10^{12}M_\odot)/3 + \ell n(M_I/10^{12}\text{GeV})$, i.e., the mass scale of interest is much smaller than the total mass within the fluctuation region, $Ht_o \simeq \beta$ and $\dot\phi(t_o) \simeq (c_2/8\pi c_1 a)H\,m_{pl}$. Then from eqn. (9) it follows that

$$(\delta\rho/\rho)_{HOR} \simeq 32\left(\frac{2\pi}{3}\right)^{\frac{1}{2}} \frac{c_1^{3/2}\,a^{\frac{1}{2}}}{c_2}\left(\frac{M_I}{m_{pl}}\right)^2 . \tag{13}$$

With $(c_1 a)^{\frac{1}{2}} \simeq O(10)$ and $M_I \simeq O(10^{12}\text{GeV})$, c_1/c_2 must be $O(10^7)$ to achieve $(\delta\rho/\rho)_{HOR} \simeq 10^{-4}$. In this model c_1/c_2 is not precisely determined; however, c_1/c_2 is naturally $\gtrsim O(10^4)$. Because $\phi >> H$ and $\dot\phi >> H^2$ during the inflation epoch the catastrophe of excessively large density inhomogeneities is avoided. Unfortunately, it is not clear whether or not sufficient reheating of the Universe occurs in this particular model.

SUMMARY

Most of the puzzles in my list involve initial data of some sort, information that we are unlikely to ever have knowledge of. The idea of inflation is extremely attractive because it naturally explains these puzzles by ensuring that the present state of the Universe is insensitive to the initial data. The epoch when the energy density of the Universe is dominated by vacuum energy cleanses the Universe of its past, and provides 'new initial data' which result in the kind of Universe which we observe today. The scenario makes only a few testable predictions: (i) $\Omega = 1$ to a very high degree of precision; (ii) due to the quantum fluctuations in the scalar field a spectrum of adiabatic, scale-free density inhomogeneities; and (iii) the only monopoles present today should be those produced in particle collisions - probably too few to detect; however, the uncertainties in calculating the number produced are in the exponent (see Turner 1982). Finally, inflation merely exploits the puzzle of the vanishing cosmological constant to explain the other 6 puzzles. It does nothing to further our understanding (i.e., complete lack of understanding!) of why Λ is so small today. A fundamental explanation of this fact may result in some surprises.

Because of the large density inhomogeneities which result from the scalar field fluctuations, the simplest model of 'new inflation', the SU(5) model with Coleman-Weinberg SSB, appears to be dead. The basic problem is that during inflation $\dot{\phi} \lesssim H^2$ (and $\phi \lesssim H$), and since $(\delta\rho/\rho)_{HOR} \simeq (2a^{-\frac{1}{2}}/\pi)\, H^2/\dot{\phi}_o$ density inhomogeneities of greater than O(1) result. Linde (in his contribution to this volume), and Ford and Vilenkin (1982) have pointed out that because $\phi \lesssim H$ during inflation the quantum fluctuations in ϕ lead to the addition of a negative mass-squared term to the effective potential which can quicken the evolution of ϕ, and possibly prevent inflation. Both difficulties can be avoided if the epoch of inflation occurs when $\phi \gg H$ and $\dot{\phi} \gg H^2$.

Although the simplest model of 'new inflation' is dead, it leaves as its legacy the prescription for 'newer inflation': a scalar potential which is: (i) flat for $\phi \gg H$ so that inflation occurs when $\phi \gg H$ and $\dot{\phi} \gg H^2$; (ii) strongly-curved near the SSB minimum to ensure efficient reheating; of course, the light fields must also couple to ϕ with sufficient strength. Supersymmetric models with potentials of the O'Raifeartaigh-Witten type can easily satisfy (i); but, it is not yet clear whether they can also satisfy (ii).

Most of the work I have described here was completed during the Nuffield Workshop, which is a testimony to the true workshop atmosphere that existed, due in large measure to the efforts of the organizers: Gary Gibbons, Stephen Hawking, and Judy Fella. It is a pleasure to thank all the members of the workshop for helpful comments, criticisms, and conversations. I especially thank Alan Guth, A. A. Starobinskii, and Stephen Hawking, all of whom also worked on this problem, for valuable discussions and for sharing their work and ideas on this topic with us. This work was supported in part by the DOE through contract DE-AC02-80ER 10773 (at Chicago), by the NSF through grant PHY 77-27084 (at ITP-UCSB), by NASA at the Aspen Center for Physics where a preliminary version of this paper was written during the 1982 Astrophysics Program, and by the Nuffield Foundation.

REFERENCES

Abbott, L., Farhi, E. & Wise, M. (1982). Phys. Lett. 117B, 29.

Albrecht, A., Dimopoulos, S., Fischler, W., Kolb, E. W., Raby, S. & Steinhardt, P. J. (1983). in preparation.
Albrecht, A. & Steinhardt, P. J. (1982). Phys. Rev. Lett., 48, 1220.
Albrecht, A., Steinhardt, P. J., Turner, M. S. & Wilczek, F. (1982). Phys. Rev. Lett., 48, 1437.
Bardeen, J. M. (1980). Phys. Rev., D22, 1882.
Bardeen, J. M., Steinhardt, P. J. & Turner, M. S. (1982). in preparation.
Coleman, S. & Weinberg, E. (1973). Phys. Rev., D7, 1888.
Dimopoulos, S. & Raby, S. (1982). Nucl. Phys. B., in press.
Guth, A. (1981). Phys. Rev., D23, 347.
Guth, A. & Pi, S.-Y. (1982). Phys. Rev. Lett. 49, 1110.
Harrison, E. R. (1970). Phys. Rev., D1, 2726.
Hawking, S. W. (1982). Phys. Lett., 115B, 295.
Hawking, S. W. & Moss, I. G. (1982). Phys. Lett., 110B, 35.
O'Raifeartaigh, L. (1975). Nucl. Phys., B96, 331.
Starobinskii, A. A. (1982). Phys. Lett. B., in press.
Turner, M. S. (1982). Phys. Lett., 115B, 95.
Vilenkin, A. & Ford, L. (1982). Phys. Rev. D26, 1231.
Witten, E. (1981). Phys. Lett., 105B, 267.
Zel'dovich, Ya. B. (1972). Mon. Not. R. Astron. Soc., 160, 1p.

GENERATION OF PRIMORDIAL PERTURBATIONS IN THE EARLY UNIVERSE

V.N. Lukash
Space Research Institute, Academy of Sciences of the USSR

I.D. Novikov
Space Research Institute, Academy of Sciences of the USSR

Abstract. The problem of the origin of the primordial cosmological perturbations which led to the formation of the large-scale structure of the observed world is considered. A theory of formation of these perturbations is proposed, with the effect of spontaneous phonon (sound wave) creation during the cosmological expansion as a basis. A spectrum of primordial sound waves due to vacuum phase transitions is derived.

0 INTRODUCTION

The following three properties of the present Universe are very important in understanding the physics of the beginning of its expansion.

I. Structure. A high degree of homogeneity and isotropy on large scales ($\delta\rho/\rho < 10^{-4}$ on the horizon scale) along with a well-developed structure on scales less than 0.01 of the horizon scale.

II. High specific entropy $\approx 10^8$. The mean baryon density is ~ 0.03 of the critical value.

III. No Λ-term. In the present state of the Universe the vacuum of all physical fields has no gravity.

Property I proves that galaxies and their clusters were formed from initially small amplitude perturbations of a homogeneous Universe since small perturbations may grow up to the order of unity only when the horizon is many times larger than their linear scales.

One of the basic implications of property II is that in the past the Universe was hot and its expansion was governed by the gravity of intensively interacting relativistic particles. At this stage the matter may, to a high degree of accuracy, be considered a hydrodynamic perfect fluid with an equation of state $p = \frac{\varepsilon}{3}$. Small perturbations of such a gravitating fluid are sound waves propagating through the matter with a constant amplitude. Mathematically, the propagation of such waves in the expanding Universe is identical to the propagation of acoustic oscillations in a non-gravitating homogeneous stationary medium in flat spacetime. It is shown below that this analogy between potential perturbations of a spatially homogeneous and isotropic non-gravitating fluid in flat space-time and in the Friedmann world whose expansion is governed by the $p = \frac{\varepsilon}{3}$ equation of state, is much more profound; no phonons, cosmological perturbations of the perfect fluid, are produced spontaneously in this case. For generation of primordial perturbations to occur on the scales of galaxies and their clusters it is necessary that there should be a stage in the evolution of the very early Universe which does not obey the $p = \frac{\varepsilon}{3}$ law of expansion. This is likely to have happened in the history of the early Universe if in the past the physical vacuum had a high density (see property III).

The physical meaning of the instability of the homogeneous gravitating fluid against small perturbations will be discussed below. Now we only mention that the mechanism has nothing to do with the classical gravitational instability. It takes place on scales much greater than the horizon scale and is a consequence of the conformal non-invariance of sound waves in the early Universe. Before going to our viewpoint, we discuss to what extent adiabatic perturbations are, in fact, interesting from the point of view of the formation of the Universe's structure. To avoid misunderstanding, we want to stress that our model of the evolution of the Universe differs from some of the other models (see for example [59,60]).

1 PERTURBATIONS OF THE HOMOGENEOUS MODEL AND THEIR ROLE IN THE FORMATION OF STRUCTURE

Observational data about the present structure of the Universe, together with Lifshitz' theory of gravitational instability [1], state that at the radiation dominated stage when $p = \frac{\varepsilon}{3}$ (say, by the end of the first second) the constant amplitude of the 'growing' mode of sound waves should be of the order of 10^{-4} within the scales of the present galaxies and clusters.

The goal of the theory is to explain the origin of fluctuations of this scale and amplitude.

Recently papers have appeared where an attempt is made to explain the origin of primordial perturbations by some physical processes that occurred in the very early Universe. In this case it is usually assumed that the initial state of cosmological expansion is highly symmetric, i.e. highly homogeneous and isotropic.

We do not know yet whether primordial perturbations formed

together with the cosmological Friedmann model at Planckian curvatures or whether they originated in the process of homogeneous and isotropic expansion which is described by the equations of classical General Relativity. In the first case we have no theory; however, it is quite evident that by no process could the Universe be born strictly homogeneous and isotropic: there always exist quantum fluctuations of the metric and physical fields; the initial fluctuations could be of statistical, random character or they may be thermal, etc. Let us show how, for instance, initial fluctuations arise in the versions often discussed recently of the origin of the expanding Universe as a result of tunnelling from the de Sitter or Minkowski vacuum state of quantum fields or some superfield (e.g. [2-18]). Usually, one derives the probability of a sub-barrier transition in the class of homogeneous metrics since it is the birth of a homogeneous mode (or a homogeneous bubble) from vacuum that is most probable. It is obvious, however, that, owing to the conformal non-invariance of potential perturbations, which is their common property, such tunnelling will inevitably occur in the state with small-amplitude chaotic fluctuations (since the transition time is finite) that is, the Universe thus born will be homogeneous and isotropic only on the average. Certainly, a calculation of the dispersion (inhomogeneity amplitude) depends on a model we may only guess about. Below in §3 we describe the most general features of initial cosmological fluctuations.

In the absence of a fundamental theory of the initial state, the idea of getting primordial cosmological perturbations at the Friedmann expansion stage when the spacetime curvatures are much smaller than at the Planck stage appeared to be very attractive. Uncertainty in what we know about the initial state has resulted in many approaches to the given problem, emphasizing one or another phenomenon responsible for the formation of perturbations - for example, quantum-gravitational effects in a one-loop approximation [13,14,19,20]; phase transitions in gauge theories with spontaneous symmetry breaking and the formation of ordered-phase bubbles, bubble walls, etc. [6-12,21-25]; the formation of strings in the Grand Unification theory [26-28]; possible anisotropy of deformation at early stages of the expansion, dissipative processes, particle creation, etc. (e.g. [29-33]). On the other hand, interest in the origin of primordial perturbations has initiated the development of Lifshitz' classical theory. It developed along the lines of deriving gauge-invariant quantities which characterize perturbations of the metric and the matter [34-39].

In 1980 the Lagrange theory of potential cosmological perturbations was developed, one of the most important consequences of which was the conformal non-invariance of sound waves [40-43]. This provides for the effect of parametric amplification of initial potential fluctuations of the metric and density during cosmological expansion. Gravitational waves in the Friedmann model have a similar amplification property [44]. Conformal invariance of the linearized equations was analysed by Grishchuk in 1980

The process of amplification of initial perturbations in the case of phase transitions near a cosmological singularity is the basis of our approach to the origin of primordial perturbations [41-43,46-49]. Further on we restrict ourselves to adiabatic perturbations - simultaneous perturbations of matter and radiation which do not violate the conditions of local equilibrium - under the assumption that all fields are intensively interacting at early stages of the expansion. The question of how the presence of gravitational waves, also born in phase transitions, affects the evolution of potential perturbations is not discussed here. Neither is the problem of the possible pumping into the entropy mode of a part of the sound wave energy. This effect is likely to be weak (see [50,51]) when initial entropy perturbations are not present because of the non-conservation of baryon number in the epoch of Grand Unification.

The next section briefly summarizes the basic aspects of the gauge-invariant theory of potential perturbations. §3 describes the general properties of initial fluctuations which are of interest for us. §4 is devoted to the effect of amplification of initial seed fluctuations for the simplest model of a phase transition and to the formation spectrum of the primordial perturbations of density.

2 GAUGE-INVARIANT THEORY OF POTENTIAL PERTURBATIONS

A field of potential perturbations is described by the real 4-scalar $q(x^i)$ (see Appendix A), which satisfies the usual formalism of a test Hamiltonian field on a classical Friedmann background. This allows the possibility of defining a local energy momentum tensor [40-43].

The 4-scalar $q(x^i)$ is a gauge-invariant linear combination of hydrodynamic and gravitational potentials describing perturbations of the matter velocity and the metric. In the short-wave approximation q is a hydrodynamic velocity potential, while for large scales q is mainly metric perturbations. The choice of the Lagrangian density $L(q)$ is connected with the masslessness of the q-field and results from expanding the action for the potential motions of a perfect gravitating fluid up to the second order in q :

$$L(q) = \tfrac{1}{2}\mathcal{D}^{ik}q_{,i}q_{,k} \qquad (2.1)$$

Here $(,)$ is a derivative in the background metric g_{ik} , $\mathcal{D}_{ik} = \xi^2(u_i u_k + \beta^2 P_{ik})$ is the background 4-tensor, u_i is the matter velocity, $P_{ik} = g_{ik} - u_i u_k$ is the projection tensor; the pressure p and energy density ε of the matter are related by equation of state:

$$p = p(\varepsilon) \; ; \qquad (2.2)$$

and $\xi = \gamma^{\frac{1}{2}}/2\beta$, $\gamma = 1 + \frac{p}{\varepsilon}$, $\beta = (dp/d\varepsilon)^{\frac{1}{2}}$ are hydrodynamic scalars.

The equation of motion of the q-field can be written in the homogeneous reference system $x^i = (t, \vec{x})$, in which all background quantities depend only on time $(8\pi G = c = \hbar = 1)$ and

$$ds^2 = dt^2 - a^2(dx^2 + dy^2 + dz^2) \; , \qquad (2.3)$$

in terms of the conformal field $\bar{q} = \xi a q$ as:

$$\Box_\beta \bar{q} = U\bar{q} \; , \qquad (2.4)$$

where $a = a(t)$ is the scale factor, $\eta = \int \frac{dt}{a}$ is conformal time, $(') = a(\dot{}) = \frac{\partial}{\partial \eta}$, $\Box_\beta = \frac{\partial^2}{\partial \eta^2} - \beta^2 \Delta$ is the acoustic d'Alambertian operator, $\Delta = \frac{\partial^2}{\partial x^2} + \frac{\partial^2}{\partial y^2} + \frac{\partial^2}{\partial z^2}$, and $U = (\xi a)''/\xi a$ is the effective potential reflecting the nonstationarity of the Universe.

A physical meaning of the conformal non-invariance of sound waves becomes transparent when the effective potentials of sound and of gravitational waves are compared: $U_S = \frac{(\xi a)''}{\xi a}$, $U_g = \frac{a''}{a}$. By 'sound wave' we mean any potential perturbation with $\beta \neq 0$. In a short-wave approximation, β is the sound velocity. Tensor perturbations do not interact with matter and their conformal non-invariance has a pure gravitational nature caused by the nonstationarity of the background ($a(\eta)$ is a function of time). Besides this effect, potential perturbations are conformally non-invariant also partly for a hydrodynamic reason. The latter reflects the play of two oppositely directed forces: the adiabatic damping of the sound wave amplitude propagating through the expanding perfect fluid and the growth of the amplitude because of the negative time derivative of the background pressure $\dot{p} < 0$. This Special Relativistic effect means that the pressure has a spatial gradient when measured in the reference system co-moving with the wavefront (due to the Lorentz transformation from homogeneous to co-moving coordinates), and this accelerates the sound wave. (This effect is similar to the increase in amplitude of an acoustic wave propagating through the upper layers of the atmosphere.)

Certainly, the gravitational and hydrodynamic effects which produce the conformal non-invariance of sound waves are unlikely to be separated in principle. However, for scales much greater than the horizon all pressure gradient effects are negligible and the nature of the conformal non-invariance of the metric perturbations as well as of the tensor types is qualitatively the same.

Second quantization of q-field over plane sound waves yields:

$$q = \int_{-\infty}^{\infty} d^3\vec{k}\,(a_{\vec{k}} q_{\vec{k}} + a^{\dagger}_{\vec{k}} q^{*}_{\vec{k}}) \,, \tag{2.5}$$

where $(q_{\vec{k}}, q_{\vec{k}'}) = [a_{\vec{k}}, a^{\dagger}_{\vec{k}'}] = \delta(k-k'), (q_{\vec{k}}, q^{*}_{\vec{k}'}) = [a_{\vec{k}}, a_{\vec{k}'}] = 0$. $a_{\vec{k}}$ and $a^{\dagger}_{\vec{k}}$ are the annihilation and creation operators of phonons (the field quanta); (.,.) is the scalar product in Hilbert space; and [.,.] is the commutator. In the coordinates of (2.3) we have:

$$q_{\vec{k}} = q_{\vec{k}}(x^i) = \frac{\nu_k}{(2\pi)^{3/2}\xi a}\, e^{i\vec{k}.\vec{x}} \,, \tag{2.6}$$

where $\nu_k = \nu_k(\eta)$ satisfies the equations

$$\nu_k'' + (\beta^2 k^2 - U)\nu_k = 0 \,, \quad \nu_k \nu_k^{*\prime} - \nu_k^{*} \nu_k' = i \,, \quad k = |\vec{k}| \,. \tag{2.7}$$

The total energy of the field of the potential perturbations on the Cauchy hypersurface t = constant (the field Hamiltonian) is given by:

$$\begin{aligned} H &= \frac{\bar{H}}{a} = a^3 \int E\, d^3\vec{x} \,, \\ E &= \frac{1}{2a^4}(\bar{q}'^2 + \beta^2 \bar{q}_{,\alpha} \bar{q}^{,\alpha} - U\bar{q}^2) \,, \end{aligned} \tag{2.8}$$

where E is the local energy density of the field q. In the non-relativistic limit, $E = \rho\frac{\vec{v}^2}{2} + \beta^2\frac{(\delta\rho)^2}{2\rho}$, where $\rho+\delta\rho,\vec{v}$ are the matter density and velocity.

An essential aspect is that in a hot Universe with $p = \frac{\varepsilon}{3}$, the q-field is conformally coupled:

$$p = \frac{\varepsilon}{3}: \qquad U = 0\ , \quad \bar{H} = \text{const.} \tag{2.9}$$

The integral of motion (2.9) of the q-field exists because phonons are neither created nor annihilated in the process of cosmological expansion. When $U \neq 0$, phonons interact with the background metric (see (2.4)) and their number is not conserved. This provides spontaneous and induced creation of phonons in the process of cosmological expansion. This effect works as a typical parametric amplification.

To solve the question quantitatively two items have to be considered: the initial seed fluctuations (before amplification) and the intermediate stage when $p \neq \frac{1}{3}\varepsilon$ (what are the most favourable parameters for the best amplification effect?). Before considering these items, we need operators for the 'growing' and 'decaying' modes at the stage when (2.9) holds:

$$p = \frac{\varepsilon}{3}: \qquad C_{g\vec{k}} = \left(\frac{\omega}{2}\right)^{1/2}\frac{a_{\vec{k}} - a^{\dagger}_{-\vec{k}}}{ia'}\ , \quad C_{d\vec{k}} = \left(\frac{\omega}{2}\right)^{1/2}\frac{a_{\vec{k}} + a^{\dagger}_{-\vec{k}}}{a'} \tag{2.10}$$

where the operators $C_{g\vec{k}}$ and $C_{d\vec{k}}$ both satisfy $C^{\dagger}_{\vec{k}} = C_{-\vec{k}}$, $|C^{\dagger}_{\vec{k}}|^2 = C^{\dagger}_{\vec{k}}C_{\vec{k}} = C_{\vec{k}}C^{\dagger}_{\vec{k}}$; $\nu_k = (2\omega)^{-1/2}\exp(-i\omega\tau)$ is the phonon representation; $\omega = k/\sqrt{3}$; $a' = \text{const}$; and $\tau = a/a' \sim \eta$ (see (4.1)).

Eqs (2.5) and (2.10) yield

$$p = \frac{\varepsilon}{3}: \qquad q = (2\pi)^{-3/2}\int_{-\infty}^{\infty} d^3\vec{k}\, e^{i\vec{k}.\vec{x}}\left(C_{g\vec{k}}\frac{\text{Sin}\ \omega\tau}{\omega\tau} + C_{d\vec{k}}\frac{\text{Cos}\ \omega\tau}{\omega\tau}\right)\ , \tag{2.11}$$

$$\bar{H} = \frac{S}{2}\int_{-\infty}^{\infty} d^3\vec{k}\left(|C_{g\vec{k}}|^2 + |C_{d\vec{k}}|^2\right)\ ,$$

$$\bar{H}_{\text{Reg}} = \int_{-\infty}^{\infty} d^3\vec{k}\ \omega\, N_{\vec{k}} \tag{2.12}$$

where $S = a'^2$ and $N_{\vec{k}} = a^{\dagger}_{\vec{k}}a_{\vec{k}}$ is the operator of the number of phonons with momentum $\vec{k}/a$. In Eq (2.12) the contribution from point-zero oscillations of the q-field is subtracted (compare (2.11)).

3 RANDOM INITIAL FLUCTUATIONS. EQUIPARTITION HYPOTHESIS

The most tempting way of getting primordial perturbations is from initial quantum metric point-zero fluctuations [41-43]. But it is not the only possibility: initial fluctuations might be of another type - statistical, thermal etc. The equipartition hypothesis displays general features of initial fluctuations.

We propose that initially (at moment t_o close to the Planck time) the energy of fluctuations should be equally distributed over all modes of perturbation [46-48,52]. This means that the initial phases of the fluctuations were random. This is exactly the situation which holds for quantum or thermal fluctuations, so they are special cases of the equipartition hypothesis. (For another approach, see [58].)

Further on we assume that at $t = t_o + 0$ relativistic particles predominated, so the equation of state was $p = \frac{\varepsilon}{3}$ and the temperature $T = T_p(t_p/t\sqrt{N})^{\frac{1}{2}}$, where N is a number of massless degrees of freedom (for three families of leptons and quarks $N \sim 10^2$).

For the case of potential perturbations the hypothesis is as follows: the density matrix ρ of the q-field depends either (weak variant) on the phonon number operators $N_{\vec{k}}$ at the stage $t = t_o + 0$, or (strong variant) on the Hamiltonian H_o at $t = t_o + 0$ (see (2.12)).

The weak variant follows from the requirement that the density matrix does not depend on the initial phases of the field of potential perturbations: $\rho = \mathrm{inv}(a_{\vec{k}} \to a_{\vec{k}} \exp i\phi_{\vec{k}})$. It leads to the equipartition of a given scale fluctuation energy over both physical modes:

$$<|C_{g\vec{k}}|^2> = <|C_{d\vec{k}}|^2> \qquad (3.1)$$

where $<\ldots> = \mathrm{Trace}(\rho\ldots)$.

For scales much greater than the horizon ($\omega\eta_o \ll 1$) it would be natural to strengthen the assumption about initial phase randomness with the condition of statistical independence (over scales) of the fluctuation field amplitudes. In this case the density matrix is factorized over representation (2.10); we call such states *statistical*.

The strong variant is not as natural as the weak one. It follows from the idea of an initially stationary state of the Universe with tunnelling into the Friedmann state; this of course needs additional exploration. (Note that the statistical distributions of the strong variant are identical to the normal ones [47].) According to this idea the initial state of the expanding Universe was characterized by a single parameter - the total energy of all the fields - and this energy was distributed over all the internal degrees of freedom in the most probable way, i.e. the distribution corresponded to the largest number of microscopic configurations. This assumption will hold if the strengths of all the interactions are unified. The notion of the space-time extent (and of the energy) arises here after the spontaneous breaking of the initial supersymmetry by tunnelling to a state that was characterized by the local Lorentz group (gravitational field) and the fields unifying all the other types of interactions. (Massive particles and the interactions of our present epoch are produced by subsequent symmetry breaking of the unified fields.) After this tunnelling transition the density matrix of any free field (q-field or gravitational waves) for $t \geq t_o$ is likely to depend only on the Hamiltonian $H_o = H(t_o)$ of this field at the initial moment $t = t_o + 0$.

It is not difficult to show that on large scales the field distribution is of the Hibbs form if, in the simplest model of tunnelling transition, the initial field state before transition corresponded to a vacuum of high symmetry. Indeed, if we neglect the effects of interactions associated with the details of the transition, then relative to the final vacuum with the broken symmetry the field state will obey the Gauss distribution with dispersion equal to the energy difference of the initial and final vacuums [53]. This means that for large scales the same energy ($\sim T_p$) will account for each interval of momenta $d^3\vec{k}$. Note that the *thermal* state of the Universe at $t = t_o + 0$ was, in principle, prepared in this model at the very beginning since the initial state (vacuum with the restored symmetry) had been already stationary with Gauss' law of distribution (see, e.g. [54]).

4 PARAMETRIC AMPLIFICATION OF INITIAL FLUCTUATIONS

Here we briefly outline the amplification effect and discuss a model.

The exact solution of Eq (2.4) in the asymptotic region $\omega\eta \ll 1$ [42]: $q = C_1(\vec{x}) + C_2(\vec{x}) \int \frac{d\eta}{(\xi a)^2}$ affords an opportunity to derive the amplification coefficient for large scales independently of the rate of cosmological expansion at the intermediate stage when $p \neq \varepsilon/3$ [46-48].

Let us consider a cosmological model with a change of state (see fig 1). The nature of this change at the b-stage is immaterial: the

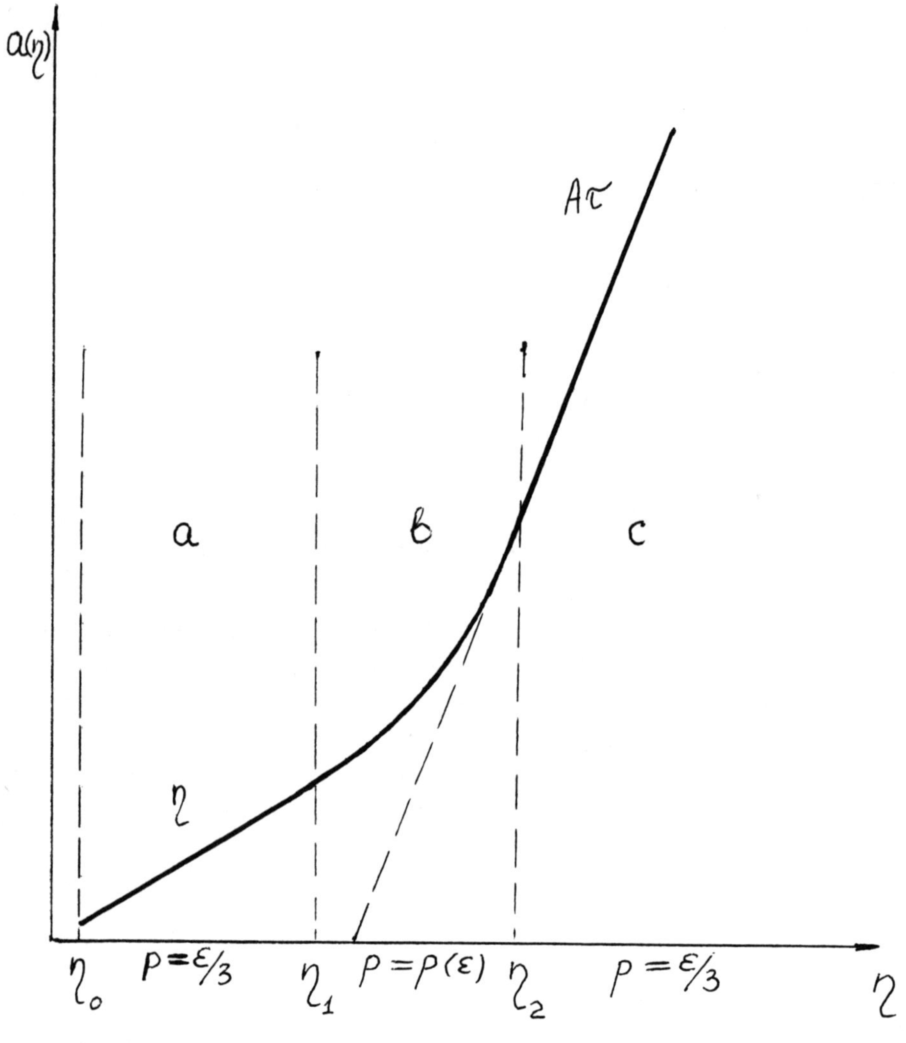

Figure 1. Evolution of the scale factor in the early Universe:
a-stage $(\eta_o < \eta < \eta_1 , p = \varepsilon/3)$,
b-stage $(\eta_1 < \eta < \eta_2 , p = p(\varepsilon))$,
c-stage $(\eta_2 < \eta , p = \varepsilon/3)$.

cause may be massive short-lived particles, phase transitions etc. [7-10, 55,56].

At the $p = \varepsilon/3$ stages, the 'growing' and 'decaying' modes of the q-field are determined by the phase $\omega\tau$, where

$$\tau = \frac{a}{a'} = \eta - \int_{\eta_1} d\eta \, \frac{a''a}{a'^2} = \eta_o + \frac{1}{2}\int_{\eta_o} d\eta \, (1 + \frac{3p}{\varepsilon}) \ . \tag{4.1}$$

We choose the following normalization of the wave-vector: $k = 1$ corresponds to the Planck length at the Planck time. This means that at the a-stage:

$$a = \eta = \tau = \sqrt{2t} \ , \tag{4.2}$$

and at the c-stage

$$a = A\tau \ , \ A = \text{const} \ , \ \tau - \eta = \text{const} \ .$$

Let $a_{\vec{k}}$ and $c_{\vec{k}}$ be the phonon representations at the stages 'a' and 'c' respectively. Then

$$\begin{aligned} c_{\vec{k}} &= \alpha_k a_{\vec{k}} + \beta_k^* a^+_{-\vec{k}} \ , \quad |\alpha_k|^2 - |\beta_k|^2 = 1 \ , \\ 2\beta_k A^{-1} &= -\frac{i\kappa}{\omega} - 1 + A^{-2} + 0(\omega\eta_1) \ , \end{aligned} \tag{4.3}$$

where

$$\kappa = \frac{1}{\eta_1} - \frac{1}{A^2\tau_2} - \int_{\eta_1}^{\eta_2} d\eta/(\xi a)^2 \ .$$

The relative energy of the field of potential perturbations at the a- and c-stages is as follows (2.12):

$$\frac{H_{\text{Reg}}}{\varepsilon a^3 \int d^3\vec{x}} = \frac{1}{24\,\pi^3}\int_{-\infty}^{\infty} d^3\vec{k}\omega \begin{cases} n_{ka} \\ A^{-2} n_{kc} \end{cases} , \tag{4.4}$$

and the phonon occupation numbers n_{kc} and n_{ka} are related by

$$n_{kc} = n_{ka} + |\beta_k|^2 (2n_{ka} + 1) \ .$$

n_{kc} is larger than n_{ka} because of induced and spontaneous creation. The factor A^{-2} takes into account the magnitude of phonon cooling at the intermediate b-stage.

A very important point is that after amplification the mutual correlation of phases of the created perturbations corresponds to the 'growing' mode:

$$C_{gc} \simeq C_{ga} + \frac{\kappa}{\omega} C_{da} ,$$

$$C_{dc} \simeq A^{-2} C_{da} , \tag{4.5}$$

$$\langle |C_{gc}|^2 \rangle \gg \langle |C_{dc}|^2 \rangle \quad \text{for} \quad \omega\eta_1 \ll 1 .$$

So, from initial random perturbations the 'growing' mode is created. If initially only the 'growing' mode were present, then no amplification would take place at all since the evolution of the 'growing' mode does not depend on the matter equation of state (metric fluctuations are constant for large scales). The possibility of amplification is directly connected with presence of the 'decaying' mode at the beginning. The equipartition of energy means that the metric perturbation amplitude on large scales is almost all in the 'decaying' mode: $q_d \sim c/\omega\eta \gg q_g \sim c$, where $c \sim k^2$ for quantum perturbations, $c \sim k^{3/2}$ for thermal and so on. To clarify the physical meaning of the amplification effect, let us consider a situation in which the duration of the intermediate stage ($p \neq \varepsilon/3$) is rather short. Then the perturbation amplitude will not change significantly after this stage but now it will be approximately the same in every mode ($q_d \sim q_g \sim c/\omega\eta_1$), so we have a large increase in the 'growing' mode. Analysis shows that for extended intermediate stages in cosmology a typical situation is the same: at the epoch close to t_1 , when first $p \neq \varepsilon/3$ and a parametric barrier (non-zero effective potential) arises, the amplitude of the 'growing' mode increases and becomes of the order of the 'decaying' mode amplitude at $t = t_1$, and then remains constant as far as metric perturbations on large scales are concerned. The amplification coefficient $\frac{\kappa}{\omega}$ is greater the earlier the intermediate stage begins and it is equal in order of magnitude to the ratio of the perturbation scale to the horizon at $\sim t_1$. The amplification effect does not practically depend on the duration of the intermediate stage and on the rate of expansion at this stage.

Taking into account that at the Planck time the amplitude of the initial fluctuations is about unity on the horizon scale, we obtain that, for the effect to work, the first change in the expansion rate must occur not later than in $\sim 10^4$ Planck times, which corresponds to an energy of $\sim 10^{17}$ Gev. This figure does not yet conflict with the estimate of the amount of gravitational waves produced in the change, that will influence the large-scale isotropy of the relict radiation.

Now the question about scales. It is necessary to have a scale that initially (at t_1) was $\sim 10^4$ times larger than the Planck length to be the scale of clusters by now. To get this we do not even have to invent something because a spontaneously broken theory prompts the answer: to stretch the scale factor we need the total duration of the supercooling inflationary stages to be about several tens of cosmological times. It does not matter when these exponential stages take place: it might be at the time of grand unification, or electroweak unification (or they might both contribute) etc. Of course, some other effects (not necessarily exponential stages) might increase the scale factor.

At a first glance the existence of such an inflationary stage seems to be enough to get the primordial fluctuations needed because it converts small scales (but $\omega\eta < 1$!) where fluctuations are always stronger, into large cosmic scales. But it is not so. First of all if the field of potential perturbations were conformally invariant the phonon vacuum would remain the vacuum independently of the cosmological expansion rate and no real perturbations would arise. If initially we have statistical or some other (non vacuum) fluctuations, again the amplification effect is very important because it amplifies the 'growing' mode near the singularity. Indeed, the initial amplitude of the 'growing' mode might be ten orders of magnitude less than is required for the primordial fluctuations.

The simplest model of the inflationary stage is considered in Appendix B. During this stage all matter effects become negligible. The interesting question arises why the conserved metric perturbations, which would have been fictitious for the pure vacuum de Sitter model, become physical 'growing' mode perturbations after the phase transition (remember that q-field is gauge-invariant!). It is just because the phase transition occurs on the hypersurface of constant temperature (or nearby). In a sense, the model remembers the hot beginning of the expansion. Note that the details of the phase transition at $t \sim t_2$ do not influence the resulting spectra (see fig 2). Indeed, parametric amplification

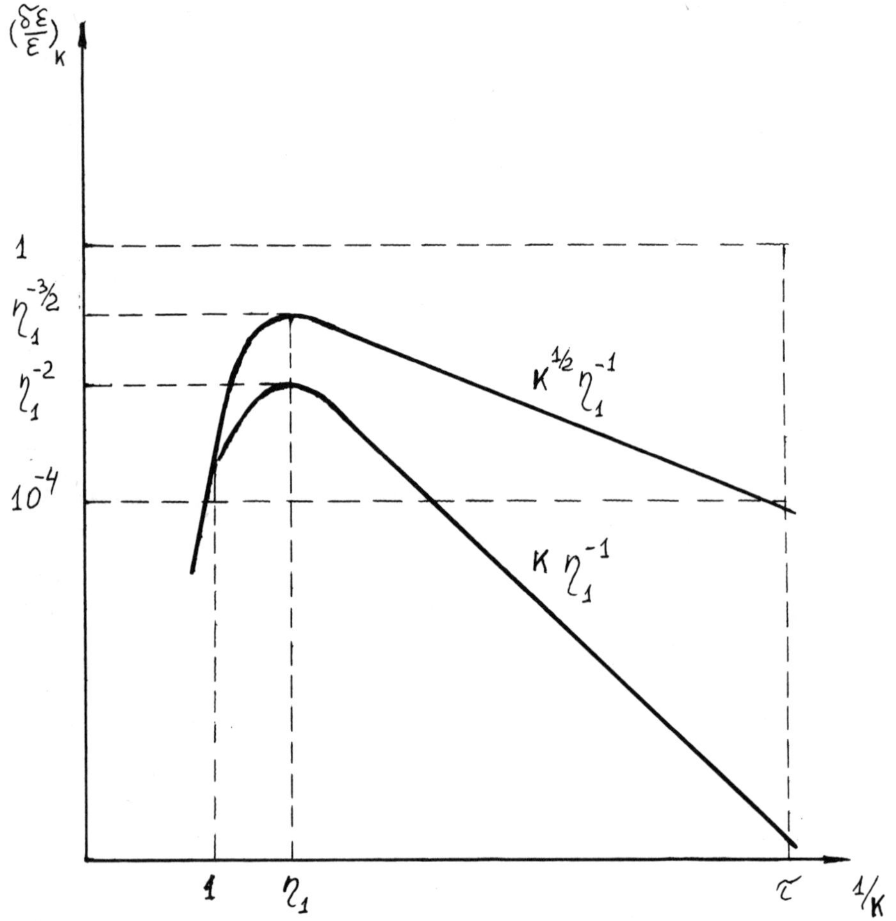

Figure 2. Spectra of the primordial matter density perturbations in the radiation-dominated epoch, created from initially quantum ($n_{ka} = 0$, bottom curve) or thermal ($n_{ka} \simeq \omega^{-1}$, top curve) fluctuations of metric. $k = \tau^{-1}$ is the horizon scale. $(\frac{\delta\varepsilon}{\varepsilon})_k = \frac{\omega\kappa}{\pi} (n_{ka} + \frac{1}{2})^{1/2}$, $\eta_1 < k^{-1} < \tau$.

By definition $\left\langle\left(\frac{d\varepsilon}{\varepsilon}\right)^2\right\rangle = \int_0^\infty \frac{dk}{k} \left(\frac{\delta\varepsilon}{\varepsilon}\right)_k^2$.

generates perturbations on scales $\lambda \geq ct_1 \exp t_2/2t_1 >> ct_1$ at the epoch of the phase transition because the barrier (non-zero effective potential) is localized near the beginning of the inflationary stage. On the other hand, phase transition details give inhomogeneities on smaller scales (since the bubbles form at $t > t_1$), that die more rapidly on larger scales than that in fig 2.

APPENDIX A

In a perturbed reference system

$$\delta u_i = P_i^k V_{,k} + \frac{1}{2} u_i u_k u_\ell h^{k\ell} ,$$

$$\frac{\delta p}{\varepsilon + p} = u^k V_{,k} - \beta u^k_{\ ;k} V - \frac{1}{2} u_k u_\ell h^{k\ell} ,$$

$$\delta g_{ik} = h_{ik} = 2(A e_{ik} + B g_{ik} + D_{;ik}) + (u_i C)_{;k} + (u_k C)_{;i} .$$

The functions A,B,C,D are the coefficients of general linear decomposition of a second-rank tensor in terms of scalars and $e_{ik} = 2u_i u_k - g_{ik}$. C and D are arbitrary functions (gauge freedom of h_{ik}); $B = 0$ for perfect fluid. $V = \zeta + C + u^i D_{,i}$ is a hydrodynamic (velocity) potential, A and ζ are gauge-invariant functions.

The q-function is

$$q = 2\sqrt{3}\, (\frac{\dot a}{a} \zeta + A) .$$

The inverse transformations are

$$A = \frac{\sqrt{3}}{4} \frac{\dot a}{a^2} \int dt\, a\gamma q , \quad \zeta = \frac{\sqrt{3}}{4} (\frac{2a}{3\dot a} q - \frac{1}{a} \int dt\, a\gamma q) .$$

The Einstein equations give:

$$\beta^2 \Delta A = \frac{\sqrt{3}}{4} \gamma \frac{a'}{a} q' .$$

On scales of interest, the background hypersurfaces $t =$ constant are assumed to be flat.

APPENDIX B

Let us consider a simple cosmological model with relativistic interacting particles ($p = \varepsilon/3$) and high vacuum density which transforms to heat at t_2 [10,23,46-48]:

$$t_o < t < t_2: \quad a = \begin{cases} (2t_1 \,\mathrm{Sh}\, \frac{t}{t_1})^{1/2} \\ (2A(t-\Delta))^{1/2} \end{cases}, \quad \varepsilon = \begin{cases} \frac{3}{4t_1^2} \mathrm{cth}^2 \frac{t}{t_1} \\ 3A^2/a^4 \end{cases}, \quad \xi = \begin{cases} 1/\mathrm{ch} \frac{t}{t_1} \\ 1 \end{cases}, \quad t_2 < t:$$

where $\Delta = t_2 - t_1$, $A = \mathrm{Sh}\ t_2/t_1 \simeq \exp \Delta/t_1 >> 1$. The vacuum density $3/4t_1^2$ predominates at $t_1 < t < t_2$.

If we neglect the details of the phase transition at t_2 then the sound effective potential has a standard form [47,48]:

$$U = -2\left(\frac{a}{t_1 \,\mathrm{ch}\, \frac{t}{t_1}}\right)^2 \simeq -\frac{4}{t_1} \begin{cases} \frac{t}{t_1} & , \quad t < t_1 \\ 2\exp(-\frac{t}{t_1}) & , \quad t_1 < t \end{cases},$$

and $\kappa \simeq 1/\eta_1$. We omit the integral in Eq (4.3) for κ since the exponential growth of the integrand at the de Sitter stage is ensured by the sharp exponential decrease of the hydrodynamic pressure with time. (This results in the growth of hydrodynamic velocity owing to the high pressure space-gradients in the co-moving reference system, while metric perturbations do not change.) Moreover, a formal application of hydrodynamic theory to this case gives the same estimate for κ . One should only take into account a jump of pressure at t_2 . By smoothing out all functions at $t \sim t_2$ it is easy to make sure that this pressure-jumping effect strictly compensates for the integral growth at the de Sitter stage, so the integral proves to be $\sim 1/\eta_1$.

For another approach see [57].

REFERENCES

[1] E.M. Lifshitz. ZhETF 16, 587 (1946).
[2] A.A. Grib, S.G. Mamaev, V.M. Mostepanenko. Quantum effects in intensive external fields. Atomizdat (1980).
[3] L. Parker (1976). In Asymptotic Structure of Space-Time, eds. F.P. Esposito and L. Witten. New York and London: Plenum.
[4] Ya.B. Zeldovich, A.A. Starobinski. ZhETF 61, 161 (1971).
[5] V.Ts. Gurovich. Preprint IFM, Academy of Sciences of Kirgizskaya SSR, Frunze (1981).
[6] S. Coleman, E. Weinberg. Phys.Rev. D7, 1888 (1973).
[7] D.A. Kirzhnitz, A.D. Linde. Ann.Phys. 101, 195 (1976).
[8] A.D. Linde. Rep.Phys. 43, 389 (1979).
[9] S. Weinberg. Phys.Rev.Lett. 42, 850 (1979).
[10] A.H. Guth, E.J. Weinberg. Phys.Lett. 45, 1131 (1980).

[11] A.H. Guth. Phys.Rev. D23, 347 (1981).
[12] S.A. Bludman (1981). Ann.N.Y.Acad.Sci. 375, 419.
[13] A.A. Starobinski. Phys.Lett., 91B, 99 (1980).
[14] V.F. Mukhanov, G.V. Chibisov. Pisma ZhETF 33, 549 (1981).
[15] S.W. Hawking. Proc. of the 2nd Seminar 'Quantum Gravity', Moscow. Oct. 13-15 (1981).
[16] L.P. Grishchuk, Ya.B. Zeldovich. Proc. of the 2nd Seminar 'Quantum Gravity', Moscow. Oct. 13-15 (1981).
[17] D. Atkatz, H. Pagels. Phys.Rev. D25, 2065 (1982).
[18] P. Van Nieuwenhuizen. Five Lectures on Supergravity. University of Toronto, Ontario. March (1981).
[19] J.B. Hartle, G.T. Horowitz. Phys.Rev. D24, 257 (1981).
[20] A.A. Starobinski. Pisma ZhETF 34, 460 (1980).
[21] A.D. Linde. Proc. of the 2nd Seminar 'Quantum Gravity', Moscow. Oct. 13-15 (1981).
[22] W.H. Press. Phys.Scripta 21, 702 (1980).
[23] K. Sato. Proc. of the Int. School of Nuclear Astrophys., Erice. 25 March - 6 April (1980).
[24] S. Coleman. Phys.Rev. D15, 2929 (1977).
[25] A.D. Linde. Decay of the false vacuum at finite temperature. Preprint Lebedev Inst. of Phys.Acad. of Sci., USSR, 265 (1981).
[26] Ya.B. Zeldovich. MNRAS 192, 663 (1980).
[27] A. Vilenkin. Phys.Lett. 46, 1169 (1981).
[28] A. Vilenkin. Phys.Rev. D24, 2082 (1981).
[29] J.B. Hartle. Phys.Rev. D22, 2091 (1980).
[30] L. Parker. Phys.Rev.Lett. 21, 562 (1968).
[31] J.D. Barrow, M.S. Turner. Nature 292, 38 (1981).
[32] D. Nanopoulous. Lecture at this workshop.
[33] C.J. Hogan. Nature 286, 360 (1980).
[34] S.W. Hawking. Astroph.J. 145, 544 (1966).
[35] E.R. Harrison. Rev.Mod.Phys. 39, 862 (1967).
[36] R.K. Sachs, A.M. Wolfe. Astroph.J. 147, 73 (1967).
[37] G.B. Field, L.C. Shepley. Astroph.Spa.Sci. 1, 309 (1968).
[38] W.H. Press, E.T. Vishniac. Astroph.J. 239, 1 (1980).
[39] J.M. Bardeen. Phys.Rev. D22, 1882 (1980).
[40] V.N. Lukash. Pisma ZhETF 31, 631 (1980).
[41] V.N. Lukash. ZhETF 79, 1601 (1980).
[42] V.N. Lukash. Generation of sound waves in the isotropic Universe. Preprint Spa.Res.Inst., Acad. of Sci., USSR, p.-559 (1980).
[43] V.N. Lukash. Generation of sound waves and galaxy formation in the expanding Universe. Preprint Spa.Res.Inst., Acad. of Sci., USSR, -300 (1980).
[44] L.P. Grishchuk. ZhETF 67, 825 (1974).
[45] L.P. Grishchuk, V.M. Yudin. J.Math.Phys. 21, 1168 (1980).
[46] D.A. Kompaneets, V.N. Lukash. Astron.Zh. 58, 482 (1981).
[47] D.A. Kompaneets, V.N. Lukash, I.D. Novikov. Origin of the initial inhomogeneities of the Universe. Astron.Zh. 59, 424 (1982).
[48] D.A. Kompaneets, V.N. Lukash, I.D. Novikov. Proc. of the 2nd 'Quantum Gravity' Seminar, Moscow. Oct. 13-15 (1981).
[49] G.S. Bisnovatyi-Kogan, V.N. Lukash, I.D. Novikov. Proc. of 5-Regional meeting (IAU/EPS), Liege, Belgium. 29-31 July (1980).
[50] E.T. Vishniac. Nonlinear effects on cosmological perturbations. Preprint 1431, Preprint 1432, Harvard-Smithsonian Center for Astrophysics (1980).
[51] W.H. Press, E.T. Vishniac. Astroph.J. 236, 323 (1980).
[52] V.N. Lukash, I.D. Novikov. ZhETF 81, 1000 (1982).

[53] D.A. Kompaneets, V.N. Lukash. Quantum massive fields in the accelerated reference system. Preprint Space Res.Inst. of the Acad. of Sci. of the USSR 570 (1981).
[54] D.W. Sciama, P. Candelas, D. Deutsch. Advances in Physics 30, 367 (1981).
[55] A.G. Polnarev , M.Ju. Khlopov. Phys.Lett. 97B, 383 (1980).
[56] B.J. Carr. Proc. of the 2nd Seminar 'Quantum Gravity', Moscow. Oct. 13-15 (1981).
[57] V.F. Mukhanov, G.V. Chibisov. Galaxy formation and phonons. Preprint Lebedev Inst. of Phys. of the Acad. of Sci. of the USSR 162 (1980).
[58] Ya.B. Zeldovich, I.D. Novikov. Astron.Zh. 46, 960 (1969).
[59] S. Hawking. Article in this volume (1982).
[60] A. Starobinski. Lecture at this workshop.

THE EFFECTIVE POTENTIAL IN DE SITTER SPACE

B. Allen

Department of Applied Mathematics and Theoretical Physics, Silver Street, Cambridge, England.

There have been many references to the "one-loop effective potential" in these seminars. I want to show what it is, and how it can be calculated for any gauge theory in the inflationary model of the early universe.

The de Sitter universe ($k = 0, \pm 1$) possesses a unique Euclidean section, which is a four-sphere S^4 of radius a, on which our quantum field theory is well defined. One obtains this Euclidean S^4 by Wick rotating the Lorentz de Sitter space, which has topology $R^1 \times S^3$. The radius a is related to other speakers' measures of curvature as follows: Hubble's constant $H = a^{-1}$; the Einstein cosmological constant $\Lambda = 3a^{-2}$; and the Ricci scalar curvature $R = 12a^{-2}$. We use units with $\hbar = c = 1$, so $mass = length^{-1}$.

The effective potential is a measure of the vacuum energy of the quantum fields. It is defined like the free energy in thermodynamics. The path integral is

$$Z = \int d[\varphi]d[g]\exp -S[g,\varphi] \qquad (1)$$

where $S[g,\varphi]$ is the action of a field configuration φ on the manifold with Euclidean metric g. The effective potential is defined for a constant background field φ_b by

$$\exp [-\Omega V_{eff}(\varphi_b)] = Z \qquad (2)$$

where Ω is the four-volume of S^4. V_{eff} is an energy density: its units are $mass \times length^{-3} = mass^4$.

We choose background fields φ_b and g_b which are solutions of the classical field equations. Here φ_b is the expectation value of the Higgs field, and g_b is the standard metric of constant curvature on S^4.

We then expand S as a Taylor series in $(\varphi-\varphi_b)$ and $(g-g_b)$. Since φ_b and g_b extremise the action, there is no linear term. In the one-loop approximation we keep only the quadratic (gaussian) term, and drop the higher orders. Then the path integral is

$$\log Z = -S[g_b,\varphi_b] + \log Det(\mu^{-2}Q)^{-1/2} \tag{3}$$

where Q is the operator for the second variation of the action, and μ is a constant with dimensions of mass introduced to define the path integral measure in (1).

For example, if the action of a real scalar field φ were

$$S = \int [\tfrac{1}{2}(\partial_\mu\varphi)(\partial^\mu\varphi) + V(\varphi)]\sqrt{g}\, d^4x \tag{4}$$

then the operator Q would be $Q = -\Box + V''(\varphi_b)$, acting on real scalar functions. In de Sitter space the operators Q can always be decomposed into five "elementary" representations for fields of different spin. The eigenfunctions of these operators form a complete set; any set of fields (of any spin) can be uniquely expanded in terms of them. Let $L = 0, 1/2, 1, 3/2, 2$ denote spin. The relevant operators are $Q = -\Box + x$ for bosons (L integer) and $Q = \gamma^\mu\partial_\mu + \sqrt{x}$ for fermions (L half-integer). The x are polynominal functions of the constant background field φ_b and the coupling constants. The eigenvalues λ_n and multiplicity g_n of Q are (n = 0, 1, 2,...)

$$\lambda_n = n^2 + (2L+1)n + x \tag{5}$$

$$g_n = dim(n+L,L) = \tfrac{1}{3}(2L+1)(n+1)(n+L+3/2)(n+2L+2) \tag{6}$$

The eigenfunctions are pure spin L representations. For example, the spin L=1 eigenfunctions A^μ satisfy $\nabla_\mu A^\mu = 0$.

Because the eigenvalues (5) increase without bound, the determinants are defined with the aid of the generalised zeta function (Hawking 1977):

$$\zeta(z) = \sum_{n=0}^{\infty} g_n\lambda_n^{-z} \tag{7}$$

The sum can be analytically continued to a function which is regular at z = 0 and has poles at z = 1 and z = 2. The determinant is defined as

$$Det\ Q = \exp -\zeta'(0) \tag{8}$$

(where $' = \frac{d}{dz}$) which is formally the product of all the λ_n. One can easily show that if we scale the eigenvalues by μ

$$Det\ (\mu Q) = \mu^{\zeta(0)}\ Det\ Q \tag{9}$$

so that $\zeta(0)$ can be interpreted as the "number of rows" of Q.

The detailed calculations appear elsewhere (Allen 1982); the result is that for a unit radius (a = 1) four-sphere

$$\zeta(0) = ax^2 + bx + c \tag{10}$$

where a, b and c are constants that depend upon the spin L. The determinant is

$$\zeta'(0) = -\frac{1}{3}(2L+1)\left[\int_{L+\frac{1}{2}}^{L+\frac{1}{2}+\sqrt{\Delta}} u(u-L-\tfrac{1}{2})(u-2L-1)\psi(u)du + \right. \tag{11}$$

$$\left. \int_{L+\frac{1}{2}}^{L+\frac{1}{2}-\sqrt{\Delta}} u(u-L-\tfrac{1}{2})(u-2L-1)\psi(u)du - \frac{1}{4}\Delta^2 + \frac{1}{2}(L^2-\frac{1}{12})\Delta + constant\right]$$

where $\Delta = (L+\frac{3}{2})^2 - x$, and $\psi(z) = \frac{d}{dz}\log\Gamma(z)$ is the ψ function. Shown in figure 1 is $\zeta'(0)$ for a spin L=1 vector field. Each time that x cancels an eigenvalue, $\zeta'(0) \to \infty$, and $Det\ Q \to 0$.

To show how to use these determinants, I will compute the one-loop effective potential for massless scalar electrodynamics. The Euclidean action is

$$S[A^\mu,\varphi] = \int [\ \frac{1}{4}F_{\mu\nu}F^{\mu\nu} + \frac{1}{2}(D_\mu\varphi)^* D^\mu\varphi + V(\varphi)]\ \sqrt{g}\,d^4x \tag{12}$$

with a tree potential $V(\varphi) = \frac{1}{2}\xi R\varphi^2 + \frac{\lambda}{4}\varphi^4$. The path integral can be calculated in Landau gauge, and gives

$$Z_{one-loop} = \exp\,[\,-\Omega V(\varphi_b)]\; Det\;\mu^{-2}(W + e^2\varphi_b^2)^{-1/2} \tag{13}$$

$$Det\;\mu^{-2}(Q + \varphi_b^{-1}\,V'(\varphi_b))^{-1/2}\; Det\;\mu^{-2}(Q + V''(\varphi_b))^{-1/2}$$

where the operator $Q = -\Box$ on scalars and the operator

Figure 1. Shown is the real part of $\zeta'(0)$ for a spin-1 field, as a function of Δ. The logarithmic singularities occur at those values of Δ which cause the determinant to vanish.

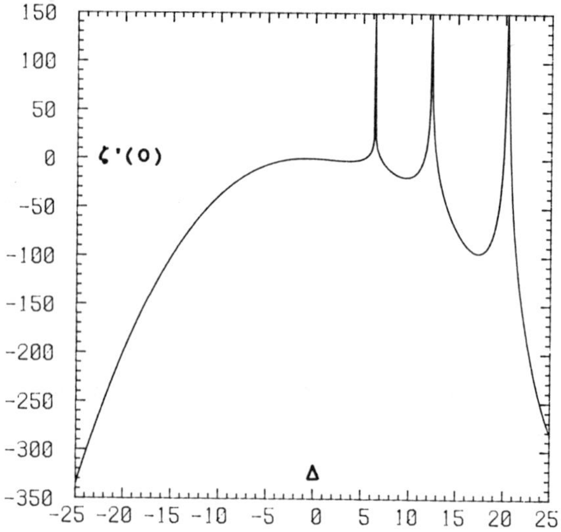

$W = -g_{\mu\nu}\Box + R_{\mu\nu}$ on vectors. The effective potential is then

$$V_{eff}(\varphi) = \frac{1}{2}\xi R\varphi^2 + \frac{\lambda}{4}\varphi^4 + \frac{1}{2\Omega}\Big[\; Det\;\mu^{-2}(W + e^2\varphi_b^2) \tag{14}$$

$$+ \log\,Det\;\mu^{-2}(Q + V''(\varphi_b)) + \log\,Det\;\mu^{-2}(Q + \varphi_b^{-1}V'(\varphi_b))\Big]$$

The first determinant is the sum of gauge field loops

$$\log\,Det\;\mu^{-2}(W + e^2\varphi_b^2) = \tag{15}$$

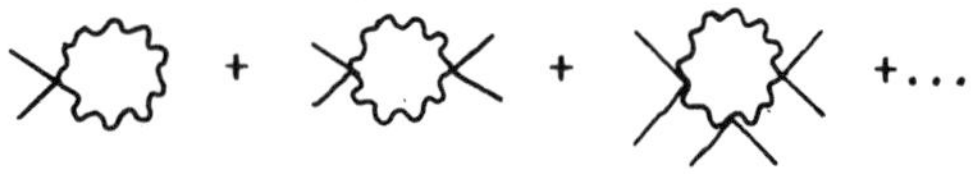

and the second and third determinants are the sum of scalar field loops.

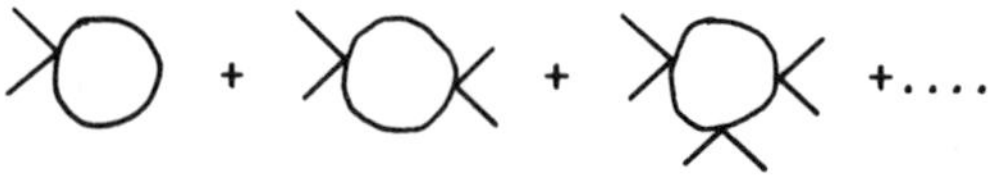

(16)

Now if we assume that the quartic coupling constant λ is of order e^4 (e is the gauge field coupling constant), then we can neglect the second and third determinants.

The shape of this potential function (14) depends upon a, and upon a single parameter which I call P.

$$P = \frac{32\pi\xi}{e^2} - \frac{8\pi^2\lambda}{9e^4} - \frac{3}{2} \tag{17}$$

For all values of λ, e, ξ, and μ, the limit as the four-sphere S^4 becomes very big, and locally flat, is

$$\lim_{a \to \infty} V_{eff}(\varphi,a) = \frac{3e^4}{64\pi^2}\varphi^4[\log \frac{e^2\varphi^2}{M^2} - \frac{1}{2}] \tag{18}$$

which is the standard flat-space result (Coleman & Weinberg 1973). This limit defines the mass M which is the mass of the vector gauge field in flat space. We will see that how this limit is approached as $a \to \infty$ depends critically on the value of P.

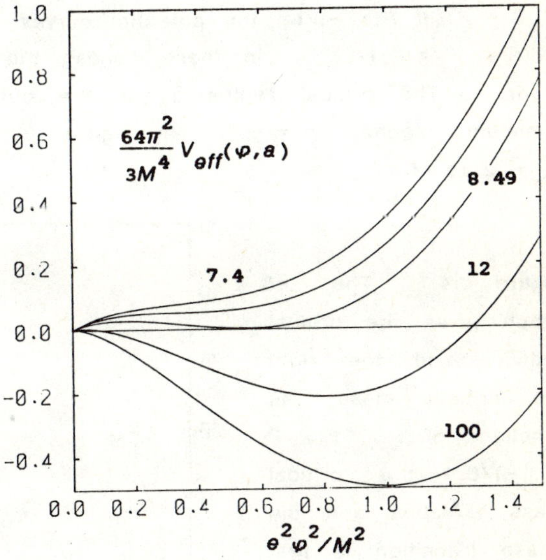

Figure 2. Shown is $V_{eff}(\varphi,a)$ as a function of the background field φ^2 for several values of the radius a. The number above a curve is the value of a in units of M^{-1}. For large radius, V_{eff} approaches the flat space potential (18).

If $P > -1/6$, the potential curves $V_{eff}(\varphi,a)$ form a family like that shown in figure 2, for the case P = 10. As the curvature increases, the broken symmetry is restored. There is a first-order phase transition at a critical radius a_c, with a critical gauge field mass M_c^2. In figure 2 the critical radius is $a_c = 8.49\ M^{-1}$ and the critical mass is $M_c^2 = 0.55\ M^2$.

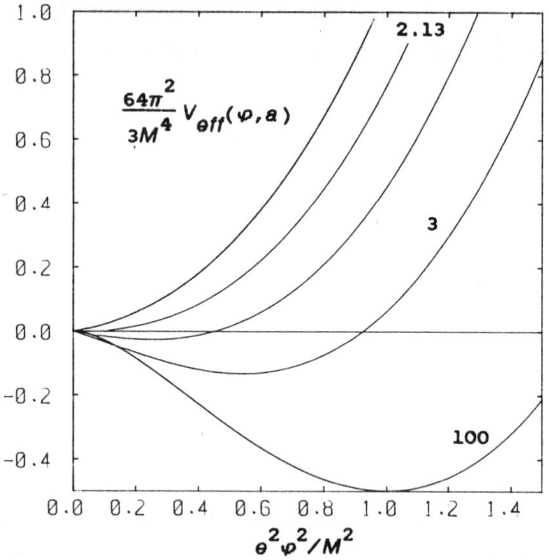

Figure 3. Shown is $V_{eff}(\varphi,a)$ for a parameter $P < -1/6$. The phase transition is second order, and so the critical mass $M_c = 0$. The number above a curve is the value of a in units of M^{-1}.

If $P < -1/6$ the potential curves $V_{eff}(\varphi,a)$ look like figure 3 (where $P = -1/2$). In these cases, the phase transition is second order. The critical radius a_c is the curvature at which the global minimum reaches $\varphi = 0$. In figure 3, the critical radius is at $a_c = 2.13\ M^{-1}$.

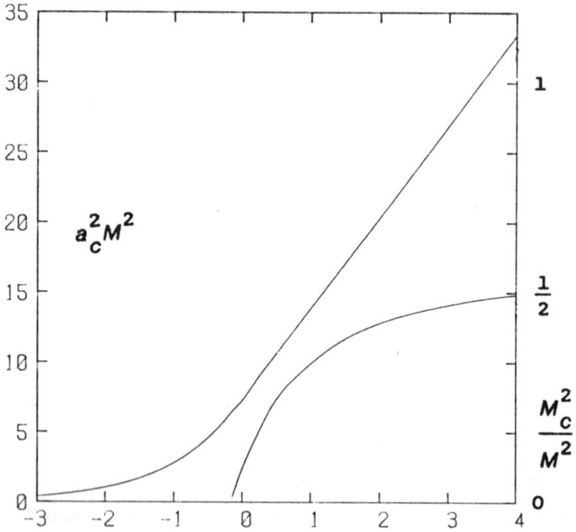

Figure 4. The left graph gives the critical radius, and the right the critical mass, as functions of P. For P < -1/6, the critical mass vanishes and the phase transition is first order.

We can work out the critical mass and curvature for each value of P; shown in figure 4 is a graph of the results. Apart from a small numerical discrepency, these critical values agree with the two special cases (conformal and minimal) considered by Shore (1980). In the Guth scenario, $V_{eff} = 0$ in the flat space broken symmetry minimum, and $a \approx M_p M_x^{-2}$ in the exponentially expanding phase. There will be a barrier between the two minima only if $\xi \gtrsim \frac{e^2}{4\pi}$.

These determinants can be used to find the one-loop effective potential for any gauge theory. For the simple SU(5) model, the phase diagram depends not on a single parameter P, but on two parameters (Allen, 1982). In the case of conformal scalar electrodynamics, we have thown away an infinite energy density, because the theory contains only bosons. There is then no reason not to put the zero of V_{eff} at $\varphi = \varphi_0$ rather than at $\varphi = 0$. However a supersymmetric theory will not have an infinite background energy (Allen & Davis, 1983). In these theories the vacuum energy that we can calculate is meaningful (i.e. finite). However there is no longer the freedom to add a constant to V_{eff}. Indeed, the natural ground state of gauged supergravity is anti de Sitter space, with a negative cosmological constant. It may be rather difficult to construct an inflationary universe model based on such theories.

REFERENCES

Allen, B. (1982). Phase Transitions in DeSitter Space, to appear in Nuclear Physics

Allen, B. (1982). SU(5) Effective Potential in DeSitter Space, DAMTP Preprint, Cambridge University

Allen, B. & Davis, S. (1983). Vacuum Energy in Gauged Extended Supergravity, to appear in Physics Letters

Coleman, S. & Weinberg E. (1973). Radiative Corrections as the Origin of Spontaneous Symmetry Breaking. Phys. Rev. D7, 1888.

Hawking, S.W. (1977). Zeta Function Regularisation of Path Integrals in Curved Spacetime. Comm. Math. Phys. 55, 133.

Shore, G.M. (1980) Radiatively Induced Spontaneous Symmetry Breaking and Phase Transitions in Curved Spacetime. Ann. Phys. 128, 376

QUANTUM FLUCTUATIONS AND PHASE TRANSITIONS IN COSMOLOGY

L.H. Ford
Physics Department, Tufts University, Medford, MA 02155 USA

Abstract. The effects of the coupling to the background gravitational field upon phase transitions in an inflationary universe model are discussed.

In much of the work which has been undertaken on the early universe, one treats the particle interactions as though one were in flat spacetime at finite temperature. The cosmological expansion then simply serves to decrease the temperature of the universe. Although this is a good approximation at sufficiently late times, it presumably fails at the Planck time and possibly at times much later than the Planck time. When this happens, the coupling to the cosmological background gravitational field can no longer be ignored. One place where such coupling might be expected to be important is in the study of phase transitions in the early universe.

As will be discussed below, quantum field fluctuations can be sensitive to the details of the coupling to the cosmological background. These field fluctuations can be crucial in either producing or inhibiting a phase transition, particularly in an inflationary universe model (Guth 1981).

In the following this effect will be illustrated in a simple model. Only the results will be quoted, and for derivations and further details the reader is referred to more complete accounts (Vilenkin 1982a, Vilenkin & Ford 1982). Let us consider the theory of two interacting scalar fields with the Lagrangian

$$\mathcal{L} = \frac{1}{2}\,[(\partial_\mu\phi)^2 - m^2\phi^2 - \xi R\phi^2 + (\partial_\mu\psi)^2 - \frac{1}{6}R\psi^2 - \frac{1}{6}\lambda\phi^4 - g^2\phi^2\psi^2] \tag{1}$$

where R is the scalar curvature. This model can be regarded as a simplification of a gauge theory in which ψ plays the role of the gauge fields and ϕ that of the Higgs field. The ϕ field may acquire a nonzero vacuum expectation value which breaks the $\phi \to -\phi$ symmetry. This phenomenon is usually studied by means of the effective potential; however, it is also possible to use the equations of motion of the theory (Ford 1980, Ford & Toms 1982). The latter approach is especially well suited to time-dependent situations where the effective potential is not well defined. In lowest order perturbation theory, the symmetric phase ($\langle\phi\rangle = 0$) will be unstable if the equation

$$(\Box + M^2)\,\Phi = 0, \tag{2}$$

with $M^2 = m^2 + \xi R + g^2\langle\psi^2\rangle + \lambda\,\langle\phi^2\rangle$, has growing solutions. [For a discussion of some of the higher order contributions in this model, see Vilenkin (1982 ab).] Here $\langle\phi^2\rangle$ and $\langle\psi^2\rangle$ are free field expectation values; their appearance in Eq. (2) represents the influence of quantum fluctuations upon the stability of the symmetric phase. For example, if $\lambda > 0$ then a positive $\langle\phi^2\rangle$ tends to stabilize this phase. The thermal contribution to $\langle\phi^2\rangle$ is positive and consequently one expects the symmetric phase to be stable at high temperatures. However, it is possible to have $\langle\phi^2\rangle < 0$ in the vacuum state due to the gravitational coupling. In this case the symmetric phase may become unstable as the universe cools. To better understand when such phase transitions can occur, it is necessary to calculate $\langle\phi^2\rangle$ and $\langle\psi^2\rangle$ in particular spacetimes.

Let us focus our attention on the behavior of $\langle\phi^2\rangle$ in an inflationary-type model for which the metric is

$$ds^2 = dt^2 - a^2(t)\,d\vec{x}^2 \tag{3}$$

where $a(t) \sim t^{\frac{1}{2}}$ for $t \lesssim H^{-1}$ and $a(t) \sim e^{Ht}$ for $t \gtrsim H^{-1}$. That is, the universe is radiation dominated at early times and has the metric of de Sitter space at later times. We are interested in $\langle\phi^2\rangle$ in the de Sitter regime. The quantum state is taken to be a thermal state in which the definition of particles is that associated with the conformal vacuum in the radiation dominated regime. We may write $\langle\phi^2\rangle = \langle\phi^2\rangle_T + \langle\phi^2\rangle_o$ where $\langle\phi^2\rangle_T$ and $\langle\phi^2\rangle_o$ are the thermal and vacuum contributions, respectively.

We assume that the thermal contribution can be described by an incoherent superposition of particles with a Bose-Einstein spectrum and write

$$\langle\phi^2\rangle_T = \pi^{-2}\int_0^\infty dk\; k^2|\psi_k|^2(e^{\varepsilon_k/\theta}-1)^{-1}, \qquad \varepsilon_k = (k^2+m^2+\xi R)^{\frac{1}{2}}. \tag{4}$$

The temperature at the beginning of the de Sitter regime ($t \simeq H^{-1}$) is θ and is assumed to be large compared to H. Here the ψ_k are the positive frequency mode functions in de Sitter space and are of the form

$$\psi_k = (\pi/4)^{\frac{1}{2}} H\eta^{3/2} [C_1 H_\nu^{(1)} (k\eta) + C_2 H_\nu^{(2)} (k\eta)] \tag{5}$$

where $\eta = -H^{-1} e^{-Ht}$, $\nu^2 = \frac{9}{4} - 12 (m^2/R + \xi)$, and C_1 and C_2 are constants such that $|C_2|^2 - |C_1|^2 = 1$. These constants determine the definition of the vacuum state. At the beginning of the exponential expansion, one has

$$\langle\phi^2\rangle_T \simeq \frac{1}{12} (\theta/a)^2, \tag{6}$$

which is the flat spacetime result for a thermal state at temperature θ/a. At late times we have, for $m^2/R + \xi << 1$,

$$\langle\phi^2\rangle_T \sim A\theta H\, e^{-8(m^2/R + \xi)Ht} \tag{7}$$

where A is a constant of order unity. Note that in the particular case $m^2 + \xi R = 0$ (as for the massless, minimally coupled theory),

$$\langle\phi^2\rangle \rightarrow A\theta H \text{ as } t \rightarrow \infty. \tag{8}$$

Thus the late time behavior is drastically modified by the coupling to the background gravitational field.

Let us now consider the vacuum contribution $\langle\phi^2\rangle_o$. If $0 < m^2/R + \xi << 1$, then the asymptotic form at late times is

$$\langle\phi^2\rangle_o \sim \langle\phi^2\rangle_{BD} - \frac{R^2}{384\pi^2} (m^2 + \xi R)^{-1} e^{-8(m^2/R + \xi)Ht} \tag{9}$$

where $\langle\phi^2\rangle_{BD}$ is the expectation value in the de Sitter invariant state defined by setting $C_1 = 0$ and $C_2 = 1$. This was first calculated by Bunch & Davies (1978) and for small $m^2/R + \xi$ is given by

$$\langle\phi^2\rangle_{BD} \sim \frac{R^2}{384\pi^2} (m^2 + \xi R)^{-1} \tag{10}$$

Thus in the limit $m^2 + \xi R \to 0$, the de Sitter invariant state ceases to be physically meaningful. However, $\langle\phi^2\rangle_o$ is still well defined and if $m^2 + \xi R = 0$, it grows linearly in time

$$\langle\phi^2\rangle_o \sim (4\pi^2)^{-1} H^3 t + \text{constant}. \tag{11}$$

These results for $\langle\phi^2\rangle_o$ were obtained independently by Linde (1982) and by Starobinsky and reported in their talks at this workshop.

The overall behavior of $\langle\phi^2\rangle$ is illustrated in Fig. 1. If $0 < m^2/R + \xi \ll 1$ (solid line), $\langle\phi^2\rangle$ is initially dominated by the thermal contribution which first decreases as in Eq. (6) and later decreases much more slowly, resulting in the plateau illustrated. Meanwhile the vacuum contribution increases and eventually dominates, with $\langle\phi^2\rangle$ asymptotically approaching $\langle\phi^2\rangle_{BD}$. If $m^2/R + \xi = 0$ (dashed line), the behavior is essentially the same, except that $\langle\phi^2\rangle$ ultimately grows without bound. This rather complicated time dependence is due to the effects of the background spacetime. The fact that $\langle\phi^2\rangle_T$ approaches a nonzero constant and $\langle\phi^2\rangle_o$ grows when $m^2/R + \xi = 0$ is due to the influence of very long wavelength modes. In this case, the de Sitter invariant vacuum does not exist because of infrared divergence problems (Ford & Parker 1977).

We are now in a position to see that the coupling to the cosmological background can strongly influence the quantum field fluctuations (i.e., the value of $\langle\phi^2\rangle$) which in turn influence the stability of the symmetric phase via Eq. (2). A detailed discussion of the destabilization

Fig. 1. The behavior of $\langle\phi^2\rangle$ as a function of time for $0 < m^2/R + \xi \ll 1$ (solid line) and for $m^2/R + \xi = 0$ (dashed line).

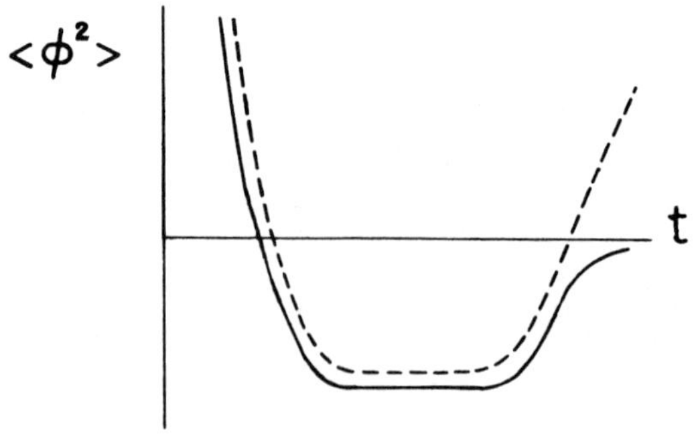

of the symmetric phase for particular choices of the parameters of the theory of Eq. (1) is given in Vilenkin & Ford (1982). The effect of the gravitational coupling can be large even when the universe is far away from the Planck regime. One reason for this is that in an inflationary model, the thermal contribution $\langle\phi^2\rangle_T$ is decreased by the exponential expansion and curvature contributions to $\langle\phi^2\rangle$ (which are usually expected to be of order R) soon become important. In the case of a massless, conformally coupled theory ($\xi = \frac{1}{6}$), the curvature contribution can be described as due to the Gibbons-Hawking temperature of de Sitter space (Gibbons & Hawking 1977); more generally, this contribution is contained in $\langle\phi^2\rangle_o$ and does not need to be added on separately. A second reason for the importance of the cosmological background is that the vacuum contribution $\langle\phi^2\rangle o$ can sometimes be much larger than R, as illustrated by Eq. (10). This is especially true when the coupling of the scalar field to the curvature is not conformal, and illustrates the fact that the gravitation influence upon field interactions is greatest for nonconformally invariant theories. Another example of this is the loss of CPT invariance in the early universe for nonconformal fields (Ford 1982).

This work was supported in part by the National Science Foundation (PHY-8106978).

REFERENCES

Bunch, T.S. & Davies, P.C.W., (1978). Quantum field theory in de Sitter space: renormalization by point splitting. Proc. R. Soc. London A, 360, 117-34.

Ford, L.H. (1980). Instabilities in interacting quantum field theories in non-Minkowskian spacetimes. Phys. Rev. D, 22, 3003-11.

Ford, L.H. (1982). Particle decays and CPT noninvariance in cosmology. Nuclear Phys. B, 204, 35-44.

Ford, L.H. & Parker, L. (1977). Infrared divergences in a class of Robertson-Walker universes. Phys. Rev. D, 16, 245-50.

Ford, L.H. & Toms, D.J. (1982). Dynamical symmetry breaking due to radiative corrections in cosmology. Phys. Rev. D, 25, 1510-8.

Gibbons, G.W. & Hawking, S.W. (1978). Cosmological event horizons, thermodynamics, and particle creation. Phys. Rev. D, 15, 2738-51.

Guth, A.H. (1981). The inflationary universe: a possible solution to the horizon and flatness problems. Phys. Rev. D, 23, 347-56.

Linde, A. (1982). Scalar field fluctuations in expanding universe and the new inflationary universe scenario. Phys. Lett. 116B, 335-339.

Vilenkin, A. (1982a). Gravitational Effects in Guth Cosmology. Phys. Lett. B, 115, 91-4.

Vilenkin, A. (1982b). Phase transitions in de Sitter space. Tufts University preprint TUTP-82-10.

Vilenkin, A. & Ford, L.H. (1982). Gravitational effects upon cosmological phase transitions. Phys. Rev. D, 26, 1231.

SYMMETRY BEHAVIOR AT FINITE TEMPERATURE IN DYNAMIC SPACETIMES

B. L. Hu
Department of Physics and Astronomy, University of Maryland
College Park, Maryland 20742 U.S.A.

Abstract. Symmetry Breaking at finite temperature (T) for a $\lambda\phi^4$ theory in Robertson-Walker spacetimes is studied by means of the quasi-adiabatic approximation. Quantum corrections to the temperature dependent mass are calculated and found to consist of three distinct parts arising from adiabatic expansion ($\sim T^2$), non-adiabatic expansion ($\sim T^{-2}$, T^{-4},..) and low frequency modes. Specialization to the de Sitter exponential expansion is discussed.

1 INTRODUCTION

The problem of symmetry breaking in curved spacetime has been investigated by a number of authors. By adding a term in the effective mass due to the scalar curvature, Coleman & DeLuccia (1981) and others (Abbott 1981; Fujii 1981; Hut & Klinkhamer 1981) considered the classical effect of curvature. However, as their treatment uses essentially flat-space field-theoretical techniques, the dynamical effects of spacetime was not considered fully. For curved but static spacetimes, Ford (1980), Toms (1980), Kennedy (1981), Denardo & Spallucci (1981a) discussed the effect of spacetime curvature including quantum corrections. For dynamic spacetimes, Ford & Toms (1982) analyzed the symmetry breaking in a Robertson-Walker universe due to radiative corrections. Hawking & Moss (1982) and others discussed the Coleman-Weinberg potential in the de Sitter universe. Most of this earlier work were carried out in the zero temperature context. Finite temperature effects on the restoration of symmetry are not too difficult to deal with for static spacetimes (Dowker & Critchley 1977, Dowker & Kennedy 1978) or spacetimes admitting Euclidean sections such as the de Sitter space (Shore 1980, Denardo & Spallucci 1981; Vilenkin & Ford 1982). In the former case, thermal equilibrium is put in "by hand" and there is no question about its being maintained while in the latter case, the thermal nature is intrinsically associated with the existence of event horizons. In both cases, finite temperature theory can directly be extended from the flat-space theory by the imaginary-time method. The case of the de Sitter spacetime which is of special interest to us here in connection with Grand Unification era phase transitions (Guth 1981; Linde 1982; Albrecht & Steinhardt 1982) is discussed extensively in the talks given by Allen, Ford, Linde, Moss and Starobinsky.

For general dynamic spacetimes, due to the prevailing non-equilibrium situation, more sophisticated techniques in curved-spacetime quantum statistical theories are needed for the full treatment of quantum statistical processes like symmetry breaking and particle production. The meaning and viability of finite temperature theories in dynamic spacetimes have been discussed in a recent paper of mine (Hu 1982a). There it was shown that under certain restrictive conditions (near conformal invariance of the field and near conformal flatness of the spacetime) a state of quasi-equilibrium can be defined and a finite temperature theory can be established up to a certain order in some nonadiabaticity parameter. In this talk I shall apply methods outlined there to the analysis of the symmetry behavior of a simple $\lambda\phi^4$ theory at finite temperature in Robertson-Walker (RW) spacetimes. (For a discussion of the thermal Green function approach, see Drummond 1981.) We start by writing down the equation obeyed by the fluctuation field ϕ of a classical solution $\hat{\phi}$. Our aim is to calculate the variance $<\phi^2>$, which enters into the equation governing $\hat{\phi}$ as vacuum $<\phi^2>_o$ and finite temperature $<\phi^2>_T$ quantum radiative corrections to the effective mass. We seek solutions to ϕ by using an adiabatic expansion for the high frequency modes and a "sudden" approximation for the low frequency modes. The former contributes two types of terms to $<\phi^2>_T$, one (AD) arising from the adiabatic part of the expansion of the background (proportional to T^2) and the other (NA) from the non-adiabatic part of the expansion ($\sim T^{-2}, T^{-4}, ..$). The low frequency contribution (LF) is proportional to ξR, where ξ is the coupling constant with the spacetime of curvature R ($\xi=0$ is conformal). It is of particular importance to rapidly expanding universes with event horizons like the de Sitter, in which case it is proportional to T_o, the temperature at the beginning of the inflationary era at which a significant amount of modes get red-shifted drastically. Our result agrees with others in the AD and LF terms but most previous work on this subject seem to have unduely neglected the non-adiabatic contributions. Our result is more general in that the RW spacetime dynamics is left unspecified and contains the de Sitter result as a special case.

2. FINITE TEMPERATURE $\lambda\phi^4$ THEORY IN ROBERTSON-WALKER SPACETIMES

Consider a massive (m) scalar field $\hat{\phi}$ with quartic self-interaction (λ) described by the Lagrangian density

$$L = -\frac{1}{2}\tilde{\phi}(x)[\Box+(1-\xi)R/6 + m^2]\tilde{\phi}(x) - \frac{\lambda}{4}\tilde{\phi}^4 \tag{2.1}$$

in a Robertson-Walker (RW) universe with metric

$$ds^2 = dt^2 - a^2 d\ell^2 \tag{2.2}$$

where a is the scale length and $d\ell^2$ is the line element of a three-dimensional space of constant curvature (κ=0,1, -1 corresponds to flat, closed and open spaces respectively). The equation of motion for $\tilde{\phi}$ derived from (2.1) is given by

$$[\Box+(1-\xi_B)R/6 + m_B^2]\tilde{\phi} + \lambda\tilde{\phi}^3/6 = 0 \tag{2.3}$$

The classical action has an extremum at $\tilde{\phi} = \hat{\phi}$ which satisfies the classical equation of motion (m_B, ξ_B, and λ_B denote the bare quantities)

$$[\Box+(1-\xi_B)R/6 + m_B^2 + \lambda_B\hat{\phi}^2/6]\hat{\phi} = 0 \tag{2.4}$$

To study the vacuum stability of this classical solution, consider quantum fluctuations $\phi = \tilde{\phi} - \hat{\phi}$ around $\hat{\phi}$ (which is assumed to have vanishing vacuum expectation value $\langle\tilde{\phi}\rangle_o = \langle 0|\phi + \hat{\phi}|0\rangle = \hat{\phi}$). The classical field equation contains now a source term arising from the variance of the fluctuation, i.e.

$$[\Box+(1-\xi_R)R/6 + m_R^2 + \lambda_R\hat{\phi}^2/6 + \lambda_R\langle\phi^2\rangle/2]\hat{\phi} = 0 \tag{2.5}$$

where m_R, ξ_R and λ_R are the renormalized quantities after divergences in $\langle\phi^2\rangle_o$ are properly removed. The fluctuations obey the equation

$$[\Box+(1-\xi_R)R/6 + m_R^2]\phi + \frac{1}{6}\lambda_R(\phi^3 + 3\phi^2\hat{\phi} + 3\phi\hat{\phi}^2 - 3\langle\phi^2\rangle\hat{\phi}) = 0$$

In the lowest order (1-loop) approximation, only linear terms in ϕ are retained. This yields,

$$[\Box+ (1-\xi_R)R/6 + m_R^2 + \frac{1}{2}\lambda_R \hat{\phi}^2]\phi = 0 \tag{2.6}$$

For a constant $\hat{\phi}$ this has the form of an equation for free scalar fields with effective mass $M^2 = m_R^2 + \frac{1}{2}\lambda_R\hat{\phi}^2$.

Symmetry behavior of this system at zero temperature can be studied by analyzing the stability of $\hat{\phi}$ in (2.5) due to $\langle\phi^2\rangle_o$. Thermal stability of the system is studied by including the finite temperature contributions $\langle\phi^2\rangle_T$, in $\langle\phi^2\rangle$, i.e. $\langle\phi^2\rangle = \langle\phi^2\rangle_o + \langle\phi^2\rangle_T$, where $\langle\ \rangle_T$ denotes the canonical ensemble average of the system in equilibrium at temperature T. In principle, the fundamental quantity of importance is the finite temperature effective action $\Gamma_T(\bar{\phi})$ or, equivalently, the free energy density f, which can be used to derive all statistical properties of the system of interest. For the study of phase transitions, suppose that at zero temperature the system possesses a symmetry such that

$\delta\Gamma_T(\bar\phi)/\delta\bar\phi = 0$ at some $\bar\phi \neq 0$, one wants to determine whether at some finite temperature the symmetry is restored, corresponding to having $\delta\Gamma_T(\bar\phi)/\delta\bar\phi=0$ at $\bar\phi = 0$. For static, homogeneous spacetimes, it sufficies to study $\Gamma_T(\bar\phi)$ for constant $\bar\phi(x) = \hat\phi$, one can then define the effective potential $V_T(\hat\phi) = -(\mathrm{Vol})^{-1}\Gamma_T(\phi)|_{\bar\phi=\hat\phi}$, where (Vol) denotes the spacetime volume. This is what is usually assumed in most calculations for flat or curved but static spacetimes. For dynamic spacetimes, where $\bar\phi(x)$ changes in time, this is no longer true in general. However, if the background spacetime and background field evolve very slowly compared to the change of the fluctuation field, i.e., $\dot a/a \sim \dot{\hat\phi}/\hat\phi << \dot\phi/\phi$, one can invoke the adiabatic method for the calculation of the effective potential. This was carried out for the RW universe in a recent paper (Hu 1982b). In the opposite limit when the background field evolves as rapidly as the fluctuations, one might have to appeal to methods developed for treating problems in dynamical critical phenomena.

3 HIGH FREQUENCY CONTRIBUTIONS AND ADIABATIC APPROXIMATION

For the study of thermal stability, a simpler alternative is to calculate the temperature dependent mass, defined by

$$m^2(T) = m^2 + (1-\xi)R/6 + \frac{\lambda}{2}(\langle\phi^2\rangle_o + \langle\phi^2\rangle_T) \tag{3.1}$$

where $\langle\phi^2\rangle_o$ is the contribution from the (regularized) vacuum fluctuations and $\langle\phi^2\rangle_T$ is from the thermal fluctuations. (Here and after m, ξ, λ are understood to be the renormalized quantities.) The behavior of $\langle\phi^2\rangle_o$ for power-law dependence of a(t) has been studied in some detail by Ford & Toms (1982) and for the exponential expansion in great depth by a number of speakers in this workshop. For this reason I shall only concentrate on the structure of $\langle\phi^2\rangle_T$. Consider the equation (2.6) governing the fluctuations ϕ. Define $\chi(\eta) = a\phi(t)$. The amplitude χ_k of the kth mode obeys the wave equation

$$\chi_k''(\eta) + (\Omega_k^2 - \xi Ra^2/6)\chi_k(\eta) = 0 \tag{3.2}$$

where $\Omega_k^2 = q^2 + M^2a^2$, $k^2=q^2-\kappa$, $M^2 = m^2 + \frac{1}{2}\lambda\hat\phi^2$. Defining $\mathcal{M}^2 = M^2 - \xi R/6$, one can rewrite the natural frequency as $W_k^2 \equiv \Omega_k^2 + Q = q^2 + \mathcal{M}^2a^2$, where $Q \equiv -\xi Ra^2/6$ denotes the curvature contribution and $\mathcal{M}^2$ contains the time-dependent parameters $\hat\phi^2$ and R. The thermal average of $\langle\phi^2\rangle_T$ is given by

$$\langle\phi^2\rangle_T = (\pi a)^{-2} \int d\nu(k)|\chi_k|^2\langle n_k\rangle \tag{3.3}$$

where $\int d\nu = \int q^2dq$ for $\kappa = 0, -1$ and $\Sigma_{q=1}^{\infty} q^2$ for $\kappa = +1$, and $\langle n_k\rangle = (e^{W_k/\Theta} - 1)^{-1}$ is the Bose-Einstein distribution. The temperature paramter Θ = Ta gives a measure of the number of ambient radiation quanta which remain constant in adiabatic expansions. Under the condition that the background spacetime and the background field vary slowly - the high frequency contributions to χ_k can be approximated by the positive-frequency generalized WKB solution,

$$\chi_k \simeq (2\tilde{W}_k)^{-1/2} \exp(-i \int^{\eta} \tilde{W}_k \, d\eta')$$

where $$\tilde{W}_k = W_k[(1 + \underline{\varepsilon}_2)(1 + \underline{\varepsilon}_4)]^{1/2} = \Omega_k(1 + \varepsilon_2 + \varepsilon_4)^{1/2} \tag{3.4}$$

with $$\underline{\varepsilon}_2 = - W^{-3/2} \partial_\eta (W^{-1}\partial_\eta W^{1/2}) \tag{3.5}$$

$$\underline{\varepsilon}_4 = - W^{-1}(1 + \underline{\varepsilon}_2)^{-3/4} \partial_\eta[W^{-1}(1 + \underline{\varepsilon}_2)^{-1/2} \partial_\eta(1 + \underline{\varepsilon}_2)^{1/4}]$$

and $$\varepsilon_2 = \underline{\varepsilon}_2 + Q\Omega^{-2}, \quad \varepsilon_4 = \underline{\varepsilon}_4 + \underline{\varepsilon}_2 \, Q\Omega^{-2}$$

These quantities were introduced and calculated in detail for RW space-times in Fulling *et al* (1974). To the fourth adiabatic order, $|\chi_k|^2$ is calculated to be (henceforth we shall drop all subscripts k):

$$|\chi|^2 \simeq (2\tilde{W})^{-1} \simeq (2\Omega)^{-1}\{1 - [\tfrac{1}{2} \varepsilon_{2(2)}] + \tfrac{3}{8} \varepsilon^2_{2(2)} - \tfrac{1}{2} \varepsilon_{2(4)} - \tfrac{1}{2} \varepsilon_{4(4)}\} \tag{3.6}$$

The number in parenthesis after ε_n indicates its adiabatic order (denoted by T^{-n}, where T is a large time scale characteristic of the change in the background). The change in Ω is related to the change in the scale factor and the background field by ($X \equiv \Omega^2$)

$$X'/X = 2C^2E, \quad X''/X = 2C^2(E'+2E^2), X'''/X = 2C^2(E''+6EE'+4E^3), \text{ etc.} \tag{3.7}$$

When these are substituted back to (3.5) for ε_n, (3.6) becomes

$$|\chi|^2 \simeq (2\Omega)^{-1}\{1 - \frac{1}{2\Omega^2}[- \frac{C^2}{2} (E'+2E^2) + \frac{5}{4} C^4E^2+Q] + \frac{1}{8\Omega^4} (Q''+3Q^2) + O(\Omega^{-6})\} \tag{3.8}$$

Here, $D \equiv a'/a$, $E = a'/a + M'/M$, $C \equiv Ma/\Omega$, $M'/M = \hat{\phi}'/\hat{\phi}/(1+2\, m^2/\lambda\hat{\Phi}^2)$. The terms are arranged in increasing adiabatic order (which can be measured by counting the order of time derivatives). The higher order terms encompass contributions from an increasingly rapid-changing background.

Let us take a closer look at (3.8). Physically, terms in (3.8) fall into two groups: those containing powers of Q are recognized to come from the binomial expansion of $\Omega/W \simeq 1 - \frac{Q}{2\Omega^2} + \frac{3}{8}\frac{Q^2}{\Omega^4} - \ldots$ (this corresponds to setting $|\chi|^2 \simeq (2W)^{-1}$). The contribution of these terms to $\langle\phi^2\rangle_T$ arises from the adiabatic variation in the spacetime (i.e. when the dynamics is constituted simply as a succession of static spacetimes). Terms of this type give ,

$$<\phi^2>_T^{AD} = \frac{1}{2\pi^2 a^2}\int \frac{q^2 dq}{W}\frac{1}{e^{W/\Theta}-1} = \frac{T^2}{2\pi^2}\int_0^\infty \frac{x^2 dx}{(x^2+\mu^2)^{1/2}}\frac{1}{e^{(x^2+\mu^2)^{1/2}}-1} \tag{3.9}$$

where we have introduced the notation , $x = q/\Theta$, $\mu = M/T$, $u = M/T$, $v = \sqrt{\xi R/6}/T$, whence $\mu^2 = u^2 - v^2$. Notice that the integral has the same form as that in flat space (*cf.* Dolan & Jackiw 1974). When $M \ll T$, which corresponds to the case of relativistic particles ($m/T \ll 1$) and small curvature or near-conformal coupling ($\xi R/6T^2 \ll 1$), one can carry out the high temperature expansion in a similar manner as in flat space and get

$$<\phi^2>_T^{AD} \simeq \frac{T^2}{12} - \frac{MT}{4\pi} + M^2 \ln\frac{M^2}{T^2} + \ldots \tag{3.10}$$

This has the same form as for a static (Einstein) spacetime (where $R/6 = \kappa/a^2$, $M^2 = M^2 - \xi\kappa$) or the flat spacetime (where $R = 0, M = M$). Like in flat space, the second term arises from the zero-mode contributions, and the $M^2 \ln M^2$ term is cancelled by the zero-temperature contribution. In spontaneous symmetry breaking, when m^2 assumes a negative value, this term becomes imaginary. However, it is to be suppressed when the higher-loop contributions are taken into consideration, according to an argument similar to that presented for the flat space case. The case of massless minimally coupled fields needs a separate treatment, as there is a discontinuity in the first derivative of the effective potential at the minimum near $\hat{\phi} = 0$. The infrared problem we encounter here is the same as that in a static Einstein universe which is disucssed in detail in a recent work of O'Connor *et al* (1982).

The remaining terms in $<\phi^2>_T$ containing C and Q' and their derivatives are contributions arising from the non-adiabatic change of the background spacetime and the induced change in the background field. Since the non-adiabaticity of the system is related to the non-conformal nature of the fields (Hu 1982a) (the spacetime being conformally flat here), parameters which break conformal invariance like m and ξ appear in these terms. Denoting them collectively by $<\phi^2>_T^{NA}$, we have

$$<\phi^2>_T^{NA} = \frac{1}{2\pi^2 a^2}\int \frac{q^2 dq}{\Omega}(J_4\,\Omega^{-4} + J_6\,\Omega^{-6} + \ldots)<n(q)> \tag{3.11}$$

where
$$J_4 = \frac{1}{4}M^2 a^2 (E' + 2E^2) + \frac{1}{8}Q''$$
$$J_6 = -\frac{5}{8}M^4 a^4 E^2 - \frac{M^2 a^2}{16}(E''' + 8E''E + 6E'^2 + 24E'E^2 + 8E^4).$$

To facilitate high temperature expansion, we have arranged terms in increasing orders of inverse frequency, and in so doing mixed terms of different adiabatic orders. Carrying out the thermal averaging we get

$$<\phi^2>_T^{NA} = \frac{J_4 I_4}{2\pi^2 a^2 T^2} + \frac{J_6 I_6}{2\pi^2 a^6 T^4} + \ldots \quad , \tag{3.12}$$

where $I_n = L_n(\mu=0)$ and $L_{n-1}(\mu) = \int \frac{x^2 dx}{(x^2+\mu^2)^{n/2}} \frac{1}{\exp[(x^2+\mu^2)^{1/2}]-1}$ are the Bose-Einstein type integrals. The leading contribution has a T^{-2} dependence, the next one T^{-4}, etc. It is seen that non-adiabatic contributions to $<\phi^2>_T$ become more pronounced at lower temperatures. The relative importance of these terms depends on J_n which in turn depends on the expansion law a(t) and $\hat{\phi}(t)$, the solution to the lowest order classical equation. Even though these terms are smaller at higher temperatures, they are in some sense more distinguished, as they are the ones which encompass the full dynamics of the system. Without them the result would be incorrect as that would amount to treating the dynamics as a sequence of instantaneously static spacetimes.

4 LOW FREQUENCY CONTRIBUTIONS AND THE SUDDEN APPROXIMATION

The WKB expansion used above for the wavefunction is in general good for treating high frequency modes ($\omega > t^{-1}$ at time t) or for gradual expansions. If the background expands rapidly, like in the inflationary behavior $a \sim e^{Ht}$ of de Sitter spacetimes, then the otherwise high-frequency modes would all be quickly reduced to low frequency by the large red-shift. Hence at late times the contribution of long wavelength modes become very pronounced and could play an important role in the symmetry behavior of the system. To take into account these low frequency modes, we can use the so-called "sudden approximation". This method was used by Hu & Parker (1978) and earlier by Zel'dovich & Starobinsky (1971) for the treatment of long wavelength modes in the quantum domain for cosmological particle production. As a formal solution to (3.2), for each mode k, χ and χ' can be replaced by two complex-valued functions $\alpha(\eta)$ and $\beta(\eta)$ defined by

$$\chi = (2\Omega)^{-1/2} (\alpha e_+ + \beta e_-)$$
$$\chi' = -i(\Omega/2)^{1/2} (\alpha e_+ - \beta e_-) \tag{4.1}$$

where $e_\pm \equiv \exp(\mp i \int_{\eta_0}^{\eta} \Omega \, d\eta')$, where η_0 is the onset time when the system undergoes rapid expansion. The conserved Wronskian condition on χ (which guarantees the validity of the canonical commutation relations in time) gives $|\alpha|^2 - |\beta|^2 = 1$. Eq. (4.1) can be reduced to two linear equations governing α and β :

$$\alpha' = \frac{1}{2}\left(\frac{\Omega'}{\Omega} - i\frac{Q}{\Omega}\right)\beta\, e_+^2 - \frac{iQ}{2\Omega}\alpha \qquad (4.2)$$

$$\beta' = \frac{1}{2}\left(\frac{\Omega'}{\Omega} + i\frac{Q}{\Omega}\right)\alpha\, e_-^2 + i\frac{Q}{2\Omega}\beta$$

For low frequency modes ω(or in short durations Δt such that $\omega\,\Delta t \ll 1$) the exponential $\exp(2i\int_{\eta_o}^{\eta}\Omega d\eta') \sim 1$; one finds that the general solution to (4.2) is

$$\alpha = c_1\left(\Omega^{1/2} - i\,\Omega^{-1/2}\int_{\eta_o}^{\eta} Qd\eta'\right) + c_2\,\Omega^{-1/2}$$

$$\beta = c_1\left(\Omega^{1/2} + i\Omega^{-1/2}\int_{\eta_o}^{\eta} Qd\eta'\right) - c_2\Omega^{-1/2} \qquad (4.3)$$

where $c_1(k)$ and $c_2(k)$ are complex numbers which by virtue of the Wronskian condition satisfies $c_1 c_2^* + c_2 c_1^* = \frac{1}{2}$. The choice of c_i is dictated by the nature of the state vector at the initial moment (at $\eta=\eta_o$). Assuming that particle production from the vacuum is absent at this time (which is approximately valid for post-Planckian era and certainly valid for processes in the GUT epoch) so that the solution (4.1) matches with the flat space solution at η_o, one can impose the initial condition $\alpha(\eta_o) = 1$, $\beta(\eta_o) = 0$ which gives $c_1 = \frac{1}{2}\Omega_o^{-1/2}$ and $c_2 = \frac{1}{2}\Omega_o^{1/2}$, where $\Omega_o(\sim\eta_o^{-1})$ is the value of Ω at time t_o. (The choice of vacuum state for de Sitter space is discussed by Vilenkin & Ford (1982)). Under these conditions, $|\chi|^2 = 2|c_1|^2 = \Omega_o^{-1}$. This gives a low frequency contribution to $\langle\phi^2\rangle_T$ equal to

$$\langle\phi^2\rangle_T^{LF} \simeq \frac{1}{2\pi^2 a^2}\int \frac{q^2 dq}{\Omega_o}\,\frac{1}{e^{W/\Theta}-1} \qquad (4.4)$$

At sufficiently high temperature (for de Sitter space $T_o \gg T_H$, the Hawking temperature), the Bose-Einstein factor can be approximated by $\Theta/W = Ta/(\Omega^2-\xi Ra^2/6)^{1/2}$. The integration limit on q should be from the lowest defined by the curvature $\sqrt{\xi R}a$ to the highest defined by the limit of validity of the sudden approximation $q_m \sim a/t$. For conformal fields (ξ=0), the integration yields approximately, (See Appendix of Hu & Parker 1978).

$$\langle\phi^2\rangle_T^{LF}(\xi=0) \simeq \frac{\Theta}{4\pi^2}\,\frac{q_m^2}{a^2} \simeq \frac{\Theta}{4\pi^2}\left(\frac{1}{t^2}\right) \qquad (4.5)$$

In near-adiabatic expansion, $\Theta \sim$ constant, the t^{-2} dependence reduces the low frequency contribution to zero at late times, as expected. For non-conformal (massless) fields, the dominant contribution to the integral comes from $q_c \simeq (\xi Ra^2/6)^{1/2}$, which yields approximately,

$$\langle\phi^2\rangle_T^{LF}(\xi\neq 0) \simeq \frac{\Theta}{4\pi^2\Omega_o}\left(\frac{q_c}{a}\right)^2 \simeq \frac{\Theta}{4\pi^2\Omega_o}\left(\frac{\xi R}{6}\right) . \tag{4.6}$$

For power-law expansions $a \sim t^n$, $R \sim t^{-2}$, the low frequency contribution also reduces to zero at late times with the same t^{-2} dependence, However, for exponential expansions $a \sim e^{Ht}$, $R = 12H^2$ = constant, this term remains to give a non-zero contribution even at late times. Assuming that the transition to inflationary epoch occurs at a relatively late time (with temperature T_o) so that $\Omega_o \sim \Omega_c$, then (4.6) reduces to

$$\langle\phi^2\rangle_T^{LF}(\xi\neq 0\ ,\ \text{de Sitter}) \simeq \frac{\sqrt{2\xi}}{4\pi^2}\ T_o H \tag{4.7}$$

for massless, non-conformal fields. This result shows that for non-conformal fields the effect of curvature in de Sitter universe can exert a strong influence on the behavior of $\langle\phi^2\rangle_T$.

5 SUMMARY

In summary, three types of terms associated with different physical sources contribute to the finite temperature quantum correction of the effective mass (3.1), i.e.,

$$\langle\phi^2\rangle_T = \langle\phi^2\rangle_T^{AD} + \langle\phi^2\rangle_T^{NA} + \langle\phi^2\rangle_T^{LF}$$

The first term $\langle\phi\rangle_T^{AD} \sim T^2$ arising from the adiabatic change of the background is most important at high temperatures. The second term $\langle\phi^2\rangle_T^{NA} \sim T^{-2}$, T^{-4} and T^{-2n} arise from the nonadiabatic change of the background. The third term $\langle\phi^2\rangle_T^{LF} \sim \xi R$ arises from the accumulation of red-shifted long-wavelength modes. It is important at late times for non-conformal fields in exponential expansions.

In this talk I have outlined an approach for treating symmetry breaking problems at finite temperature in curved spacetimes, and applied it to the Robertson-Walker universe as an example. The details for specific functional dependence of the scale factor - say, power law behavior for the Friedmann solutions or the exponential behavior for the de Sitter solutions - can be worked out by explicit substitution into the given results. Once the quantum corrections are calculated, one can follow the standard procedure in deducing the critical temperature and related quantities of interest. The conditions of validity for the various approximations used should also be examined more closely in specific cases (a case in question is the equilibrium condition for low frequency modes). The method outlined here can be applied to the calculation of the finite temperature effective action, from which one can also infer the symmetry behavior of the systems concerned. Calculations using these methods for more general spacetimes of cosmological interest are also in progress.

Acknowledgements

I would like to thank Professors Stephen Hawking, Gary Gibbons and Stephen Siklos for their unfailing effort in providing a stimulating atmosphere throughout the workshop and for their warm hospitality during my stay at Cambridge. I have benefited from conversations with Professors Larry Ford and Alex Vilenkin on this problem during the workshop. This work was supported in part by the National Science Foundation under Grant PHY81-07387.

References

Abbott, L. F. (1981) Nucl. Phys. B185, 233.
Albrecht, A. & Steinhart, P. J. (1982). Phys. Rev. Lett. 48, 1220.
Allen, B. (1982). Phase Transitions in de Sitter Space. Univ. of Cambridge DAMTP preprint.
Coleman, S. & De Luccia, F. (1980). Phys. Rev. D21, 3305.
Denardo, G. & Spallucci, E. (1981a). Nuovo Cimento 64A, 27.
Denardo, G. & Spallucci, E. (1981b). Nuovo Cimento 64A, 15.
Dolan, L. & Jackiw, R. (1974). Phys. Rev. D9, 3357.
Dowker, J. S. & Critchley, R. (1977). Phys. Rev. D15, 1484.
Dowker, J. S. & Kennedy, G. (1978). J. Phys. A11, 895.
Drummond, I. T. (1981). Nucl. Phys. B190 [FS3] 93.
Ford, L. H. (1980). Phys. Rev. D22, 3003.
Ford, L. H. & Toms, D. J. (1982). Phys. Rev. D25, 1510
Fujii, Y. (1981). Phys. Lett. 103B, 29.
Fulling, S. A., Parker, L. & Hu, B. L. (1974). Phys. Rev. D10, 3905.
Guth, A. H. (1981). Phys. Rev. D23, 347.
Hawking, S. W. & Moss, I. G. (1982). Phys. Lett. 110B, 35.
Hu, B. L. (1982a). Phys. Lett. 108B, 19.
Hu, B. L. (1982b). Finite Temperature Effective Action for $\lambda\phi^4$ Theory in Robertson-Walker Universes. Univ. of Maryland preprint.
Hu, B. L. & Parker, L. (1978). Phys. Rev. D17, 933.
Hut, P. & Klinkhamer, F. R. (1981). Phys. Lett. 104B, 439.
Kennedy, G. (1981). Phys. Rev. D23, 2884.
Linde, A. D. (1982a) Phys. Lett. 108B, 389.
Linde, A. D. (1982b). Scalar Field Fluctuations in Expanding Universe and the New Inflationary Universe Scenario. Phys. Lett. 116B, 335.
Moss, I. G. (1982). Talk presented at the Nuffield Workshop on the Very Early Universe.
O'Connor, D. J., Hu, B. L. & Shen, T. C. (1982). Symmetry Restoration in Curved Spacetime I. $\lambda\phi^4$ Theory in Einstein Universe with Arbitrary Coupling. Univ. of Maryland preprint.
Shore, G. M. (1980). Ann. Phys. (N.Y.) 128, 376.
Starobinsky, A. A. (1982). Talk presented at the Nuffield Workshop on the Very Early Universe.
Toms, D. J. (1980). Phys. Rev. D21, 2805.
Vilenkin, A. & Ford, L. H. (1982). Phys. Rev. D 26, 1231.
Zel'dovich, Ya. B. & Starobinsky, A. A. (1971). Zh. Eksp. Teor. Fiz. 61, 2161. [Sov. Phys. - JETP 34, 1159 (1972)].

SOME REMARKS ON THE PROBLEM OF VERY EARLY UNIVERSE

M.A. Markov
Institute for Nuclear Research, Academy of Sciences of the USSR

Abstract. Specific features of a perpetually oscillating universe are considered. In a perpetually oscillating universe there is no problem of origination of the world, which in other versions of the universe may appear to be unsolvable.
In the first section, the solution of the singularity problem is associated with the possible existence of asymptotic freedom for gravitational interactions. This is formulated in such a way that at small times $t < t_o$ the gravitational constant κ or the product of κ and the mass of the universe M is time dependent:

$$\kappa M = \kappa_o M_o \left(\frac{t}{t_o}\right)^n .$$

If, for example,

$$R = (9/2\ \kappa_o M_o)^{1/3} \left(\frac{t}{t_o}\right)^{n/3} t^{2/3} ,$$

then for $n > 1$ the acceleration $\ddot{R}$ in the universe expansion is positive, which testifies to a state with violated energy dominance. At small $t < t_o$ the ratio $ct/R(t)$ for $n > 2$ tends to infinity, not to zero, so the horizon problem is avoided. It is shown that quantum corrections of the type $R_{\mu\nu}\ \ell n(\Box R^{\mu\nu})$ imply asymptotic freedom in the form

$$R \sim (t/t_o)^{1/2} t^{1/2} .$$

The possibility of the existence of the limiting matter density $\rho_q = c^5/\hbar\kappa^2 \sim 10^{94}$ gm cm^{-3} is discussed. The condition $\rho \leq \rho_q$ leads to a new equation of the type:

$$R^\nu_\mu - \frac{1}{2} R\delta^\nu_\mu = \frac{8\pi}{c^5} \kappa T^\nu_\mu \left(1 - \left(\frac{T}{\rho_q}\right)^2\right)^n - \Lambda\left(\frac{T}{\rho_q}\right)^n \delta^\nu_\mu$$

which describes the de Sitter world when $T \to \rho_q$ and the classical Einstein equations if $\hbar \to 0$.
The de Sitter stages in the universe evolution must play the role of 'purgatory': in these stages the entropy must always have the same value.
In the second section, the final state of a black hole is assumed to be a stable particle (maximon) of Planck mass and dimensions. In attempts to conform the stability of maximons with astrophysical data attention is paid to the effect of an intensive annihilation (transformation into radiation) of stable particles of large mass in the formation and in the

decay of black holes and 'blinis' (pancakes) in a two-component medium consisting of radiation (relativistic matter) and maximons.

0 INTRODUCTION

The present paper, concerning the problems of the very early universe, is aimed not at listing what is known about the physics of extreme matter states typical of the 'very early universe', but at investigating the 'reserves of arbitrariness', which can be utilized for building different theories of the very initial stage of the history of the universe.

We should recall that the matter density in our universe at the present time is such that it is not excluded that its metric is closed. Thus, its collapse in future is not excluded. Let us call a theory 'complete' or 'perfect' if in its framework a consistent description can be given to the origin of the universe, to the period corresponding to the classical singularity, to the subsequent history of the universe and its possible collapse.

The most complicated version, which could help verify the efficiency of such a 'perfect theory', is a perpetually oscillating universe. This is much richer in its difficulties than, for example, an open universe: the problem of entropy increase both in the course of expansion and contraction (Tolman 1934), the description of the collapse etc. But it should be especially emphasized that this version does not deal with the origin of the universe, which may appear to be the most difficult problem concerning the universe.

1 ASYMPTOTIC FREEDOM OF GRAVITATIONAL INTERACTIONS AND THE PROBLEM OF THE UNIVERSE SINGULARITY

As is known, the viewpoint was expressed long ago that the classical description of the universe is inapplicable near the classical singularity. General hopes were cherished for the replacement in this region of the classical description by a quantum description. But a consistent quantum theory of gravitational fields has not yet been created. The promising ideas of grand unification of all interactions have not been concretely realized either.

In recent years some models have been considered in which the gravitational field is left classical, but the state of matter has violated energy dominance

$$p = -\varepsilon \tag{1}$$

in regions close to the singularity under conditions of super-high matter density.

In other words, if the principle of energy dominance ($p < \varepsilon$) was regarded earlier as a firm principle leading inevitably to the appearance of a singularity in the history of the universe, now some hopes appeared that under conditions of superdense matter, more precisely, near the classical singularity, there may appear a state with violated energy dominance. As is known, energy dominance violation is typical of all de Sitter type or steady-state universes (Hawking & Ellis 1973). To say it differently, theories are being sought at present in which at initial moments the Friedman universe turns out to be de Sitter or steady-state.

The de Sitter version is developed in a number of papers taking into account the one-loop approximation in the energy-momentum tensor (Mamayev & Mostepanenko 1980; Starobinski 1979, 1980), i.e. the corrections leading to vacuum polarization and to particle production by a strong gravitational field. The vacuum-like state of matter seems also to appear in a natural way in phase transitions of matter at superhigh temperatures in the early universe (Linde 1981, 1982).

In macroscopic physics a state with violated energy dominance has analogs in the states of superheated liquid or supercooled vapour. The high instability of such states is a very favourable property since there must evidently be a rapid transition from the vacuum-like state with $p = -\varepsilon$ into the ordinary state $p < \varepsilon$ typical of our universe.

In the above-mentioned attempts to consider the effects of polarization and particle production in a strong gravitational field, the latter is not quantized. The applicability of the theory is limited to the following condition

$$R_{\mu\nu\lambda\delta} R^{\mu\nu\lambda\delta} < \frac{1}{\ell_{p\ell}^4} \qquad (2)$$

where $\ell_{p\ell}$ is the Planck length.

If, for the solution of fundamental problems of cosmology and field theory, one can really limit oneself to the condition (2), such a situation would be a most valuable present to theorists, a favour from nature.

In (Markov 1981a) an attempt is made to interpret the appearance of the vacuum-like state as asymptotic freedom of all interactions, including gravitational ones, in the case when particles of matter appear to be at some super-small distances from one another.

Formally, we are dealing with the transformation of the right-hand side of the Einstein's equation $\kappa T_{\mu\nu}$ into the Λ-term of de Sitter's universe, when in the process of collapse the universe approaches the classical singularity

$$\kappa T_{\mu\nu} \to \Lambda g_{\mu\nu} \quad \text{as} \quad R \to 0 . \tag{3}$$

It was suggested (Markov 1981b) that this asymptotic freedom effect appears when account in $T_{\mu\nu}$ is taken of quantum corrections for matter fields (vacuum polarization, particle production) in the one-loop approximation. It seems that this consideration can be illustrated even in classical language.

Indeed, the well-known classical model of an expanding universe with $p = 0$ has an expansion law of the form

$$R = (9/2\ M\kappa)^{1/3} t^{2/3} , \tag{4}$$

where M is the total mass of the universe.

Formally, asymptotic freedom could arise either from a decrease of the gravitational constant at small distances, i.e. at small t ; for example,

$$\kappa(t) = \kappa_o (\frac{t}{t_o})^n , \tag{5}$$

or from time dependence of the mass M

$$M = M_o (\frac{t}{t_o})^n , \tag{6}$$

which can naturally be interpreted as the creation of matter. Or else, the factor $(\frac{t}{t_o})^n$ could be referred to the combination

$$\kappa M = \kappa_o M_o (\frac{t}{t_o})^n . \tag{7}$$

In other words, at small times $0 < t < t_o$, where t_o is, for instance, the Planck time

$$t_o = t_{p\ell} = (\frac{\hbar\kappa}{c^5})^{1/2} \sim 10^{-43} \text{ sec}$$

the law (4) takes the form

$$R = (9/2\ M_o\kappa_o)^{1/3}\left(\frac{t}{t_{p\ell}}\right)^{n/3} t^{2/3} \tag{8}$$

and for $t > t_{p\ell}$ it goes back to its original form (4).

It is noteworthy that in the case (4) the acceleration is negative:

$$\ddot{R} = -2/9(9/2\ M_o\kappa_o)^{1/3} t^{-4/3} \tag{9}$$

but in the case of the law (8), if $n > 1$, the acceleration is positive. If pressure is also taken into account the acceleration is

$$\ddot{R} = -\frac{4\pi\kappa}{3c^2} R(\varepsilon + 3p)\ . \tag{10}$$

For radiation $(p = \frac{1}{3}\varepsilon)$, R is proportional to $t^{1/2}$:

$$R = \left(\frac{32\pi}{3c^2}\bar{k}\kappa\right)^{1/4} t^{1/2}\ , \tag{11}$$

where $\bar{k} = \varepsilon R^4$. If now the time dependence in (11) is altered:

$$t^{1/2} \rightarrow \left(\frac{t}{t_o}\right)^{n/4} t^{1/2}$$

we get positive acceleration for $n > 2$. Eq (10) shows that this implies a new equation of state with $\varepsilon + 3p < 0$. Thus we see that asymptotic freedom of the form (7) actually causes energy dominance violation.

In their preprint 'Quantum Gravity Removes Classical Singularities' Frolov and Vilkoviski (1979) consider the problem of collapse in a one-loop approximation proceeding from a Lagrangian which takes into account terms of the type $R_{\mu\nu}\ \ell n(\Box R^{\mu\nu})$. They show that the singularity of a collapsing object vanishes: there appears a 'potential' of the type

$$\phi = \frac{A}{r}\left(1 - \exp(-r/\ell_{p\ell})\right)\ .$$

For small r,

$$\phi \sim A/\ell_{p\ell}\ . \tag{12}$$

If we assume without any contemplations that the potential has the form (12) when $0 < t < t_p$, then from equality of the kinetic and the potential energies $\frac{1}{2}\dot{r}^2 = \frac{\kappa M}{\ell_{p\ell}}$, we would obtain

$$r \sim (2\kappa M)^{1/2}\left(\frac{t}{\ell_{p\ell}}\right)^{1/2} t^{1/2}$$

instead of the expansion law (11).

It is natural that an expansion law of the type (8) could be obtained as a result of the appearance of a corresponding 'static potential'. We have used the words 'static potential' in inverted commas since, in fact, we are dealing with a dynamic picture, involving equations with higher derivatives. In this sense the power n in the formulation of asymptotic freedom probably arises specifically from the higher derivatives in the quantum corrections. (Static potentials of this form in the framework of a second-order equation were considered in detail already in that remote time when there were attempts to solve the problem of divergences in electrodynamics by introducing ad hoc formfactors. Unfortunately, the author spent much time searching for such possibilities until he came to the conclusion that the formfactor providing finiteness of diverging quantities cannot be static, but must in some sense be dynamically deformable, otherwise there appears contradiction with Special Relativity, with finiteness of causal signal propagation (Markov 1956). Already in those remote times the hopes of solving the problem of divergences were associated with the specific nature of General Relativity (Markov 1947).)

1.1 The horizon problem

It is known that there is a difficulty in understanding how one should explain, within the universe expansion laws discussed at the present time $(R \sim t^{2/3},\ p = 0$ or $R \sim t^{1/2},\ p = \frac{1}{3}\varepsilon)$, the homogeneity of the universe in the absence of causality among different regions in the whole course of the universe evolution (Rindler 1956; Guth 1981). Formally this difficulty is expressed by the fact that the ratio of the length of the causal signal $\ell = ct$ to the scale factor R tends to zero, not to infinity, as is required by causality. The ratio

$$\frac{\ell}{R} \sim \frac{ct}{at^n} \to \sim t^{1-n}$$

tends to zero if $n < 1$. If there exists an asymptotic freedom of the form (7) with $n > 1$, the ratio ℓ/R tends to infinity.

Frolov and Vilkoviski (1969) analyse the limiting case $n = 1$, $\frac{\ell}{R} \to \text{const}$ but it is possible that here we also deal with the case of $n > 1$ if we bear in mind the approximations made.

The duration of causality depends both on the value of the power in (7) and on the value of t_o. From some viewpoints it is even reasonable not to associate t_o with $t_{p\ell}$ since times earlier than $t_{p\ell}$ may turn out to have no physical meaning at all.

In a closed universe another length may also have a physical meaning. This is

$$\ell_o = (\ell_{p\ell} R_o)^{1/2} ,$$

where R_o is the radius of the universe at the moment of its maximal expansion. This length is associated with the time

$$\tau_o = \frac{\ell_o}{c} .$$

For our universe $(R \sim 10^{28}\text{ cm})$:

$$\ell_o \sim 10^{-3}\text{ cm} .$$

ℓ_o is just the dimension acquired by our universe when it expands by the law (11) for the Planck time $t_{p\ell} \sim 10^{-43}$.

1.2 Matter density ρ in the time interval $0 < t < t_{p\ell}$

According to the formula for the gravitating matter density at early times, $\rho \sim \frac{1}{\kappa t^2}$, which is independent of the equation of state, the density ρ at the Planck time $t_{p\ell} \sim (\frac{\hbar\kappa}{c^5})^{1/2}$ takes the so-called quantum value ρ_q :

$$\rho_q \sim \frac{c^5}{\hbar\kappa^2} \sim 10^{94}\text{ gr cm}^{-3}$$

Of course, variation of the density at $t < t_{p\ell}$ is also one of the uncertainty factors. Among various suggestions, the one that is distinguished is that it should be constant for $t < t_{p\ell}$ and equal to ρ_q. If matter creation takes place in this case, the universe must be steady-state-like for $0 < t < t_{p\ell}$. As is known, such a universe, the same as the de Sitter universe, is characterized by a state with violation of the energy dominance condition (Hawking & Ellis 1973). The assumption that matter density cannot be more than ρ_q is attractive in the sense that it is an essential limitation, the appearance of which in our physics testifies to great progress in our understanding of nature. Suffice it to recall the limitations imposed by the knowledge that signal velocities cannot exceed the velocity of light and that the quantum of action cannot be less than $\hbar$. Of course, one cannot exclude a priori the possibility of $\rho > \rho_q$.

But the existence of the limiting density can play an important role in preventing unlimited increase of density perturbations at the end of a contracting universe. The dependence of the gravitational constant on density ρ can be formulated explicitly by rewriting the Einstein equations in the form

$$R_\mu^\nu - \frac{1}{2} R\delta_\mu^\nu = \frac{8\pi\kappa}{c^5} T_\mu^\nu (1 - (\frac{\rho}{\rho_q})^2)^n - \Lambda(\frac{\rho}{\rho_q})^n \delta_\mu^\nu .$$

If $\rho = \rho_q$, we have the de Sitter world: $R_\mu^\nu - \frac{1}{2} R\delta_\mu^\nu - \Lambda\delta_\mu = 0$. If $\hbar \to 0$ we have Einstein's equations. Instead of ρ one can put R/c^2 or T/c^2, or even $R_{\mu\nu\sigma\lambda}R^{\mu\nu\sigma\lambda}/c^2$. Using the property of G_μ^ν : $G_{\mu;\nu}^\nu = 0$ we can get a condition for $T_{\mu;\nu}^\nu$ which describes the disappearance of matter during the process of collapse. This version of the theory will be considered elsewhere.

It follows from the formulation of the problem that we are considering a closed model of the universe. This has in out opinion a great advantage over the open universe: in a closed universe the total energy is equal to zero, and both the production and the disappearance of matter particles are admissible without violation of the energy conservation law as a whole. If the phenomenon of asymptotic freedom in the formulation (7) can be considered as associated with the process of matter creation at the initial moment of the universe it seems that an inverse process, a collapse, must be accompanied by the 'disappearance' of matter, provided we want to have an exactly repeated version of a perpetually oscillating universe. If such a possibility is excluded, if after collapse the initial state cannot be repeated exactly, the same problem of the origin of the initial state, which we face in the open universe version, remains unsolved. The requirement of asymptotic freedom in the form (7) admits an interpretation which implies formally a decrease of the gravitational constant. It is only the non-zero gravitational constant that enables physical manifestation of gravitating matter, provided that, as in this case, no other forces are present. It can be said in this case (collapse, ($\kappa(t)$) that the 'weight' of the matter becomes increasingly smaller. One can say, of course, that matter transforms into another 'less weighty' state. But the question arises how far this 'decrease' of matter may go. If gravitational properties of matter vanish completely, we again face the problem of the origin of the universe. If we assume that the mass density cannot exceed the quantum value ρ_q, our universe would reach this value at the radius $R \geq 10^{-13}$ cm.

The universe could stop contracting, and from this radius $R \sim 10^{-13}$ cm a new phase of expansion would begin. But the final radius depends on the amount of matter created both in the process of expansion and contraction. This amount would change (and entropy too) in any new cycle of expansion and contraction, and we would again face an unsolved problem of the universe origin. But at a further decrease of the universe radius the mass density can remain constant only if $\kappa(t)$ decreases correspondingly at smaller distances.

In order not to contradict the assumption that ρ_q remains constant in the course of further contraction, we must put forward a certain new hypothesis about the minimum radius which can be reached in the collapse. It is essential in this case that this radius must have the same numerical value in each expansion and contraction cycle. We can point out one such possibility, when the discussed character of collapse is realized automatically. We mean a model of the universe which is almost closed, with a stable external mass. In the outer space, this world has all the properties of a black hole. If this is the question of the stability of a black hole, this may be the stability of its lower state only, the stability of a black hole with the Planck mass, i.e. the maximon stability. (A maximon is an elementary black hole with mass $(\hbar c/\kappa)^{1/2}$.) The mass conservation law would regulate the final stage of universe contraction. However, there exist very weighty arguments against the stability of the final state of a black hole (Khlopov & Polnarev 1981; Hawking 1976).

The main objection to the stability of maximons is a general objection to the production in the early universe of stable particles of large masses, specifically, massive monopoles, maximons, etc. The density of massive particles falls with the expansion of the universe as $\sim 1/R^3$, and the radiation density falls as $\sim 1/R^4$. So when particles of large masses are created in the early universe, the state of matter with $p = 0$ appears rapidly with all the ensuing astrophysical consequences, in particular, for the production of He^4 etc. (Khlopov & Polnarev 1981).

The stable maximon hypothesis could be brought into agreement with the available astrophysical facts if some process of stable maximon vanishing could be found. Markov (1978,1981c) discusses such a process. Indeed, two elementary holes (maximons) can form one, unstable, black hole of large mass. This black hole, radiating in the usual matter, transforms into one maximon. So, a reservoir filled with an arbitrarily large number of stable maximons may in principle turn out in some time to be in a state, when only a single maximon remains in the reservoir (Hawking 1976). But if there exists a large probability of maximon formation from the products of their decay, i.e. from ordinary matter, then the situation in cosmology becomes again more complicated.

It should be noted that in the cosmological problem it is required, on the one hand, that the number of primary maximons should be decreased by many orders of magnitude, and on the other hand, that some factors exist that would suppress the probability of new maximon production from the decay products of primary maximons.

2 IS THE FINAL STATE OF THE BLACK HOLE STABLE OR UNSTABLE?

2.1 Creation of miniholes and the conservation of entropy

The entropy of a black hole of mass M_{BH} is expressed as

$$S_{BH} = 4\pi\left(\frac{M_{BH}}{m_{p\ell}}\right)^2 k \sim 10^{-5} M_{BH}^2 ,$$

where $k = 1.3\times10^{-16}$ erg deg^{-1} is the Boltzmann constant. If black holes are formed in a medium with $p = \frac{1}{3}\varepsilon$ at temperature T, the entropy of a volume V of such a medium is

$$S_T = \frac{16\sigma}{3c} VT^3 ,$$

where $\sigma = 5.67\times10^{-5}$ gr sec^{-3} deg^{-4} is the Stefan-Boltzmann constant.

Both the mass density ρ and the temperature T are determined by the cosmic time

$$\rho = \frac{4.5\times10^5}{t^2} = \frac{3}{32\pi\kappa t^2}$$

$$T(°K) = 1.5\times10^{10}\ \alpha^{-1/4}t^{-1/2}\ ,$$

where α is a dimensionless coefficient larger than unity, that takes into account that different sorts of particles are in equilibrium. For simplicity we assume $\alpha = 1$.

If a black hole is formed of a matter volume $V = M_{BH}/\rho$ then the expression for the volume entropy can be written in the form

$$S_T = \frac{16\pi}{3c}\frac{M_{BH}}{\rho}T^3$$

or, expressing S_T in terms of t :

$$S_t \sim 10^{11}t^{1/2}M_{BH}\ .$$

The equality $S_{BH} = S_t$ holds provided that

$$M_{BH} \approx 10^{16}\ t^{1/2}\ .$$

Thus, at $t \sim 10^{-24}$ sec the entropies S_t and S_{BH} are equal only for $M_{BH} \sim 10^5$ gr . In this period it is only for masses $M_{BH} > 10^4$ gr that black holes are formed with an increase of entropy.

So, the entropy of a black hole of mass 10 gr is given by

$$S_{BH} \sim 10^{-5}\ \text{erg deg}^{-1}$$

while the entropy of a corresponding mass of relativistic medium at the moment $t \sim 10^{-30}$ sec is equal to

$$S_t \sim 10^{-3}\ \text{erg deg}^{-1}$$

i.e. larger by two orders of magnitude.

One can believe that in these cases black holes and maximons are not formed at all (Sciama 1976).

2.2 Processes responsible for the disappearance of stable maximons

In a medium with a relatively high maximon density, such mini-maximon systems as maximonium, maxitritium, maxihelium, etc. are likely to be formed (Markov 1978). These systems are unstable: emitting gravitational waves (if the maximons are not electrically charged), the systems collapse in the end. Only one maximon remains after a complete collapse of each such system.

The time of such a collapse

$$t \sim \frac{c^5 r_o^4\ 100^{-3}}{\kappa^3 m_{p\ell}^3} \sim 10^{87}\, r_o^4 \ .$$

At the initial value $r_o \sim 10^{-17}$ cm the maximonium lifetime is of the order of that of our universe:

$$t \sim 10^9 \text{ years} \ .$$

Apparently, the most important and intensive process of maximon vanishing may proceed not through mini-maximon system formation, but through the same black miniholes, when the maximon component of a cosmic medium considerably increases, but the maximon density is still several orders of magnitude lower than the radiation mass density. Indeed, suppose the maximon formation process leads to the appearance of a total maximon mass $M \sim 10^{55}$ gr at $t \sim 1$ sec when the radiation density $\rho_1 \sim 10^5/t^2 = 10^5$ gr cm^{-3} and the universe dimensions, according to (11), are

$$R_1 \sim 10^{17} t^{1/2} \sim 10^{17} \text{ cm} \ , \quad V_1 \sim R^3 \approx 10^{51} \text{ cm}^3 \ .$$

Then the total radiation mass (i.e. the mass of relativistic matter, Wheeler's geons, etc.) $\rho_1 V_1 \sim 10^{56}$ gr at this moment is an order of magnitude larger than the total maximon mass and, accordingly, the ratio of the densities $\rho_1/\rho_m \sim 10$.

So, we are now dealing with the case

$$p = \gamma\varepsilon \ , \quad \text{where} \quad \gamma \sim 1/3 \ .$$

Suppose at this moment a black hole of mass $\sim 10^5$ gm is formed. Then this black hole must contain 10^4 gr maximons, and its decay must lead to the disappearance (transformation back to radiation) of n maximons, where

$$n \sim 10^9 \text{ particles} \ .$$

This scenario requires, of course, further consideration. The previous remarks were intended to draw attention to the existence of the situation when there is an intensive process of vanishing of stable maximons.

On the other hand, the formation of very small miniblackholes (pbh - primordial black holes) may be suppressed by the law of the conservation of the entropy and the formation of larger pbh seems to be strongly limited (Carr 1981; Khlopov & Polnarev 1981).

It may turn out as well that the final number of stable maximons is relatively small, that at the present time the maximum mass density $\rho_m < 10^{-29}$ gr cm^{-3}, and that our universe is open. But in this case the missing mass could be made up of heavy neutrinos which can form not-directly observable cosmic bodies of huge masses and dimensions (Markov 1964).

Let us now consider other scenarios of the early universe, which can be admitted at the present time by the lack of our knowledge of the structure of the primary matter itself at a time $t \sim t_{p\ell}$.

2.3 Possible structures of primary matter

We mean possible structures of matter originating at $t \sim t_{p\ell}$. According to some authors, this matter may arise in the form of very heavy elementary particles. At any rate, in extremely strong gravitational fields the probability of particle creation is proportional to some power of their mass (Grib et al. 1980), which seems natural in the case of gravitational forces. Assume that an elementary particle of maximum possible mass is a maximon. On the one hand, a maximon may appear to be an elementary black hole of Planck mass and dimensions. On the other hand, it may be a particle of the same Planck dimensions, but with a structure essentially different from a black hole. Such a maximon could also be referred to as papapetron. (One possibility, pointed out by Papapetron (1947), is an electrically charged particle, in which the forces of electric repulsion are balanced by gravitational forces. See Markov (1966) and Markov & Frolov (1972) for further discussion.)

These elementary particles are distinguished from the series of elementary particles in that their gravitational radius coincides with their Compton length ℓ_c:

$$\ell_c = \frac{\hbar}{m_p c} \sim \frac{m_p}{c^2} \sim r_{gr} \sim 10^{-33} \text{ cm} .$$

The relation $\ell_c >> r_{gr}$ may be considered as one of the conditions defining the concept of an elementary particle (Markov 1981c). We introduced the term 'maximon' for particles of mass $m = (\hbar c/\kappa)^{1/2}$ (Markov 1965, 1966) if these particles could arise from collapsing regions of matter in a superdense medium in the very early universe. In the present paper we discuss another situation, when these particles (maximons) are not created from some other matter, but appear already produced in strong gravitational fields, like all other elementary particles.

Black holes of large masses may be created from existing black holes. As this process proceeds, entropy increases. The increase due to the confluence of two black holes of masses M_1 and M_2 is

$$\Delta S \sim \frac{M_1 M_2}{m_{p\ell}} k ,$$

where k is the Boltzmann constant.

In the phase of maximum expansion, black holes occupy the entire volume within the Schwarzschild sphere. These volumes are proportional to M_1^3 and M_2^3 , and the volume occupies by the two black holes is

$$V_o \sim M_1^3 + M_2^3 .$$

The maximum volume of a black hole arising from confluence of the two previous black holes is

$$V_f \sim (M_1 + M_2)^3 ,$$

and the ratio of the volumes V_f and V_o in the case $M_1 = M_2$ is

$$\frac{V_f}{V_o} = 4 .$$

For the confluence of N particles this ratio increases like N^2 . The matter within the Schwarzschild sphere is not static: it either collapses or anticollapses. If a black hole has internal Planck dimensions and an external mass equal to the Planck mass, the matter density in it is quantum (ρ_q). If it is not decaying, such a black hole represents some degenerate case: it can neither collapse, nor anticollapse if one assumes that the mass density cannot exceed ρ_q . In other words, the requirement of the existence of a limiting density is very strong and leads to nontrivial consequences.

If the primary matter is actually a maximon 'liquid' comprising stable black holes in the above-mentioned state, then the entire volume of the maximon 'liquid' seems to be some stable volume of matter. If this volume begins, nevertheless, expanding, like in the Friedman case, there must be some physical reasons for the initial start leading to such an expansion. The question arises, whether the confluence of maximons in the indicated metastable liquid can offer such a possibility. (From a topological point of view the maximon liquid is a model of a quasi-isotropical space.)

Indeed, in a black hole formed even of two maximons, there appears a possibility, or rather a necessity, of its anticollapse. (It cannot collapse if ρ cannot be greater than ρ_q. Initially, the matter cluster is more like a white hole.) The volumes of larger 'black holes' thus formed and, accordingly, of the entire maximon liquid would also expand correspondingly.

The described process of the appearance of large holes out of stable maximons resembles boiling of a superheated liquid and the formation of gas bubbles. Here the new phase started from 'molecules' (embryos) of the 'liquid' itself. As in this case, the resulting bubbles of maximon liquid can grow.

If a 'future' black hole of such a mass is produced in a medium with a density $\rho_q \sim 10^{94}$ gr cm^{-3} this cluster of matter has initial dimensions

$$r_i \sim 10^{-28} \text{ cm} .$$

The final dimension of an anticollapsing cluster of mass $M \sim 10^{10}$ gr (i.e. its gravitational radius) is

$$r_{gr} \sim 10^{-18} \text{ cm} .$$

A cluster of matter with mass $\sim 10^{10}$ gr could appear in a medium with density $\rho = \rho_q$ for time $t \sim r_i/c \sim 10^{-39}$ sec. If we assume that the mass densities exceed ρ_q, the time of formation of such massive 'future' black holes could be arbitrarily small. Here we again suffer from lack of knowledge.

We are also ignorant of such important things as how in these small times the horizon problem is solved in nature; what is the duration of the initial causal interaction in the regions occupied by primary matter from the very start of the Friedman phase or even earlier, in the 'steady-state' phase, the necessity of which is connected with the assumption of the initial creation of matter with a constant quantum density ρ_q . This problem is also connected with the appearance of initial inhomogeneities necessary for galaxy formation.

We see that the assumption of the existence of a limiting final mass density (ρ_q) has far-going consequences. But at the same time this assumption is in no way grounded - it is merely not inconsistent in itself and is for the moment admissible.

On the basis of these considerations we can say nothing about the transition phase from the steady state to a clearly expressed Friedman era, how long this transition phase lasts and what the expansion law of the resulting Friedman universe is. But it is true that if our scenario is realistic, then a radiation era must begin at some time. The decay of black miniholes created must result in the appearance of ultrarelativistic matter.

We shall not evidently encounter any difficulties with this scenario if this process, for example, by $t \sim 10^{10}$ sec , leads to a radiation density $\rho_r = \varepsilon_r/c^2$ still higher than that of the remaining maximons.

Unlike the previous standard scenario, this one involves the vanishing of primary stable maximons throughout the history of the universe. Maximon annihilation occurs at the beginning due to mini-maximon system formation, black hole and 'blini' (pancake) formation in a maximon medium. The formation of blinis leads to local increases in maximon density and hence to the higher probability of their disappearance. At subsequent moments of the radiation era, maximons are annihilated by black hole formation out of a two-component (radiation and maximons) medium.

With our ignorance of the structure of matter at the initial moment of the universe, we could construct numerous possibilities for its different forms; we could assume that the matter from the very beginning was, for example, a set of white holes. This possibility was discussed in connection with Ambartsumyan's hypothesis about the creation of stellar matter due to a superdense matter outflow from some centres. Novikov (1964) and Ne'eman (1965) tried to interpret this hypothesis, associating the phenomenon with the appearance of white holes in the very early universe.

White holes are clusters of anticollapsing matter concentrated at the initial moment in regions of smaller radius than the gravitational radius of the cluster. For instance, in the above example, the anticollapsing matter concentrated in the region 10^{-28} cm with a total mass of 10^{10} gr could be just as well a white hole if in the course of further development of its anticollapse, part of this matter could get out from under the Schwarzschild radius. White holes in the very early universe could play an especially important role if we give up for the moment the arbitrary hypothesis about the existence of the limiting matter density. The hypothesis of white holes expanding by their own laws, for example, with a considerable delay in the rates of the universe expansion as a whole, would change essentially the picture of the physics of a very early universe. If one does not associate the white hole expansion with Ambartsumyan's hypothesis, delays in the expansion of initial white holes could be within the first minute of the existence of the universe.

Thus, the topology of the universe from the very start can be very complicated. The universe could also be composed of a large number of semi-closed worlds, each of which transforms in the end into a stable maximon. These objects can be referred to as second-kind black holes (Markov 1974). They were first discovered theoretically by O. Klein (1961) and called by him 'nonstationary objects with external world' (mit Aussenwelt).

Apparently, the manifestation of these objects in the external space does not differ from the behaviour of black holes. After complete evaporation, this object is transformed into a closed Friedman universe and vanishes as an object of observation in the external space. In the process of evaporation, when the mass of this object becomes equal to the maximon mass, the object either vanishes immediately or remains a stable maximon. We have discussed above the arguments in favour of representing the universe as such an almost closed Friedman universe, a stable maximon enclosed in the external space, which can also be an internal space of a 'larger' maximon etc. In our universe there can also exist maximons, which can also be almost closed huge universes with galaxies and civilizations. For such a conception of an infinite universe, the appearance of a new relativity is typical, relativity of the concept 'micro' and 'macro'.

CONCLUSION

As is seen from the text, if the question of the universe structure could depend on the author's wish, the picture of the development of a perpetually oscillating universe would be as follows: an almost closed universe with a total mass equal to the maximon mass is at present anticollapsing. After it reaches the maximum possible expansion, it will recollapse until the mass density reaches a quantum value. In the course of further contraction the mass density remains constant, but the total mass decreases owing to the weakening of gravitational interactions, which results from asymptotic freedom. The universe enters a peculiar steady-state like phase, and its final vanishing is forbidden by conservation of its external mass. In the final state of the contraction phase it can be interpreted as a de Sitter universe of Planck dimensions, whose metric joins the external Schwarzschild type-metric. Thus, the de Sitter stages must play the role of the so-called 'purgatory', each time purifying the oscillations the universe from entropy and energy accumulated in its Friedman phase. Only under these conditions is a perpetually oscillating universe possible (Tolman 1934).

In conclusion I would like to express my acknowledgements to V.A. Berezin and V.P. Frolov for numerous fruitful discussions.

REFERENCES

Carr, B.J. (1981). In Proc. 2nd Moscow Seminar on Quantum Gravity.
Frolov, V.P. & Vilkoviski, G.A. (1979). Quantum gravity removes classical singularities. Preprint IC/79/69, International Centre for Theoretical Physics, Trieste.
Grib, A.A., Mamayev, S.G. & Mostepanenko, V.M. (1980). Quantum effects in strong gravitational fields. Preprint, Atomizdat, Moscow.
Guth, A.H. (1981). Phys.Rev. D23, 347.
Hawking, S.W. & Ellis, G.F.R. (1973). The large scale structure of space-time. Cambridge: Cambridge University Press.
Hawking, S.W. (1976). Phys.Rev. D13, 191.
Khlopov, M.Yu. & Polnarev, A.G. (1981). The stage of superheavy particle dominance in the universe and big bang nucleosynthesis. Preprint Pr-634, Space Res.Inst., Acad.Sci. U.S.S.R., Moscow.
Klein, O. (1961). In Werner Heisenberg und die Physik unserer Zeit, ed. Fritz Bopp. Braunschweig: Vieweg.
Linde, A. (1981). Preprint N229, P.N. Lebedev Phys.Inst., Moscow.
Linde, A. (1982). Phys.Lett. B108, 389.
Mamayev, S.G. & Mostepanenko, V.M. (1980). ZhETF 78, 20.
Markov, M.A. (1947). ZhETF 17, 848.
Markov, M.A. (1956). Nuovo Cimento Suppl. N4, 3, 760.
Markov, M.A. (1964). Phys.Lett. 10, 122.
Markov, M.A. (1965). Progr.Theor.Phys. Sup.Ext., 85.
Markov, M.A. (1966). ZhETF 51, 878.

Markov, M.A. (1974). In Gravitational radiation and gravitational collapse. ed. C. de Witt-Morette. I.A.U. Symp. No 64. Dortrecht, Holland and Boston, USA: D. Reidel.
Markov, M.A. (1978). Maximon-type scenario of the universe. Preprint P-0207, Inst.Nucl.Res., Acad.Sci. USSR, Moscow.
Markov, M.A. (1981a). In Proc. 2nd Moscow Seminar on Quantum Gravity.
Markov, M.A. (1981b). Is perpetually oscillating Friedman universe possible? Preprint P-0227, Inst.Nucl.Res., Acad.Sci. USSR, Moscow.
Markov, M.A. (1981c). On the maximon and the concept of elementary particle. Preprint P-0208, Inst.Nucl.Res., Acad.Sci. USSR, Moscow.
Markov, M.A. & Frolov, V.P. (1972). Theoret. and Mathem.Fiz. 13, 41.
Ne'eman, Y. (1965). Ap.J. 141, 1303.
Novikov, I.D. (1964). Astron.Zh. 41, 1075.
Papapetrou, A. (1947). Proc.Roy. Irish Acad. A51, 191.
Rindler, W. (1956). MNRAS 116, 663.
Sciama, D.W. (1976). Vistas in Astronomy 19. 385.
Starobinski, A.A. (1979). Pisma V. ZhETF 30, 719.
Starobinski, A.A. (1980). Phys.Lett. 91B, 99.
Tolman, R.C. (1934). Relativity, Thermodynamics and Cosmology. Oxford: Clarendon Press.
Zeldovich, Ya.B. (1977). Usp.Fiz.Nauk. 123, 478.
Zeldovich, Ya.B. & Novikov, I.D. (1975). Structure and Evolution of the Universe. Moscow: Nauka (in Russian).

COSMOLOGICAL CONSTRAINTS ON GUTs

G. Lazarides
The Rockefeller University, New York, New York 10021

Abstract: We construct a class of axion models with no cosmological domain wall problem. We also discuss the behaviour of SUSY GUTs in the very early universe. We find that all the non trivial models, which are consistent with cosmology belong to the O'Raifeartaigh-Witten class.

1. AXION MODELS WITH NO DOMAIN WALL PROBLEM

The QCD action must in general include the CP violating term

$$\theta . \frac{1}{32\pi^2} \int d^4x \; F^a_{\mu\nu} \; \tilde{F}^a_{\mu\nu} \; . \qquad (1)$$

The integrand in this equation is a total derivative and, thus, perturbation theory is θ-independent to all orders. However, non-perturbative field configurations (Belavin et al 1975), characterized by a non-zero Pontryagin index q, give a non-trivial contribution to (1). Due to the fact that q is an integer, θ is a periodic variable with period 2π. The absence of CP violation in strong interactions requires that θ is zero or very close to zero. Since θ is a priori arbitrary, we must find some explanation for its smallness. This is the strong CP problem.

The Peccei and Quinn (1977) (PQ) mechanism is certainly one of the most attractive solutions to the strong CP problem. One introduces an anomalous global $U(1)_{PQ}$ symmetry which is broken explicitly at the QCD scale $\Lambda \sim 100MeV$ by the gluon anomaly,

$$\partial_\mu j^{PQ}_\mu \propto F^a_{\mu\nu} \; \tilde{F}^a_{\mu\nu} \quad . \qquad (2)$$

Due to this anomaly relation, any $U(1)_{PQ}$ transformation implies a change in θ. $U(1)_{PQ}$ is also broken spontaneously at the GUT scale. (We consider only the Dine, Fischler and Srednicki(1981) (DFS) variant of the PQ scheme, where the resulting axion is invisible.) It has been argued that, in the PQ scheme, there is a range of parameters so that minimization of the effective potential gives $\theta \simeq 0$.

Very recently, Sikivie (1982) has shown that $U(1)_{PQ}$ contains a discrete subgroup which is left unbroken by the QCD gluon anomaly. Spontaneous breaking of this discrete symmetry leads to an unacceptable domain wall production in the early universe.

We will construct a class of models (Lazarides and Shafi, 1982a) which do not suffer from this cosmological problem. The spontaneously broken "discrete" subgroup of $U(1)_{PQ}$ is embedded in the spontaneously broken continuous symmetry of the theory and, thus, ceases to be a genuine discrete symmetry.

Let us first discuss Sikivie's problem. For definiteness, we will restrict our discussion to the minimally extended SU(5) GUT model of Wise, Georgi and Glashow(1981) (WGG). The Higgs sector of this model contains a complex $\underline{24}$ ϕ_{24} and two $\underline{5}$'s ϕ_5 and ϕ_5'. The fermions are in left-handed $\underline{\bar{5}}$'s $\psi^{\alpha}_{\bar{5}}$ and $\underline{10}$'s ψ^{α}_{10} (α=1,2,...,f). The global anomalous $U(1)_{PQ}$ is defined as:

$$\psi^{\alpha}_{\bar{5}} \rightarrow e^{i\theta}\psi^{\alpha}_{\bar{5}} \, , \, \psi^{\alpha}_{10} \rightarrow e^{i\theta}\psi^{\alpha}_{10} \, , \, \phi_{24} \rightarrow e^{2i\theta}\phi_{24} \, ,$$

$$\phi_5 \rightarrow e^{-2i\theta}\phi_5 \, , \, \phi_5' \rightarrow e^{2i\theta}\phi_5' \text{ for } e^{i\theta} \varepsilon \, U(1)_{PQ} \, . \tag{3}$$

This symmetry is broken by the QCD anomaly to a Z_{4f} subgroup generated by $\exp(i\pi/2f)\,\varepsilon U(1)_{PQ}$. This Z_{4f} subgroup is, in turn, spontaneously broken to Z_2 = (1,-1), by $\langle\phi_{24}\rangle$, $\langle\phi_5\rangle$ and $\langle\phi_5'\rangle$. This leads to a catastrophic production of domain walls separating 2f degenerate vacua:

$$\langle\phi_{24}\rangle = v_o e^{\frac{i\pi n}{f}} \, , \, \langle\phi_5\rangle = v_1 e^{\frac{-i\pi n}{f}} \, ,$$

$$\langle\phi_5'\rangle = v_2 e^{\frac{i\pi n}{f}} \quad (n = 0,1,..., 2f-1). \tag{4}$$

We will now discuss in some detail the scenario for the wall production. At a critical temperature $T_c \sim 10^{15}$GeV, SU(5) breaks down to SU(3) x SU(2) x U(1) and $U(1)_{PQ}$ to Z_2 = (1,-1) by $\langle\phi_{24}\rangle$. The spontaneous breaking of $U(1)_{PQ}$ leads to the production of a network of superheavy strings (Vilenkin and Everett, 1982). As one moves around a minimal string, $\langle\phi_{24}\rangle$ gets transformed by a $U(1)_{PQ}$ transformation that ranges between 1 and -1. The cosmological evolution of such "global" strings is very similar to the evolution of the more conventional "gauge" strings (Vilenkin and Everett 1982). In particular, the scale of the string network becomes of the order of the cosmic time t soon after production (Kibble 1976; Everett 1981). At $t \sim m_a^{-1} \sim 10^{-6}$ sec. ($m_a \sim 10^{-8}$ eV is the axion mass) non-perturbative QCD effects break $U(1)_{PQ}$ explicitly down to Z_{4f}. Then a domain structure appears with domains characterized by (4) and walls that separate these domains. The walls have thickness $\sim m_a^{-1}$. The phases of the Higgs fields change by $\pm\pi/f$ as we cross a minimal wall. Thus any string (around which the phases change by 2π) becomes the boundary of 2f walls that terminate on it. Such a system of walls and strings decays before dominating the universe only in the case 2f = 1 (Vilenkin and Everett 1982). In this case, the spontaneously broken discrete symmetry is trivial. In the realistic case $2f \neq 1$, however, the system of walls and strings persists and leads into a cosmological catastrophe.

This catastrophe can be avoided in the minimal Spin(10) model:

$$\text{Spin}(10) \xrightarrow[\phi_{16}]{M_s \gtrsim 10^{17}\text{GeV}} SU(5) \xrightarrow[\phi_{45}]{M_x \sim 10^{15}\text{GeV}}$$
$$[SU(3) \times SU(2)] \otimes U(1) \xrightarrow[\phi_{10}]{M_W \sim 100 \text{ GeV}} SU(3) \otimes U(1) , \qquad (5)$$

with a single family of fermions ψ belonging to a left-handed 16 of Spin(10). There are two ways to introduce a global anomalous $U(1)_{PQ}$ in this model:

$$\psi \to e^{i\theta}\psi , \ \phi_{16} \to e^{i\theta}\phi_{16} , \ \phi_{10} \to e^{-2i\theta}\phi_{10} ,$$

$$\phi_{45} \to e^{iq\theta} \phi_{45}, \text{ for } e^{i\theta} \in U(1)_{PQ} , \qquad (6)$$

with q = 2 or -4. QCD anomalies break explicitly $U(1)_{PQ}$ down to a Z_4 generated by $\exp(i\pi/2) \in U(1)_{PQ}$ (remember we have a single family). The centre of Spin(10) is a Z_4 subgroup generated by $i\Gamma^o \in$ Spin(10), where $\Gamma^o = i^5\Gamma^1\Gamma^2 \dots \Gamma^{10}$ is the "chirality" operator (Γ's are the generalized Dirac matrices in 10 dimensions). One can show that under $i\Gamma^o$,

$$\psi \to i\psi, \ \phi_{16} \to i\phi_{16} , \ \phi_{10} \to -\phi_{10} , \ \phi_{45} \to \phi_{45} . \qquad (7)$$

For q = -4, $i\Gamma^o \in$ Spin(10) and $\exp(i\pi/2) \in U(1)_{PQ}$ have identical action on all the fields of the model. They are indistinguishable. Thus, the spontaneously broken Z_4 subgroup of $U(1)_{PQ}$ is identical with the centre of Spin(10), and, therefore, ceases to be a genuine discrete symmetry. No domain wall catastrophe is expected in this case.

Let us see how this works in some detail. At a critical temperature $T_c \gtrsim 10^{17}$GeV, ϕ_{16} and ϕ_{45} develop vacuum expectation values (VEVs) and Spin(10) $\otimes$ $U(1)_{PQ}$ breaks to SU(5). A network of strings appears, since

$$\pi_1[\text{Spin}(10) \otimes U(1)_{PQ}/SU(5)] = Z. \tag{8}$$

To construct this fundamental group, we must remember that Spin(10) $\otimes$ $U(1)_{PQ}$ is not a global direct product of Spin(10) and $U(1)_{PQ}$. It is, rather, defined as Spin(10) x $U(1)_{PQ}$ with the three pairs of (Spin(10), $U(1)_{PQ}$) transformations

$$(\ (i\Gamma^{o})^{-k},\ e^{i\pi k/2})\quad (k = 1,2,3) \tag{9}$$

identified with (1,1). The fundamental group in (8) is then generated by the homotopy class of closed curves in (Spin(10), $U(1)_{PQ}$):

$$\begin{aligned} &(\ g(t),\ h(t)\ ),\quad 0 \leqslant t \leqslant 1\ , \\ &g(t) \in \text{Spin}(10)\ ,\quad h(t) \in U(1)_{PQ}\ , \\ &g(0) = 1,\quad g(1) = (i\Gamma^{o})^{-1} = -i\Gamma^{o} \\ &h(0) = 1,\quad h(1) = e^{i\pi/2},\quad 0 \leqslant \arg h(t) \leqslant \pi/2. \end{aligned} \tag{10}$$

Thus the minimal strings of the network have the following property. As one goes around such a string, $\langle\phi_{16}\rangle$ and $\langle\phi_{45}\rangle$ get transformed by a pair of transformations. One transformation belongs to the gauge group SU(5) and interpolates between 1 and $-i\Gamma^{o}$. The other transformation belongs to the global group $U(1)_{PQ}$ and interpolates, between 1 and $\exp(i\pi/2)$ with its phase between 0 and $\pi/2$. Such strings are called hybrid strings. At $t \sim m_a^{-1} \gtrsim 10^{-4}$sec ($m_a \lesssim 10^{-10}$eV), $U(1)_{PQ}$ breaks explicitly down to Z_4. A wall structure is formed at this stage. Across a minimal wall the VEVs get transformed by a $U(1)_{PQ}$ transformation that continuously interpolates between 1 and $\exp(i\pi/2)$ with its phase between 0 and $\pi/2$. Thus each hybrid string becomes boundary of a single wall that terminates on it. The system of walls and strings, then, decays rapidly (Vilenkin and Everett 1982) and the cosmological catastrophe is avoided.

In the case of three families, the problem cannot be avoided in the context of GUTs with plain family repetition. A possible way out is to extend the Spin(10) model discussed here, so that it contains 2 more families with PQ charge 1 and a horizontal group G_H which acts in the family space. G_H can be global local or mixed. It can be simple, semi-simple, containing U(1)'s or even consisting of U(1)'s exclusively. In any case, the following requirements must be met:

(i) The centre of G_H must contain a Z_3 subgroup that coincides with the Z_3 subgroup of $U(1)_{PQ}$ generated by $\exp(2\pi i/3)$.

(ii) Z_3 must be contained in a spontaneously broken continuous subgroup of G_H which is broken neither explicitly nor by the QCD anomaly.

Then, the spontaneously broken $Z_{12} \varepsilon U(1)_{PQ}$ is embedded in a continuous subgroup of Spin(10) x G_H and the wall problem is solved. Models that satisfy these requirements have been constructed (Lazarides and Shafi 1982a).

2. SUSY GUTs AND COSMOLOGY

We will employ cosmological considerations to impose constraints on SUSY GUTs. We find (Lazarides and Shafi 1982b) only two classes of realistic SUSY GUTs:

(i) Models with SUSY softly broken at or near the superheavy GUT scale M. These theories cannot be distinguished from ordinary GUTs.

(ii) Models based on Witten (1981) mechanism for generating M from the SUSY breaking scale $\sqrt{F}$. These models may or may not have inflation (Guth 1981). The two cases are perhaps experimentally distinguishable, since the former predicts absence of monopoles while the latter predicts observable monopole density.

It should be mentioned that in our analysis we do not consider exotic or unnatural possibilities; we restrict ourselves to generic theories.

Let us first consider the simplest SUSY GUT model based on SU(5) with SUSY broken softly at $\sqrt{F} << M$ (Georgi and Dimopoulos 1981; Sakai 1982). We assume throughout that $10^{15}\text{GeV} \lesssim M < M_p$, the Planck mass. We will argue that such a model is not compatible with cosmology. At temperature T = 0 (and for $\sqrt{F} = M_w = 0$) the theory possesses (Georgi and Dimopoulos 1981) three degenerate minima corresponding to the gauge symmetries G = SU(5), F = SU(4) x U(1) and H = SU(3) x SU(2) x U(1). SUSY is not broken in any of these minima and the height of the potential barriers separating these phases are all of order M^4. Weinberg (1982) has shown that gravitational effects split the degeneracy between these phases by an amount of order $GM^6 = M^4(M/M_p)^2 << M^4$ (G is the Newton's constant). The H phase has zero energy density whereas the G and F phases have negative energy densities. Weinberg's splitting is much greater than the splitting of order F^2 due to SUSY breaking, if $\sqrt{F} << 10^{13}$GeV.

We will now discuss the cosmology of this model under the assumption that $\sqrt{F} << 10^{13}$GeV. For $T \gtrsim 10^{13}$GeV, the thermal splitting of the degeneracy between the G, F and H phase is $\sim T^4$ and dominates over the gravitational splitting. The G phase has the lowest free energy and the universe is in this phase. For $T \lesssim 10^{13}$GeV, gravitational effects take over but the universe still remains in the G phase (see above). SUSY breaking at $T_c \sim \sqrt{F}$ introduces a splitting $\sim F^2 << GM^6$. Thus, the phase G remains favoured even for $T \lesssim \sqrt{F}$ and the universe is trapped in the wrong phase forever. Clearly, such models are in conflict with cosmology. Various ways out of this problem may be tried but none of them works.

(i) One possibility is to take $\sqrt{F} \sim 10^{13}$GeV. Then, the SUSY splitting is of the same order as the gravitational splitting and one can arrange parameters so that, for $T \lesssim \sqrt{F} \sim 10^{13}$GeV, the H phase is favoured. The splitting between the minima is of order F^2 whereas the height of the barriers that separate them are of order M^4. The tunnelling probability per unit time from the G to the H phase is then given (Coleman 1977; Callan and Coleman 1977; Linde 1977) by

$$P \sim M^4 \exp\left[-\left(\frac{M}{\sqrt{F}}\right)^9 \frac{\sqrt{F}}{T}\right],$$

for $$\sqrt{F} \gtrsim T \gtrsim \sqrt{F}\left(\frac{\sqrt{F}}{M}\right)^3$$

and $$P \sim M^4 \exp\left[-\left(\frac{M}{\sqrt{F}}\right)^{12}\right], \text{ for } T \lesssim \sqrt{F}\left(\frac{\sqrt{F}}{M}\right)^3 . \qquad (11)$$

Since $M \gtrsim 10^{15}$GeV and $\sqrt{F} \sim 10^{13}$GeV, this tunnelling probability is utterly negligible and the phase transition is never completed.

(ii)Another possibility is to assume that, at a critical temperature $T_c \sim \sqrt{F} \sim 10^{13}$GeV, the G phase becomes locally unstable and the transition to the H phase is rapidly completed. This leads to too many magnetic monopoles produced by the Kibble (1976) mechanism.

(iii)Yet another possibility is to arrange parameters so that the universe is in the H phase for all $T \gtrsim 100$GeV. $\sqrt{F}$ can be as low as 10^2GeV in this case. Unfortunately, such models produce too many monopoles because of thermal fluctuations.

We now turn to the discussion of SUSY GUTs of the O'Raifeartaigh (1975) - Witten (1981) (OW) variety. In such models the GUT scale M is related to the SUSY breaking scale $\sqrt{F}$ by $M = \sqrt{F}\exp(c/e^2)$, where e is the gauge coupling and $c > 0$. It has been pointed out (Dimopoulos and Raby 1982) that these models contain a third mass scale $M_W \sim (\alpha/\pi)F/M$ $(\alpha = g^2/4\pi)$. Then $\sqrt{F} \sim \sqrt{\pi M M_W/\alpha} \gtrsim 10^9$GeV, since $M \gtrsim 10^{15}$GeV.

We will concentrate on the simplest SU(5) model of the OW class. One has two complex Higgs 24-plets A and Y and a singlet Higgs field X. The superpotential is

$$W(A,X,Y) = \lambda \mathrm{tr}A^2Y + gX(\mathrm{tr}A^2-F^2). \tag{12}$$

SUSY is broken spontaneously at a scale $\sqrt{F}$. Minimization of the tree potential yields,

$$\langle A\rangle = g\sqrt{F}(\lambda^2+30g^2)^{-1/2}\ \mathrm{diag}\ (2,2,2,-3,-3) \tag{13}$$

and

$$\langle Y\rangle = \left(\frac{g}{\lambda}\right)\langle X\rangle \mathrm{diag}\ (2,2,2,-3,-3)\ . \tag{14}$$

Including one loop corrections (and for some range of parameters), one finds

$$\langle X\rangle \sim \langle A\rangle \exp(d/e^2)\ ,\ d > o\ . \tag{15}$$

SU(5) is then broken down to $H = SU(3) \times SU(2) \times U(1)$ at a superheavy scale $\langle X\rangle \sim M \gtrsim 10^{15}$GeV. It has been shown (Ginsparg 1982; Pi 1982) that the vacuum expectation value (VEV) of the X field becomes non zero only at temperatures below $\langle A\rangle \sim \sqrt{F}$. In this case, the gravitational contribution to the vacuum energies, being only of order GF^3, is much smaller than the contribution from SUSY breaking which is of order F^2. Thus, Weinberg's splitting can be ignored.

Let us discuss the cosmology of a typical OW GUT model. For $T \gtrsim M$, the universe is in the SU(5) phase. This phase persists till $T \sim \sqrt{F}$ where SUSY is spontaneously broken, and A acquires a VEV $\sim \sqrt{F}$ (we will not consider the case where $\langle A\rangle$ remains zero, even though SUSY is spontaneously broken). We will assume that the transition from $\langle A\rangle = 0$ to $\langle A\rangle \sim \sqrt{F}$ is rapidly completed. SU(5) is now broken to $SU(3) \times SU(2) \times U(1)$ but only by an expectation value $\sim \sqrt{F} \ll M$. At this point magnetic monopoles of mass $\sim \sqrt{F}/\alpha$ are produced.

We now turn to the subsequent transition from $\langle X\rangle = 0$ to $\langle X\rangle \sim M$. For the moment we will assume that, after "tunnelling" away from $\langle X\rangle = 0$, the X field rolls down to its vacuum value $\sim M$ very quickly.

So no inflation is associated with this rollover. We can then distinguish the following three cases:

(i) The T=0 scalar potential has a local minimum at X=0. For any reasonable values of the parameters the tunnelling away from $\langle X \rangle = 0$ is highly suppressed since the height of the barrier remains fixed at F^2 for $T \lesssim \sqrt{F}$.

(ii) The T=0 potential has a local maximum at X=0. In this case, the local minimum at X=0 will disappear at a temperature $T_x \sim \sqrt{F}$ and the transition will be rapidly completed. The number density of monopoles produced in this transition is, then, unacceptably large.

(iii) Finally, the T=0 potential has zero curvature at X=0. In this case, the tunnelling from $\langle X \rangle = 0$ to $\langle X \rangle \sim M$ is going to take place only after some supercooling and the monopole production is suppressed. Let us see this in some detail. The barrier, for $T \lesssim \sqrt{F}$, is of order T^4. Gravitational (and thermal) effects are going to destabilize the X=0 phase at a temperature $T_x \sim F/M_p \sim$ the Hawking temperature of this phase, and the transition to $\langle X \rangle = M$ will take place. The mass of the monopoles increases now to M/α and the universe is reheated back to a temperature $T_R \sim \sqrt{F}$. There is no extra thermal production of monopoles since $M/\alpha >> T_R \sim \sqrt{F}$. Also, there are no extra monopoles due to the Kibble mechanism since $\langle Y \rangle$ is parallel to $\langle A \rangle$ at any given point in space. The relative monopole density after annihilation (Preskill 1979) is

$$r \sim 10^{-6} \frac{\sqrt{F}}{M_p} \; . \tag{16}$$

After reheating from $T_x \sim F/M_p$ to $T_R \sim \sqrt{F}$, we have an extra suppression of r. Thus finally

$$r_f \sim 10^{-6} \left(\frac{F}{M_p^2}\right)^2 \; . \tag{17}$$

Requiring that the bound on r from nucleosynthesis is satisfied, we get

$$\sqrt{F} \lesssim 10^{16} \mathrm{GeV} \; . \tag{18}$$

If the rollover of X takes place with inflation, (Linde 1982; Albrecht and Steinhardt 1982) the precise shape of the potential at X=0 is irrelevant. Even a single fluctuating region can expand to include our universe. In this case, however, there are no monopoles at all in contrast to the previous non-inflationary scenarios where we had an observable monopole density.

This work is partially supported by the Department of Energy Grant No. DE AC02-82ER 40033. B000

References:

Albrecht, A. and Steinhardt, P. (1982). Phys. Rev. Lett. 48, 1220.
Belavin, A. et al. (1975). Phys. Lett. 59B, 85.
Callan, C. and Coleman, S. (1977). Phys. Rev. D16, 1762.
Coleman, S. (1977). Phys. Rev. D15, 2929.
Dimopoulos, S. and Raby, S.(1982). Los Alamos preprint LA-UR-82-1282.
Dine, M., Fischler, W. and Srednicki, A. (1981). Phys. Lett. 104B, 199.
Everett, A. (1981). Phys. Rev. D24, 858.
Georgi, G. and Dimopoulos, S. (1981). Nucl. Phys. B193, 150.
Ginsparg, P.(1982). Phys. Lett. 112B, 45.
Guth, A. (1981). Phys. Rev. D23, 347.
Kibble, T.W.B. (1976), J. Phys. A9, 1387.
Lazarides, G. and Shafi, Q. (1982a). Phys. Lett. 115B, 21.
Lazarides, G. and Shafi, Q. (1982b). ICTP preprint IC/82/114.
Linde, A.D. (1977). Phys. Lett. 70B, 306.
Linde, A.D. (1982). Phys. Lett. 114B, 431.
O'Raifeartaigh, L. (1975). Nucl. Phys. B96, 331.
Peccei, R.D. and Quinn, H. (1977). Phys. Rev. Lett. 38, 1440.
Pi, S.Y. (1982). Harvard preprint HUTP-81/A055.
Preskill, J. (1979). Phys. Rev. Lett. 43, 1365.
Sakai, N. (1982). Zeit. fur. Phys. C11, 153.
Sikivie, P. (1982). Phys. Rev. Lett. 48, 1156.
Vilenkin, A. and Everett, A. (1982). Tufts Univ. preprint TUTP-82-3.
Weinberg, S. (1982). Univ. of Texas preprint.
Wise, M., Georgi, H. and Glashow, S.L. (1981). Phys. Rev. Lett. 47,402.
Witten, E. (1981). Phys. Lett. 105B, 267.

FUN WITH MONOPOLES AND AXIONS

F. Wilczek
Institute for Theoretical Physics
University of California, Santa Barbara
Santa Barbara, California 93106 USA

Abstract. Some recent developments in the theory of magnetic monopoles and in the theory of axions are summarized briefly. Wildly speculative ideas concerning the capture of monopoles and the possibility of automatically cancelling the cosmological constant are discussed briefly.

Very recently there have been dramatic new developments in the theory of magnetic monopoles, and in the theory of axions. Here these developments are summarized and supplemented with a few new observations. My own work on these subjects is mainly contained in Dimopolous *et al.* (1982 a,b) and Wilczek (1982 a,b) (monopoles) and Preskill *et al.* (1982) and Wilczek (1982 c) (axions); other work is cited in these papers and below.

MONOPOLES

1. The modern era in the theory of magnetic monopoles began with the demonstration by 'tHooft (1974) and Polyakov (1974) that such particles arise as regular solutions of the field equations in many unified gauge theories. In the simplest phenomenologically acceptable unified gauge theory, based on SU(5) (Georgi & Glashow 1974), the mass of such monopoles is expected to be $\gtrsim 10^{16}$GeV.

2. There is no question of producing such heavy particles in the laboratory or even in cosmic rays; if monopoles exist they must be cosmological relics. Calculations of the density of monopoles expected to persist after being produced in the early stages of the Big Bang, based on simple extrapolation of Friedmann cosmology, gave an unacceptably large result (Zeldovich & Khlopov 1979; Preskill 1979). This was one main motivation for the proposal (Guth 1981) that the Universe underwent a period of rapid inflation -- the monopole density produced before inflation is

then diluted.

3. In a "canonical" inflationary Universe scenario (Linde 1982; Albrecht & Steinhardt 1982) the Universe inflates by an enormous factor while at the Hawking temperature $\lesssim 10^{10}$GeV., and then reheats (Albrecht *et al*. 1982) to something like 10^{14}GeV. Reheating to a temperature close to the unification scale is necessary, in order that baryon-number violating processes may proceed rapidly and create the matter-antimatter asymmetry of the Universe. The monopole density produced prior to inflation is diluted to practically zero, and the Hawking temperature is too low to allow much production during inflation. Any presently observable monopole density must be produced in the reheating. One's crude expectation is that monopoles are produced thermally (Turner 1982; Preskill 1982); in this case their density relative to photons would be supressed by the Boltzmann factor of exp $(- M_{monopole}/T_{reheat})$. This is a very small quantity for the parameters quoted above, but the estimate is very soft quantitatively (as well as being based on a highly speculative cosmological scenario!).

I think it is fair to summarize the situation by saying that the idea of inflation is a very attractive way to avoid the glaring over-abundance of monopoles predicted in straightforward Friedmann cosmology; and that present models implementing this idea suggest, but do not convincingly predict, an unobservable small monopole density.

4. In view of the uncertainties in cosmological expectations for monopole density it is certainly appropriate to pursue more phenomenological arguments. The best constraints (Parker 1971; Turner *et al*. 1982) on the galactic monopole flux for $M \approx 10^{16}$GeV., and velocity $v = 10^{-3}c$ are based on the persistence of galactic magnetic fields, and constrain the flux to be $F \lesssim 10^{-15}/\text{cm}^2$.sec.ster. So small a flux would be very difficult to detect -- an apparatus of cross-section $(10^4 \text{cm.})^2$ is required to see roughly one event per year! It is, of course, much smaller than the flux inferred from Cabrera's (1982) recent observations, $F \approx 10^{-10}/\text{cm}^2$.sec.ster.

5. Might there be a local enhancement of the monopole flux, above the galactic value? A possibility (Dimopolous *et al*. 1982 a) is that there is a cloud of monopoles orbiting the sun, not dissimilar to the cloud of meteorites and comets. Monopoles would diffuse out of the cloud

as they are accelerated by the variable magnetic fields in the heliosphere. The cloud could be replenished by monopoles in the general galactic flux which are gravitationally captured by the sun (after losing energy by passing through solar material, and being perturbed by magnetic fields). If this machanism really works, one can have a local enhancement of the monopole flux by a factor $\sim 10^8$ times the galactic flux.

The hypothesis that monopoles are orbiting around the sun can of course be tested by observing their distribution in angle and velocity. An important consequence is that their speeds would be $\sim 10^{-4}c$, as opposed to $10^{-3}c$ for galactic monopoles. The smaller velocity means that the monopoles deposit less energy in passing through matter, so that experiments based on detecting their ionization are correspondingly less sensitive.

6. In general, the enormous inertia of 10^{16}GeV. monopoles makes it utterly absurd to think of stopping them so that we could study their properties at leisure. If they are really in essentially inertial orbits around the sun, however, it might just be possible with great effort to collect some. We would need an apparatus in space which could determine the orbital elements of incident monopoles, and send small craft out after them. It could then be arranged that these craft are brought to essentially exactly the velocity of the monopole. The monopole could then be trapped in an electric bottle (analogous to plasma in a magnetic bottle) and gently brought back to an orbiting laboratory.

7. Recent studies of the interactions between fermions and monopoles have lead to some startling results.

One is that monopoles come in nearly degenerate families with different values of fermionic quantum numbers, like electron number (Jackiw & Rebbi 1976; Wilczek 1982 b). That is, there are very nearly degenerate states of the monopole with electron number $N_e = \pm 1/2$. There is a process, with cross-section of order $\alpha(1/M_e)^2$, whereby the $N_e = \pm 1/2$ monopole absorbs a positron and goes into the $N_e = - 1/2$ state, emitting a photon.

Related to this, monopoles will have quark quantum numbers and participate in the strong interactions with typical hadronic strength (Wilczek 1982 b).

8. Another interesting process is the catalysis of baryon-number violation by monopoles (Wilczek 1982 a; Rubakov 1981, 1982; Callan 1982).

As remarked by 'tHooft (1976), the standard SU(3) × SU(2) × U(1) gauge theory of particle interactions, even without unification, does not conserve baryon number. Although in the standard theory baryon number is conserved at the classical level, there are quantum corrections which violate baryon number. The effect arises from large fluctuations of the gauge fields in the weak-coupling, broken symmetry sector, i.e. in the W- and Z-boson fields. Large quantum fluctuations in these fields are very rare and accordingly the rate of B-violations calculated from this process is unobservably small.

In the presence of monopoles, we so to speak begin already "half-way there." There are of course large magnetic fields associated with the monopole itself. Within a core of size $\sim 1/m_W$ there are large magnetic components of the fields associated with the W- and Z-bosons, which are relevant to B-violation. The additional quantum fluctuations needed to give B-violation are now much smaller. Instead of a cross-section of order $1/m^2_X$, with $m_X \sim 10^{15}$GeV. the unification scale, as one obtains for B-violating scattering off ordinary targets, one gets a cross-section governed by the scale $m_W \sim 10^2$GeV.

A semi-quantitative calculation along these lines is attempted in Wilczek (1982 b). The B-violating cross-section for scattering off monopoles as estimated there is probably unobservably small, although it is very much larger for monopoles than for ordinary targets.

9. Rubakov (1981, 1982) and Callan (1982) have presented arguments that monopoles catalyze B-violation even more efficiently, in fact with roughly strong-interaction cross-sections. I cannot vouch for these calculations, which I do not think are conclusive. If it is really true that monopoles do catalyze B-violation at such rates, then I'm afraid the prospects for ever observing a monopole are slim indeed for the following reason (Dimopolous et al. 1982 b; Kolb et al. 1982).

Monopoles are readily accreted by neutron stars: the gravitational cross-section of a neutron star is $\gtrsim 10^9$cm., and monopoles passing through the very dense material of a neutron star will lose enough energy to be captured. If Rubakov and Callan are correct, then once inside the star monopoles will catalyze B-decay at a furious rate, releasing large amounts of energy. A galactic flux as small as $10^{-26}/cm^2.sec.ster.$, ten orders of magnitude below the Parker bound, would lead by these processes to an unacceptably large contribution to the luminosity of neutron stars. Even allowing for local enhancements, it would seem that such a tiny flux is unobservable for the forseeable future.

AXIONS

1. Quantum chromodynamics (QCD), the modern theory of the strong interaction, enjoys with one exception a perfect match to the observed symmetries of the strong interaction (Wilczek 1982 d). All these symmetries -- P, C, T, the chiral symmetries SU(3) × SU(3), baryon number conservation, asymptotic scale invariance -- follow almost automatically from the principles of gauge invariance and renormalizability.

The exception is that QCD contains a parameter -- best thought of as the overall phase of the quark mass matrix -- which in general breaks P and T invariance. This parameter must be very small -- the phase of the determinant of the quark mass matrix close to zero -- in order to agree with the observed P and T invariance of the strong interaction, as indicated for instance by the smallness of the neutron electric dipole moment.

It might be thought, and was until corrected in 1976 ('tHooft 1976; Callan et al. 1976 and Jackiw & Rebbi 1976 b), that such a phase could simply be defined away by renaming the quark field variables. Classically indeed this can be done, but here is another case where quantum effects spoil an apparent invariance of the classical Lagrangian. As shown by Fujikawa (1979), renaming the quark field variables changes the measure in path integrals. So we must face up to the question of why the phase of the quark mass-matrix, or θ-parameter, is so tiny.

2. Peccei and Quinn (1977) proposed that the θ-parameter be made a dynamical variable, a dynamical scalar field, rather than a non-dynamical coupling constant as it is in the standard model. This can be accomplished by adding extra Higgs fields into the theory. Different entries in the quark mass matrix can then be proportional to the vacuum expectation values of different Higgs fields. The determinant of the mass matrix is determined by minimizing the potential of the Higgs fields; in particular the potential is to be minimized with respect to different possible relative phases between the Higgs fields. If the potential has a minimum near $\theta = 0$, then we have a theory where the smallness of P and T

violations in the strong interactions is explained.

3. Shortly after this ingenious proposal was made by Peccei and Quinn, Weinberg (1978) and I (1978) noticed an important consequence. The quanta associated with the dynamical θ-field would be quasi-Goldstone bosons -- now known as axions. If the symmetry of the classical Lagrangian which allowed redefinition of the phase of the quark mass matrix were exact axions would be Nambu-Goldstone bosons, precisely massless particles. Instead they acquire a small finite mass through the quantum effects which break this symmetry.

The mass and coupling strength of the axion depend inversely on the energy scale at which the Peccei-Quinn symmetry is broken -- or, in concrete terms, the size of the vacuum expectation values of the Higgs fields which determine the phase of the quark mass matrix.

The mass of the axion is roughly (1 GeV./F) 1 MeV., where F is the above-mentioned scale. The coupling strength is likewise proportional to 1/F. The original idea (Peccei & Quinn 1977; Weinberg 1978; Wilczek 1978) was to break the Peccei-Quinn symmetry at the weak scale, $F \gtrsim 300$GeV. However a variety of experiments searching for the axion failed to find it, and it is generally agreed that the idea in its original form does not work. A very powerful constraint comes from astrophysical considerations (Dicus <u>et al</u>. 1978): axions could provide an overly efficient means of energy transfer for stars. A newer idea is to associate F with the scale of unification, $F \sim 10^{15}$GeV. (Kim 1979; Dine <u>et al</u>. 1981). The axion then couples so weakly that no practical laboratory experiment can hope to detect it. It is easy to construct unified models with this sort of "invisible" axion.

4. Very recently (partly as a result of this conference!) it was realized that the invisible axion scheme with $F \sim 10^{15}$GeV. runs into a serious cosmological problem (Preskill <u>et al</u>. 1982; Abbott & Sikivie 1982; Dine & Fischler 1982).

In fact, there are two not unrelated sorts of problem. One is that typically there are several distinct values of the axion field at which the minimum of the potential energy is achieved. This will generally lead to the formation of domain walls, with disastrous consequences (Sikivie 1982). This problem may be avoided in models where the different minima are continuously related, so that the potential domain walls can smoothly relax away (Lazarides 1982). It may also be avoided in an inflationary universe, if the inflation takes place after the Peccei-Quinn symmetry has broken. In this case, the whole presently observed Universe will lie in a single domain.

The other problem is that the axion field has a very difficult time relaxing to its minimum, even locally. At high temperatures the axion mass and potential generally vanishes rapidly, since the coherent quantum fluctuations which generate its mass are wiped out (Gross et al. 1982). After the mass does turn on, the axion field loses significant energy essentially only due to the expansion of the Universe, and only as R^{-3} like nonrelativistic matter. Whether the axion field is initially chaotic, or (as expected in the inflationary scenario) homogeneous but not at the potential minimum, there is a lot of energy stored in coherent oscillations of this field around its minimum. Calculations show that generically a million times the critical density is stored in the oscillating axion field, which of course is unacceptable.

It is consistent, although rather knavish, to suppose in the inflationary scenario that the axion field "just happened" to start at the value which would later become the minimum of the potential. This requires a coincidence of part in 10^3. Since the whole purpose of introducing the axion field was to understand the smallness of the θ-parameter, it would be quite disappointing to be reduced to invoking a coincidence in the end.

5. Fortunately, a slight but perhaps profoundly significant modification of the invisible axion scheme can remove its cosmological problem. If the scale of Peccei-Quinn symmetry breaking is not the unification scale $F \sim 10^{15}$GeV. but rather $F \sim 10^{12}$GeV., the problem is eased [One must require $F \gtrsim 10^{9}$GeV., for astrophysical reasons (Dicus et al. 1978)]. With a smaller value of F the axion couples more strongly; therefore its mass turns on earlier in the history of the Universe and the field energy gets reduced by expansion of the universe for a longer time. For $F \sim 10^{12}$GeV., the axion field energy becomes comparable to the critical density necessary to close the Universe.

This gives us a new candidate for the missing mass in the Universe -- surely the weirdest yet! For the analysis of fluctuations leading to galaxy formation, axion field energy behaves like dust on scales larger than ~ 1 cm. (the Compton wavelength). Since the axions interact very feebly with ordinary matter, there is essentially no drag on this "dust" and the fluctuations begin to grow as soon as the axions dominate the mass density, at T ∿ 10eV.

6. What is happening at the new scale? The unification symmetry is broken at a significantly larger scale. The straightforward interpretation is that there are gauge singlet fields which are not singlets under the Peccei-Quinn symmetry, and which acquire vacuum expectation values ∿ 10^{12}GeV.

It is natural, and may be very fruitful, to take a wider view here. The idea of axions is to make the phase of the quark mass matrix a dynamical variable. Might other components, or even the entire matrix, be dynamical variables? In other words, might the extra chiral flavor symmetries which the theory without mass terms enjoys be spontaneously broken?

There are several considerations which make this idea attractive (Wilczek 1982 c). The Peccei-Quinn quasi-symmetry is a funny one to postulate since it is broken by quantum effects. It could result from real symmetry considerations if the real symmetries "accidentally" forbade symmetry-breaking interactions. The domain wall problem previously discussed is avoided. Perhaps by imposing symmetries we can make a start on understanding the quark and lepton mass matrices, which at present seem completely mysterious.

7. The idea that flavor or family symmetries are spontaneously broken is subject to an important phenomenological consequence (Wilczek 1982 c). Along with the axion there should be a whole zoo of true Nambu-Goldstone bosons, coupled for instance to strangeness-changing or muon-number changing currents. I am calling these particles familons. Familons should be emitted in decays like $\mu \rightarrow ef$ or $K^+ \rightarrow \pi^+ f$. The coupling strength of familons is determined by the scale F at which the associated symmetries are broken. For F ∿ 10^{12}GeV., one predicts rates for $\mu \rightarrow ef$ and $K^+ \rightarrow \pi^+ f$ which are not far from current experimental limits.

It certainly would be thrilling if this long chain of reasoning and speculation were completed by the discovery of new kinds of massless particles in rare decays!

8. Having made the quark and lepton mass matrix a dynamical variable, one is certainly tempted to consider whether the cosmological

"constant" could similarly be made a dynamical variable [Not unrelated thoughts have also occurred to S. Weinberg (private communication)]. We might then try to understand its smallness dynamically, similarly to the way we address the smallness of the θ-parameter. There would be a cosmion field, with the minimum of its potential occurring very nearly where the effective cosmological constant vanished.

Here is a table of how I envision the analogy:

parameter observed to be "unnaturally" small	θ	Λ
field and particle corresponding to dynamical version of parameter	a, axion	λ, cosmion
coupling of field	$ma\bar{\psi}\gamma_5\psi \overset{\sim}{=} \partial_\mu a(\bar{\psi}\gamma_5\gamma_\mu\psi)$	$\lambda T^\mu_\mu = \partial_\mu \lambda(x^\nu T^\mu_\nu)$
associated quasi-symmetry	axial baryon number (Peccei-Quinn)	scale invariance

Here T^μ_ν is the energy momentum tensor, $\bar{\psi}\gamma_5\psi$ is shorthand for the current generating the Peccei-Quinn symmetry, and $\overset{\sim}{=}$ means equals classically, with quantum corrections required.

9. There are several serious difficulties in constructing an honest theory of cosmions, but I think there is at least a special class of models in which it can be done. In a wide variety of supersymmetric models, the divergences of the dilation current and of a certain axial current (associated with R-invariance) lie in a single supersymmetry multiplet (Wess & Zumino 1975). If we can construct in a supersymmetric manner an axion coupled to this axial current, it will have a cosmion as its supersymmetry partner. Close study of such models is certainly called for.

REFERENCES

Abbott, L. & Sikivie, P. (1982). Phys. Lett. 120B, 133.
Albrecht, A. & Steinhardt, P. (1982). Phys. Rev. Lett. 48, 1220.
Albrecht, A. et al. (1982). Phys. Rev. Lett. 48, 1437.
Cabrera, B. (1982). Phys. Rev. Lett. 48 , 1378.
Callan, C. (1982). Phys. Rev. D25, 2141; Phys. Rev. D (to appear); l'Ecole Normale Superiere preprint LPTENS 82120.
Callan, C. et al. (1976). Phys. Lett. 63B, 334.
Dicus, D. et al. (1978). Phys. Rev. D18, 1920.
Dimopolous, S. et al. (1982 a). Nature 298, 824.
Dimopolous, S. et al. (1982 b). ITP preprint 82-91 (to appear in Phys. Lett. B).
Dine, M. & Fischler, W. (1982). Penn. preprint UPR-0201T.
Dine, M. et al. (1981). Phys. Lett. 104B, 199.
Fujikawa, K. (1979). Phys. Rev. Lett. 42, 1195.
Georgi, H. & Glashow, S. (1974). Phys. Rev. Lett. 32, 438.
Gross, D. et al. (1982). Rev. Mod. Phys. 53, 43.
Guth, A. (1981). Phys. Rev. D23, 347.
'tHooft, G. (1974). Nucl. Phys. B79, 276.
'tHooft, G. (1976). Phys. Rev. Lett. 37, 8; Phys. Rev. D14, 3432.
Jackiw, R. & Rebbi, C. (1976 a). Phys. Rev. D13, 3398.
Jackiw, R. & Rebbi, C. (1976 b). Phys. Rev. Lett. 37, 172.
Kim, J. (1979). Phys. Rev. Lett. 43, 103.
Kolb, E. et al. (1982). (to appear in Phys. Rev. Lett.).
Lazarides, G. & Shafi, Q. (1982). Phys. Lett. 115B, 21.
Linde, A. (1982). Phys. Lett. 108B, 389.
Parker, E. (1971). Ap. J. 163, 225; ibid., 166, 295.
Peccei, R. & Quinn, H. (1977). Phys. Rev. Lett. 38, 1440; Phys. Rev. D16, 1791.
Polyakov, A. (1974). JETP Lett. 20, 194.
Preskill, J. (1979). Phys. Rev. Lett. 43, 1365.
Preskill, J. (1982). This volume.
Preskill, J. et al. (1982). Phys. Lett. 120B, 127.
Rubakov, V. (1981). JETP Lett. 33, 664; (1982). Nucl. Phys. B203, 331.
Sikivie, P. (1982). Phys. Rev. Lett. 48, 1156.
Turner, M. (1982). Phys. Lett. 115B, 95.
Turner, M. et al. (1982). Enrico Fermi Institute preprint 82-18.
Weinberg, S. (1978). Phys. Rev. Lett. 40, 223.
Wess, J. & Zumino, B. (1975). Nucl. Phys. B87, 207.
Wilczek, F. (1978). Phys. Rev. Lett. 40, 279.
Wilczek, F. (1982 a). Phys. Rev. Lett. 48, 1146.
Wilczek, F. (1982 b). "Monopole-Fermion Interactions" (to appear in the 1982 SLAC Summer School).
Wilczek, F. (1982 c). ITP preprint 82-100 (submitted to Phys. Rev. Lett.).
Wilczek, F. (1982 d). Ann. Rev. Nucl. Sci. 32.
Zeldovich, Ya. & Khlopov, M. (1979). Phys. Lett. 79B, 239.

IS INFLATION NEEDED TO SUPPRESS MONOPOLES?

Don N. Page
Department of Physics, Pennsylvania State University,
University Park, Pennsylvania 17802, USA

Abstract. Inflationary models may not be successful or necessary to explain the cosmological flatness, isotropy, homogeneity, horizon, galaxy formation and domain wall problems. Alternative solutions to several of these are suggested. They lead one to question whether inflation is necessary to solve the primordial monopole problem. Three-body nonradiative recombination is shown to be effective in getting nearly all monopoles to bind into pairs. However, the pairs would apparently not annihilate fast enough to avoid either too many monopoles or too much entropy today in noninflationary models.

NON-INFLATIONARY EXPLANATIONS IN COSMOLOGY

The inflationary model of the early universe (Guth 1981) has been suggested as a solution to the cosmological flatness, isotropy, homogeneity, horizon, galaxy formation, domain wall and monopole problems. However, the continuing difficulties with this model, as highlighted by the results of this workshop, lead one to ask whether there might not be other solutions to these outstanding cosmological problems. Indeed, for many of these problems one can invoke other solutions.

For example, the flatness problem (why the universe has expanded to at least 10^{183} Planck volumes without gravity becoming unimportant) might be explained by the weak anthropic principle applied to the quantum state of the universe (Carter 1974). That is, life might be possible only in those components of the universal wavefunction in which space grows large enough and gravitational forces remain strong enough for galaxies, stars and planets to form. Then our observation of the very close balance between expansion and gravity would be merely a selection effect conditioned by our existence.

The observed large-scale isotropy and homogeneity of the universe does not seem to be directly explained by the weak anthropic principle, since it is hard to see how our existence is directly dependent upon conditions far away. However, if one formulates the second law of thermodynamics as the statement that the universal wavefunction lacked long-range spatial correlations when the universe was very small, the large-scale isotropy and homogeneity might automatically follow. This is because large-scale anisotropies and inhomogeneities represent nontrivial gravitational fields which would seem to require long-range spatial correlations in the quantum state for the metric. If such correlations were absent initially, the large-scale anisotropies and inhomogeneities could only have developed causally during the expansion and so should be small today. Of course, this suggested solution merely reformulates the mystery of the second law (which is also unexplained by the inflationary models), but it shows how the second law and the large-scale isotropy and homogeneity might all be consequences of the single hypothesis of no long-range spatial correlations in the initial wavefunction for the universe. Note that this modification of the principle of ignorance (Hawking 1976) is not subject to the criticism that anisotropic gravitational fields (e.g. black holes) have higher entropy (Penrose 1979, 1981), for the quantum state is not assumed here to maximize the entropy.

The horizon problem is not really an independent problem but rather an emphasis on the mystery of the large-scale isotropy and homogeneity if the distant regions that appear remarkably similar were never in causal contact in their past evolution. As demonstrated by the solution to the isotropy and homogeneity problem suggested above, the particle horizons do not necessarily have to be eliminated. Indeed in classical general relativity or even in quantum field theory in a classical space-time metric, the initial conditions are given over an acausal hypersurface, so particle horizons are always present in some form or other. The situation is less clear in quantum gravity, since the quantum uncertainties in the light cones make the whole concept of horizons rather fuzzy.

The problem of getting the right perturbations for galaxy formation has as yet no satisfactory solution in any model. The inflationary models are very beautiful for being among the first to give concrete predictions for the perturbations. Unfortunately, as shown by many other participants at this workshop, even the new inflationary models (Linde 1982; Hawking & Moss 1982; Albrecht & Steinhardt 1982) predict perturbations far larger than those observed. Unless the effective potential for the matter fields has a shape that at present seems highly ad hoc, the inflationary models are actually inconsistent with the large-scale isotropy and inhomogeneity it was hoped they would explain. Thus one needs to look further for a solution to the galaxy formation problem.

The domain wall problem arises in quantum field theories with a broken discrete symmetry so that the universe tends to freeze into domains differing by the discrete symmetry and separated by domain walls. The inflationary models might explain why no such walls are seen, by predicting that the domain we are in has expanded beyond the limits of what we can now see. However, a simpler explanation might be that the fundamental theory merely has no broken discrete symmetry.

Finally we come to the monopole problem, for which inflation has appeared to be one of the most attractive solutions. The grand unified theories in which electromagnetism is embedded in a compact semisimple gauge group in order to explain the discreteness of charge have magnetic monopoles (Polyakov 1974; 't Hooft 1974). These would tend to be produced in large numbers cosmologically when the semisimple group spontaneously breaks down to its subgroup containing a U(1) factor, and a large fraction of these monopoles would apparently survive annihilation (Zeldovich & Khlopov 1978; Preskill 1979). Unless the universe later inflates sufficiently to dilute this monopole density by an enormous factor, it would be far too high to be compatible with the present cosmological mass density. If one is to find an acceptable cosmological model without inflation, this monopole problem is a key difficulty to be addressed.

MONOPOLE ANNIHILATION BY THREE-BODY RECOMBINATION

Here I shall report on an investigation with Dicus et al. (1982) of monopole annihilation resulting from previously unconsidered three-body recombination. One might think that this would be much slower than the two-body recombination that has been calculated, but since it does not require the emission of radiation it is actually faster once the monopoles cool sufficiently. However, starting with the abundances given by Preskill (1979) and using a noninflationary model, the three-body recombination does not occur at a significant rate compared with the expansion until the monopoles become decoupled from the plasma. After that the monopoles could indeed nearly all bind, but the resulting pairs would not spiral together and annihilate fast enough to get rid of the excess monopoles without producing far too much entropy. Thus we have not succeeded in suppressing the density of monopoles without inflation.

Results assuming homogeneity

If the monopoles have mass $m \sim 10^{16}$ GeV $\sim 10^{-3}\ m_{Planck}$, magnetic charge $g \sim \alpha^{-1/2} \sim 10$, temperature T_M and number density $n_M = \varepsilon s$ where $s \sim T_p^3$ is the entropy density of the relativistic plasma with temperature T_p, then the two-body radiative recombination rate per monopole is

$$\tau_2^{-1} \sim m^{-3/2}\ \alpha^{-3}\ T_M^{-1/2}\ \varepsilon\, T_p^3\ y\ , \qquad (1)$$

where $y \equiv n_f/n_M$ is the fraction of monopoles that are free. This applies for $T_M \leq m\alpha^{5/2} \sim 10^{11}$ GeV in which case discreteness effects for the bremsstrahlung photons make the capture cross-section lower than a classical calculation would indicate. The three-body recombination rate (where a third monopole carries off the excess energy in a Coulombic collision) is

$$\tau_3^{-1} \sim m^{-1/2}\ \alpha^{-5}\ T_M^{-9/2}\ \varepsilon^2\ T_p^6\ y^2\ . \qquad (2)$$

Starting with Preskill's result of $\varepsilon \sim G^{1/2} m\alpha^3 \sim 10^{-9}$, at $T_p \sim m\alpha^2 \sim 10^{12}$ GeV, τ_3^{-1} is faster than τ_2^{-1} for $T_M \sim T_p \leq m\alpha^{-2}\varepsilon \sim 10^{11}$ GeV but is slower than the expansion rate

$$H \sim G^{1/2}(m\varepsilon T_p^3 + T_p^4)^{1/2} \qquad (3)$$

roughly until the monopoles begin dominating the mass density at $T_p \sim m\varepsilon \sim 10^7$ GeV. Then a fraction less than but of order unity of the monopoles bind into pairs, and the plasma drag causes them to spiral together and annihilate. This increases the plasma entropy by a factor of roughly $G^{-1/2} m^{-1} \alpha^{-1/3} \sim 10^{-4}$ and decreases ε to $\sim G m^2 \alpha^{10/3} \sim 10^{-13}$ while the plasma drag is still effective in leading to annihilation.

However, at $T_d \sim G^2 m^5 \alpha^{10/3} \sim 1$ MeV, the monopole-plasma scattering rate

$$\tau_s^{-1} \sim m^{-1}\ T_p^2 \qquad (4)$$

becomes less than the expansion rate H. Then the plasma begins to decouple from the monopoles and no longer causes them to spiral in and annihilate once they are bound into pairs. As nonrelativistic particles, the unbound monopoles begin cooling faster than the relativistic plasma as the universe expands. This allows the recombination rate τ_3^{-1} to exceed the expansion rate H, and our results show that the fraction of free monopoles decreases as

$$y \sim (1 + G^7 m^{31/2} \alpha^{10} t^{3/2})^{-1/2} . \tag{5}$$

All but a fraction $y \sim 10^{-13}$ would be bound by $t \sim 10^{10}$ yr., but we estimated that after collisions the bound pairs would be left with orbital quantum numbers of order $G^{7/6} m^{-7/3} \alpha^{-5/2} \sim 10^{12}$ and so would take a time of order $G^7 m^{-15} \alpha^{-10} \sim 10^{15}$ yr. to spiral together by the emission of magnetic dipole radiation. Thus a homogeneous distribution of monopoles with the initial density given by Preskill (1979) would have nearly all formed bound pairs by today but would not have annihilated, in violent disagreement with the limits implied by the density of the universe.

Results assuming clumping

We also examined the effects of gravitational clumping on the three-body recombination. The enhancement in the two-body recombination had already been shown to be insufficient to produce enough monopole annihilation (Goldman et al. 1981; Fry 1981). We concurred that the annihilation could not be effective for the unstable clumps that might form before the monopoles decouple from the plasma. For quasi-equilibrium clumps of monopoles that might form after this decoupling, we found that monopole-monopole scattering would cause the free monopoles to evaporate from the clump faster than they formed bound pairs unless the number of monopoles in the clump was less than $N_0 \sim (Gm^2\alpha)^{-7/4} \sim 10^{14}$. In order that a sufficient fraction of monopoles bind and annihilate rather than escape, we needed nearly all of them to go into clumps with less than $\sim 10^{10}$ monopoles each. But by the time that the Jeans mass is small enough for such clumps to form, the monopole energy density would exceed the plasma density by a factor of at least 10^{14} , so even if nearly all the monopoles could annihilate then, far too much entropy would be produced to be consistent with the observed entropy-to-baryon ratio. In our universe the entropy attributable to monopoles is much less and may be limited to our unsuccessful efforts to understand why.

ACKNOWLEDGMENTS

This paper is largely based upon a collaboration with Duane Dicus and Vigdor Teplitz which was supported by grants from the U.S. Department of Energy and the National Science Foundation.

REFERENCES

Albrecht A. & Steinhardt P.J. (1982). Cosmology for grand unified theories with radiatively induced symmetry breaking. Phys.Rev.Lett. 48, no.17, 1220-3

Carter B. (1974). Large number coincidences and the anthropic principle in cosmology. In IAU Symposium No.63: Confrontation of Cosmological Theories with Observational Data, ed. M.S. Longair, pp.291-8. Dordrecht: D. Reidel Publishing Company.

Dicus, D.A., Page D.N. & Teplitz V.L. (1982). Two- and three-body contributions to cosmological monopole annihilation. Phys. Rev. D26, no. 6, 1306-16.

Fry J.N. (1981). Gravitational correlations and the monopole problem. Astrophys.J. 246, no.3, L93-7

Goldman T., Kolb E.W. & Toussaint D. (1981). Gravitational clumping and the annihilation of monopoles. Phys.Rev. D23, no.4, 867-75

Guth A.H. (1981). Inflationary universe: a possible solution to the horizon and flatness problems. Phys.Rev. D23, no.2, 347-56

Hawking S.W. (1976). Breakdown of predictability in gravitational collapse. Phys.Rev. D14, no.10, 2460-73

Hawking S.W. & Moss I.G. (1982). Supercooled phase transitions in the very early universe. Phys.Lett. 110B, no 1, 35-8

Linde A.D. (1982). A new inflationary universe scenario: a possible solution of the horizon, flatness, homogeneity, isotropy and primordial monopole problems. Phys.Lett. 108B, no.6, 389-93

Penrose R. (1979). Singularities and time-asymmetry. In General Relativity: An Einstein Centenary Survey, eds. S.W. Hawking & W. Israel, pp.581-638. Cambridge: Cambridge University Press

Penrose R. (1981). Time-asymmetry and quantum gravity. In Quantum Gravity 2: A Second Oxford Symposium, eds. C.J. Isham, R. Penrose & D.W. Sciama, pp.244-72. Oxford: Clarendon Press

Polyakov A. (1974). Pis'ma Eksp.Teor.Fiz. 20, 430 [JETP Lett. 20, 194]

Preskill J.P. (1979). Cosmological production of superheavy magnetic monopoles. Phys.Rev.Lett. 43, no.19, 1365-8

't Hooft G. (1974). Magnetic monopoles in unified gauge theories. Nucl. Phys. B79, 276-84

Zeldovich Ya.B. & Khlopov M.Yu. (1978). On the concentration of relic magnetic monopoles in the universe. Phys.Lett. 79B, no.3, 239-41

MASSIVE NEUTRINOS AND PHOTINOS IN COSMOLOGY AND GALACTIC ASTRONOMY

D.W. Sciama
Department of Astrophysics, Oxford, U.K.
International School for Advanced Studies, Trieste, Italy
International Centre for Theoretical Physics, Trieste, Italy

Abstract. The possibility is discussed that massive neutrinos or photinos may provide the critical density for the universe and the dark matter needed in galactic halos. They may also be a source of ultra-violet photons with observable ionising effects. All these possibilities could be realized with a particle mass ~ 100 electron volts and a radiative lifetime $\sim 10^{27}$ seconds. Reasons are given for preferring photinos to neutrinos with these properties.

THE CRITICAL DENSITY

We shall take as the starting-point of our discussion the prediction of the new inflationary scheme that the present density of the universe is essentially equal to the critical density. Since the age of the universe is observed to exceed 10^{10} years it follows that, if the cosmical constant is zero, the Hubble constant cannot exceed 50 km.sec^{-1} Mpc^{-1}. For definiteness we shall adopt this value, which falls within its observed range.

Since the critical density exceeds the baryon density (both observed and inferred from the formation of D, He^3 , He^4 and Li^7 in the big bang) by a factor ~100, we have to identify the "missing matter". We first examine the hypothesis that it consists of massive neutrinos. It has been known for a long time that conventional (left-handed Majorana) neutrinos thermally excited in the early stages of the universe would have decoupled at a temperature 1-3 Mev. After this decoupling electron pairs annihilated (at $T \sim \frac{1}{2}$ Mev), with the resulting photons feeding the 3°K radiation field but not the neutrinos. As a result of this (and the Fermi-Dirac statistics of the neutrinos)their present concentration n_ν is related to that of the 3°K photons n_γ by

$$n_\nu \sim \frac{1}{4} n_\gamma$$
$$\sim 100 \text{ cm.}^{-3}$$

The contribution of one neutrino type ρ_ν to the present density of the universe is given by

$$\rho_\nu \sim \frac{m_\nu}{100 h_o^2} \cdot \rho_{crit},$$

where h_o is related to the Hubble constant H_o by

$$H_o = 100 h_o \text{ km.sec}^{-1} \text{ Mpc}^{-1}.$$

and m_ν is the rest-mass of the neutrino in electron volts. We now assume that $\rho_\nu = \rho_{crit}$ and $h_o = 0{\cdot}5$. Accordingly we require that

$$m_\nu \sim 25 \text{ ev.}$$

(If several neutrino flavours are involved in this argument we would have $\Sigma m_\nu \sim 25$ ev).

GALACTIC DYNAMICS

We now compare this value of m_ν with that deduced from the further hypothesis that massive neutrinos provide the "missing matter" now believed to exist in extended galactic halos. This belief stems from the observation that many galaxies (including our own) possess flat (or even slowly rising) rotation curves out to great distances from their centres, in some cases as much as about 50 kiloparsecs. If the required gravitating material is in a spherical halo, the density distribution of this material ρ_h would be given by

$$\rho_h \propto \frac{1}{r^2},$$

for flat rotation curves. The increase of ρ_h as r decreases would however have to cut off at small r, where stars and gas begin to make a significant contribution to the gravitational field, but where the rotation curve remains flat.

For example, in the detailed model of our Galaxy constructed by Caldwell and Ostriker (1981) it is assumed that

$$\rho_h = \frac{\rho_o}{1 + \left(\frac{r}{a}\right)^2}$$

They then find that

$$a = 7 \cdot 8 \text{ kpc} \quad (\sim r_\odot)$$

$$\rho_h(\odot) = 4 \cdot 5\text{x}10^{-25} \text{ gm.cm}^{-3}$$

$$\rho_o = 10^{-24} \text{ gm.cm}^{-3} ,$$

with an uncertainty in ρ_o of about a factor 2.

It is tempting to identify the cut-off in ρ_h at $r \sim a$ with the existence of a maximum density for the neutrino distribution in accordance with the Liouville theorem, which would apply to decoupled neutrinos (Tremaine and Gunn 1979). One finds that

$$\rho \leqslant \rho_Q = \left(\frac{2\pi}{3}\right)^{3/2} \frac{m_\nu^4 \; v_o^3}{h^3} ,$$

if the neutrinos have a 3-dimensional Gaussian velocity distribution with dispersion v_o. Phase mixing could reduce the effective phase space density if violent relaxation occurs, but according to the numerical simulations of Melott (1982 a, b) this effect would be small in the central regions of the collapsing neutrino distribution. It is just in these regions that the velocity dispersion of the neutrinos remains low enough ($\sim$ 200 km. sec.$^{-1}$) for them to be able to be bound in galactic halos.

The quantity ρ_Q has been called the quasi-degenerate density (it differs by a numerical factor $\sim$1 from the true degenerate density associated with Fermi-Dirac statistics). Its general order of magnitude can be inferred from the simple estimate that if $v_o \sim 100$ km.sec.$^{-1}$ and $m_\nu \sim 100$ ev, then $\rho_Q \sim 10^{-24}$ gm.cm.$^{-3}$, which is indeed a galactic order of magnitude.

We now consider the hypothesis that

$$\rho_o \sim \rho_Q .$$

To estimate v_o we assume provisionally that it is independent of r (quasi -isothermal distribution). Then from the consideration of the outer halo we would have

$$v_o = \sqrt{\frac{3}{2}}\; v_c ,$$

where v_c is the (constant) rotation velocity. Since v_c is observed to be ~ 220 km.sec.$^{-1}$ we would have

$$v_o \sim 270 \text{ km.sec}^{-1}$$

Accordingly, our hypothesis that $\rho_o \sim \rho_Q$ would imply that

$$m_\nu \sim 45 \text{ ev.}$$

At first sight this value for m_ν is in remarkable agreement with our independent estimate of 25ev based on the hypothesis that neutrinos possess the critical density. However, any reduction in the phase-space density due to phase mixing would increase the value of m_ν derived from the analysis of galactic halos. In addition, we might wish to identify with neutrinos the further "missing matter" near the sun, derived by Oort from his analysis of the motions of stars perpendicular to the galactic plane. This could easily increase m_ν to about 100 ev, which would definitely be discrepant with our original estimate of 25 ev.

It seems at least worth exploring whether we could have such a large rest-mess for the dominating particle while retaining the hypothesis that its cosmological density has the critical value. To avoid an age problem for the universe we could (a) introduce a λ term, (b) invoke population III sources which produce a substantial fraction of the 3°K photons at a redshift ~ 100, thereby suppressing the neutrino concentration n_ν or (c) use massive photinos ($\tilde{\gamma}$) with $n_{\tilde{\gamma}} < n_\nu$ instead of neutrinos. We here examine the photino possibility.

We can obtain the required suppression of the photino concentration $n_{\tilde{\gamma}}$ by supposing that photinos decouple before pions and muons annihilate (Sciama 1982a). For example, if the decoupling temperature $T_d \sim 200$ Mev, then $n_{\tilde{\gamma}} \sim \frac{1}{4} n_\nu$. This would permit $m_{\tilde{\gamma}} \sim 100$ ev, with $\rho_{\tilde{\gamma}} \sim \rho_{crit}$. It would also have implications for the coupling constants of photinos and goldstinos and for the capacity of supersymmetry to solve the hierarchy problem, which are explored elsewhere (Sciama 1982 a, b). These implications are numerically self-consistent and in particular would lead to a definite value for the supersymmetry breaking parameter d, namely,

$$d \sim 10^8 \text{ Gev}^2$$

This suppression of $n_{\tilde{\gamma}}$ would also fit in with the recent tightening of the restriction on the number of particle types N which were relativistic at $T \sim 1$ Mev, derived by comparing the results of nucleosynthesis calculations with observation. One now tends to find that $N_\nu < 4$ (e.g. Barrow and Morgan 1982). If one supposes that the mass of the tau neutrino is less than 1 Mev, then already $N_\nu = 3$, so allowing less than one further "neutrino type". Thus, one would require some suppression of photinos relative to neutrinos (if they exist and have a rest-mass less than 1 Mev). Since there are no annihilating particles lying in rest-mass between electrons and pions/muons, this would immediately require that T_d for photinos jumps to at least 100-200 Mev, with the consequent implications for the supersymmetric coupling constants and the mass of the photino if $\rho_{\tilde{\gamma}} = \rho_{crit}$.

ULTRA-VIOLET ASTRONOMY

Massive neutrinos and photinos may be able to emit photons, and these photons might be detectable by ultra-violet astronomers (de Rujula and Glashow (1980), Cabibbo et al (1981)). The reactions involved are

$$\nu_1 \rightarrow \gamma + \nu_2$$

$$\tilde{\gamma} \rightarrow \gamma + \tilde{g}$$

where ν_2 is a second neutrino and $\tilde{g}$ is a massless goldstino (or massive gravitino in locally supersymmetric or supergravity theories).

If the parent particle is at rest, the energy E_γ of the photon is given by

$$E_\gamma = \frac{m_1^2 - m_2^2}{2m_1}$$
$$\sim \tfrac{1}{2} m_1 \quad \text{if } m_1 >> m_2 ,$$

which we shall assume for simplicity here. Thus, if $m_1 \sim$ 20-100 ev, $E_\gamma \sim$ 10-50 ev, corresponding to an extreme ultra-violet photon.

The radiative lifetimes τ have been calculated and are found to be given by

$$\tau_\nu \sim \frac{10^{36}}{\sin^2 2\beta_1} \left(\frac{30 \text{ ev}}{m_1}\right)^5 \text{sec} \qquad \text{with GIM suppression}$$

where β_1 is the oscillation angle between ν_1 and ν_2 :

$$\tau_\nu \sim \frac{10^{30}}{\sin^2 2\beta_2} \left(\frac{30 \text{ ev}}{m_1}\right)^5 \text{sec} \qquad \text{without GIM suppression}$$

and

$$\tau_{\tilde{\gamma}} \sim 7 \times 10^{22} \left[\frac{d}{(100 \text{ Gev})^2}\right] \left(\frac{30 \text{ ev}}{m_{\tilde{\gamma}}}\right)^5 \text{sec.}$$

Thus eliminating GIM suppression would reduce the radiative lifetime of the neutrino a millionfold (if $\beta_2 \sim \beta_1$) while with our suggested value of $\sim 10^8 \text{Gev}^2$ for d, the photino lifetime would be of the same order as the smaller neutrino one.

I have summarized the observational situation elsewhere (Sciama 1982c). Summarizing this summary, one can say that the non-observation of any effect attributable to photons from neutrinos/photinos dominating the universe/Galaxy clusters/Galaxy halos implies that

$$\tau > 10^{24} - 10^{25} \text{ sec.}$$

A radical suggestion (Sciama and Melott 1982, Sciama 1982a) is that these photons produce the high ionisation (Si|V and C|V) observed in galactic halos, including our own. This would require

$$m_{\nu,\tilde{\gamma}} \sim 100 \text{ ev}$$

$$\tau \sim 10^{27} \text{ secs.}$$

The high mass required suggests that photinos rather than neutrinos would have to be involved. It is encouraging that the required value of τ agrees well with its theoretical value for $m_{\tilde{\gamma}} \sim 100$ ev and $d \sim 10^8 \text{Gev}^2$, and so with the possible requirement of the photino suppression. A test of these ideas would be to search for a photon line at ~ 50 ev in the galactic background from a space experiment. We hope to carry out such an experiment in the next two years.

REFERENCES

Barrow, J.D. and Morgan, J. (1982), to be published.
Cabibbo, N., Farrar, G.R. and Maiani, L. (1981). Phys. Lett. 105B, 155-158.
Caldwell, J.A.R. and Ostriker, J.P. (1981). Ap. J. 251, 61-87.
Melott, A.L. (1982a). Phys. Rev. Lett. 48, 894-896.
Melott, A.L. (1982b). Nature 296, 721-723.
de Rujula, A. and Glashow, S. (1980). Phys. Rev. Lett. 45, 942-944.
Sciama, D.W. and Melott, A.L. (1982). Phys. Rev. D. 25, 2214-2217.
Sciama, D.W. (1982a). Phys. Lett. 114B, 19-22.
Sciama, D.W. (1982b). Phys. Lett. 118B, 327.
Sciama, D.W. (1982c). I.A.U. Symposium No. 104.
Tremaine, S. and Gunn, J.E. (1979). Phys. Rev. Lett. 42, 407-410.

SUPERHEAVY PARTICLES IN COSMOLOGY AND EVOLUTION OF INHOMOGENEITIES IN THE EARLY UNIVERSE

M.Yu. Khlopov
Institute of Applied Mathematics, Moscow.

A.G. Polnarev
Space Research Institute, Moscow.

ABSTRACT

The stages of dominance of superheavy metastable particles, predicted by GUTs, are shown to result in the formation of primordial black holes (PBH) in the course of evolution of small initial inhomogeneities. The minimal probability of PBH formation is estimated. The relationship between the spectrum of these PBH, the spectrum of initial metric fluctuations and the parameters of the GUTs is established. Observational astrophysical restrictions on the PBH spectrum then provide a number of restrictions on the parameters of the GUTs depending on the amplitude of initial metric perturbations.

INTRODUCTION

The fundamental relationship between particle physics and the evolution of the early Universe acquires more and more attention both from physicists, working in particle physics, and from astrophysicists, involved in the problems of cosmology. Detailed analysis of this relationship makes it possible to evolve the cosmology of the early Universe. On the other hand, analysis of cosmological consequences of particle theories provides unique information on the properties of particles and their interactions, unobtainable by modern laboratory means (Dolgov & Zeldovich 1981). Indeed the observational data on the dynamics of cosmological expansion, on the isotropy and on the spectrum of electromagnetic background, the data on the light element abundances, the observed large scale structure of the Universe make it possible to evolve a definite qualitative picture of the evolution of the Universe one second after the beginning of its expansion (Zeldovich & Novikov 1975; Weinberg 1977). This picture is based on physical laws, proved in laboratories. For a description of earlier stages of the evolution modern particle theory (Okun 1980) is needed. In the framework of such theories, unifying various particle interactions, a number of non-trivial scenarios of the evolution of the early Universe (Linde 1979) have been evolved (yet qualitatively), which provide an explanation of the observed entropy and baryon asymmetry of the Universe (Dolgov & Zeldovich 1981). The mechanisms of spontaneous symmetry breaking involved in such theories imply phase transitions (Linde 1979), inducing the formation of superheavy magnetic monopoles (Polyakov 1974; t'Hooft 1974; Kibble 1976; Zeldovich & Khlopov 1978; Preskill 1979), and in some cases of superdense walls (Kibble 1976; Zeldovich *et al* 1974) or strings (Kibble 1976). Besides the magnetic monopoles, these theories predict a number of new particles. Some of them are stable and in the framework of the above-mentioned scenarios their residual concentration may be estimated. Analysis of the influence of these particles on processes taking place in the Universe after 1 s, such as nucleosynthesis, the formation of the spectrum of relic radiation, the formation of the observed large-scale structure of the Universe and the dynamics of the Universe as a whole, provides restrictions on the number of types of such particles (Schvartzman 1969), their masses (Gerstein & Zeldovich 1966) and residual concentrations (Dolgov & Zeldovich 1981). This in turn results in restrictions on the parameters of the theories which predict these particles. The possible relationship of massive stable particles to the evolution of inhomogenities in the Universe, disclosed in such analysis, is to be mentioned especially. Effects of stable particles in the evolution of modern large scale structures were widely discussed (Szalay & Marx 1976; Bond *et al* 1980; Peebles 1980; Bisnovatyi-Kogan & Novikov 1980; Doreshkevich *et al* 1981; Zeldovich & Khlopov 1981) in connection with the possible existence of neutrino mass (Lyubimov *et al* 1980). Theoretical study of the spatial distribution of new particles in observed inhomogenities of the modern Universe provide stronger restrictions (Zeldovich *et al* 1980) on the parameters of these particles, as compared to ones (Vysotsky *et al* 1977; Lee & Weinberg 1977), obtained from analysis of the possible effect of these particles on the evolution of the Universe as a whole. On the other hand, such studies show the way to search for these particles (Khlopov 1980a,b; Blinnikov & Khlopov 1982).

However, most of the new particles predicted by the theories are unstable and their lifetime is substantially smaller than 1 s. Furthermore, for most of such particles the predicted masses are so large that formation of the particles in the late Universe, i.e. after 1 s from the beginning of expansion, is practically excluded by energetical reasons, and their existence was possible in the very early Universe only.

In the period when the thermal energy of particles, kT, exceeded their rest energy, mc^2, these particles did not differ from other species of relativistic particles from the point of view of their influence on the dynamics of early Universe as a whole and on the evolution of inhomogeneities in this period. If the lifetime of particles exceeds essentially the cosmological timescale in the period when $kT \sim mc^2$, the role of such longlived particles in the evolution of inhomogenieties in the early Universe may be similar to the role of stable particles in the evolution of inhomogeneities in the late Universe. It is natural to ask whether it is possible, by means of astrophysical data, to obtain information on these small scale inhomogeneities and on the effects of longlived particles on their development, thereby performing an indirect test of the theories which predict the existence of these longlived particles. In the present communication we shall demonstrate the principal possibility of obtaining such information on the basis of astrophysical data and we shall give a number of quantitative restrictions on the parameters of grand unified models which predict the existence of longlived particles.

Various models of grand unified theory predict several types of superheavy metastable particles with different interactions, masses and lifetimes. The frozen concentration of metastable particles may be so large that non-relativistic particles start to dominate in the early Universe, leading to early dust-like stages in the cosmological expansion.

At such stages, growth of small initial fluctuations is possible, resulting in a nonlinear stage in the formation of configurations separated from the general cosmological expansion. A small part of the configuration may, in the course of successive contractions be converted into black holes, which will be retained in the Universe after the metastable particles, from which they originated, cease to exist. So, as it is shown in Chapter 2, a principally new mechanism of primordial black hole (PBH) formation arises, which is determined by the parameters of gauge theories. For given initial fluctuations, an estimate based on this mechanism of the minimal probability of PBH formation is possible.

In Chapter 3 this minimal probability is compared with observational astrophysical restrictions on the PBH spectrum. This provides restrictions on the parameters of grand unified models, which substantially depend on the magnitude of initial inhomogeniety.

1. SUPERHEAVY METASTABLE PARTICLES IN GRAND UNIFIED THEORIES

1.1 MASSES AND LIFETIMES OF SUPERHEAVY PARTICLES

The principles of local gauge invariance enabled weak and electromagnetic interactions to be described in the framework of unified theory (Okun 1980; Weinberg 1967; Salam 1968). This approach appeared to be fruitful also in the development of the strong interaction theory - quantum chromodynamics (QCD)(Okun 1980; Gross & Wilczek 1973). The theory of unified electro-weak interaction (EWI) predicts weak logarithmic growth of the coupling constant with energy (Okun 1980). QCD predicts a logarithmic decrease of the QCD coupling constant with energy which is consistent with the accelerator data (Okun 1980; Gross & Wilczek 1973). Extrapolation of these tendencies towards energies inaccessible even in modern laboratory conditions, gives the energy scale of order $\Lambda \sim 10^{14} - 10^{16}$ GeV at which the QCD and EWI constants are equal. Just that very point is the base of modern models of grand unified theory (GUT)(Okun 1981), unifying strong, weak and electro-magnetic interaction. These models predict the magnitude of the EWI parameter $\sin^2\theta w$ which is consistent with experimental data. Other predictions of these models which are being tested experimentally involve the proton instability (Learned *et al* 1979; Cherry *et al* 1981; Fiornini 1982) and neutron-antineutron oscillations (Kuzmin 1970). It should be stressed that all these effects, which are related to an energy scale much smaller than the unification scale, have very small probability, since they are induced by transitions with superheavy virtual particles with the masses $m \sim \Lambda$). (Similar transitions may induce in some models small neutrino mass (Zeldovich & Khlopov 1981; Witten 1979).) The characteristic lifetime of such particles, predicted by the theory, is determined by the probability of decay of a boson into two light fermions or of a fermion into a boson and a fermion and is very small

$$W_o = 1/\tau = k_o m \qquad (1.1)$$

where the numerical factor $k_o \sim \alpha \div \alpha^2$ depends on the properties of concrete particles. However, besides such shortliving particles the theories predict several possible types of metastable particles, whose lifetimes substantially exceeds the "normal" timescale for shortliving particles, i.e. for which $k_o \ll \alpha^2$. Since, as we mentioned in the Introduction, we are interested in longliving particles, it is natural to restrict our consideration to such metastable particles. Metastability (or even absolute stability) of a particle in the theory may arise for two reasons. First, there may be a new quantum number in the theory. Magnetic monopoles (Polyakov 1974; t'Hooft 1974; Bisnovatyi-Kogan & Novikov 1980) whose absolute stability relative to decay ($k_o \equiv 0$) follows from magnetic charge conservation, provide a good example of this possibility. Generation (and conservation) of magnetic charge is due to the topological structure of the gauge group unifying electromagnetism with other interactions. All the existing models are based on compact groups of GUT symmetry and inevitably predict the existence of stable magnetic monopoles. However, besides monopoles, antimonopoles are predicted. The effect of monopole-antimonopole annihilation makes their lifetime finite.

Secondly, there may be particles in the theory for which direct (two-body) transitions into light particle states may be forbidden for some reason. Decays of such particles may arise only due to virtual boson exchange, with mass $\sim\Lambda$. It is easy to get an estimate of such three-body decay from dimensional consideration. In units with $\hbar = c = 1$, probability is measured in the units of mass, so that

$$W_1 = 1/\tau = k_1 . \frac{1}{\Lambda^4} . m^5 \qquad (1.2)$$

where k_1 is a numerical factor determined by the statistical weight of particles, by the magnitude of mixing angles, by the number of different decay modes etc. Examples of such particles are: a) superheavy neutral fermions, in particular, right handed neutrinos in the O(10) models (Witten 1979; Wetterich 1981) and b) superheavy quarks from the so called B multiplet in the SU(5) model with asymptotic freedom (Kalashnikov 1980). These quarks have the colour properties of gluons, so that their direct transitions into quarks are forbidden.

In the supersymmetric GUT models gravitinos are predicted (Deser & Zumino 1977; Fayet 1982; Weinberg 1982) with a mass determined by the scale of supersymmetry breaking. Estimates (Weinberg 1982) give for the probability of their decay

$$W_2 = k_2 \frac{m_G^3}{m_{pl}^2} \qquad (1.3)$$

where the numerical factor $k_2 < 1$. It will be clear from the following discussion that particles with a lifetime given by (1.2) are of interest to us, though in the case of right-handed neutrinos and gravitinos relations (1.1) and (1.3) will be of importance as well.

1.2 THE FROZEN CONCENTRATION OF METASTABLE SUPERHEAVY PARTICLES IN THE EARLY UNIVERSE

It has been already mentioned in the Introduction, that in the early Universe the conditions for the creation of superheavy and, in particular, metastable particles may be realised. Creation of particles depends on their properties. It may bear both equilibrium and nonequilibrium character. Independently of whether their creation was equilibrium or nonequilibrium in the later periods, when $kT < mc^2$, metastable particle concentration in the expanding Universe inevitably exceeds the equilibrium one (corresponding to the given temperature T).

This residual (frozen) concentration n_m of metastable particles corresponds to the relative concentration ν, given by

$$\nu = n_m/n_r = n_m/(\kappa n_\gamma) \qquad (1.4)$$

where n_r is the total concentration of relativistic particles, κ is the number of different species and n_γ is the photon concentration. The magnitude ν is determined by the cross section of annihilation of metastable particles with their respective antiparticles, by their interaction with other particles and by the details of their formation, if metastable particles are formed in a non-equilibrium way (as it may be the case for monopoles (Dolgov & Zeldovich 1981; Zeldovich & Khlopov 1978)).

Let us consider now the estimates of residual concentration of particles, discussed in the preceding Section 1.1.

Monopoles. The theory predicts (Polyakov 1974; t'Hooft 1974; Schwartz 1976) the existence of monopoles and antimonopoles with magnetic charges $g = \pm\hbar c/2e$. An estimate of the frozen concentration of monopoles and of the annihilation time scale may be obtained (Zeldovich & Khlopov 1978; Preskill 1979; Dicus *et al* 1982) on the basis of the following physical considerations:

a) In a relativistic plasma with temperature $T < m$ the main mechanism for monopole interaction with the surrounding medium is multiple scattering of plasma particles on the monopole, with effective cross section (in the units with $\hbar = c = k = 1$)

$$\sigma_s \approx 1/Tm \tag{1.5}$$

b) Annihilation of monopoles and antimonopoles is determined by their magnetic "Coulomb" attraction

c) In a dilute plasma, when effective decceleration of monopoles due to scattering on light particles is unimportant, the annihilation rate is determined by the capture (Zeldovich & Khlopov 1978) of monopoles by antimonopoles due to classical dipole bremsstrahlung

$$\langle\sigma v\rangle \sim \pi a^2 v_T \sim \left(\frac{T}{m}\right)^{11/10} \left(\frac{g^2}{T}\right)^2 \tag{1.6}$$

d) In a dense plasma the annihilation rate is determined by the rate of diffusion of monopoles towards antimonopoles through the relativistic plasma.

e) The diffusion approximation in the annihilation rate calculation is valid if the decceleration length of the monopoles (antimonopoles) determined by monopole-plasma interaction $\lambda_s = 1/n_r\sigma_s$ is much smaller than the characteristic "Coulomb" radius of a monopole (antimonopole) $r_c = g^2/T$.

f) Monopoles (and antimonopoles) are formed in the early Universe in the course of the phase transition from the symmetric GUT vacuum to the asymmetric vacuum.

The concentration of generated monopoles depends essentially on the duration and on the details of the phase transition. However, if (as we shall assume) this initial concentration is sufficiently high, so that the mechanism of diffusion annihilation is effective, the frozen concentration is independent of the initial concentration and is determined by the mass and magnetic charge of the monopole.

Taking account of a)-f), the frozen concentration is of order (Zeldovich & Khlopov 1978; Preskill 1979)

$$\nu \sim \frac{m}{m_{pl} g^5} \sim 10^{-9} \left(\frac{m}{10^{16}\,\text{GeV}}\right) \left(\frac{137}{g}\right)^5 \tag{1.7}$$

Superheavy neutral fermions (right-handed) neutrinos). In GUT models (Witten 1979; Wetterich 1981) based on higher than SU(5) gauge symmetry, superheavy neutral leptons arise, which may be identified with right handed neutrinos, L_R. Interaction of these particles with each other and with other particles may proceed via exchange of virtual X-bosons whose mass is of order Λ and exceeds essentially the mass of particles. The timescale of this interaction exceeds the cosmological one soon after the temperature T drops below Λ $(T \sim \Lambda >> m)$. This means that L_R decouple from other particles, being relativistic. Hence, their relative frozen concentration is

$$\nu_m \sim 1/_\kappa \sim 1/100 \tag{1.8}$$

where κ is the number of species of relativistic particles which are in equilibrium at $T \sim \Lambda$. Eq. (1.8) is valid for gravitinos as well, with a little bit larger κ, related to $T \sim m_{pl}$.

Superheavy quarks. Superheavy quarks (Kalashnikov 1980) G with exotic (octet) colour properties may interact with light quarks q, or gluons g and annihilate with respective antiparticles $\bar{G}$. Thermodynamical equilibrium of superheavy quarks with other particles is maintained by reactions $G\bar{G} \rightleftarrows q\bar{q};\ gg$. In QCD calculations, the analogy with quantum electrodynamics is widely used. In this sense, the considered reactions of colour interactions are similar to the processes $e^+e^- \rightleftarrows \gamma\gamma$; $\mu^+\mu^- \rightleftarrows e^+e^-$, whose rate is known (Akhiezer & Berestetsky 1969). This analogy provides an estimate of the rate of $G\bar{G}$ annihilation at $T < m$

$$\langle Gv \rangle \sim \frac{\alpha^2}{m^2} N_c \tag{1.9}$$

where N_c is the number of possible channels of the considered reaction.

Detailed analysis of the kinetics of freezing of G and $\bar{G}$ (see the review by Dolgov & Zeldovich 1981) gives for the annihilation rate (1.9) the frozen concentration of heavy quarks as

$$\nu \approx \frac{m}{m_{pl}} \frac{1}{\alpha^2 N_c} K \tag{1.10}$$

where in units with $\hbar = c = 1$ the numerical factor

$K = 10/\kappa^{\frac{1}{2}} [42 + \ln(\langle\sigma v\rangle m_p m) - \frac{1}{2}\ln\kappa] \sim 1$, m_p is the mass of proton and $\langle\sigma v\rangle$ is given by (1.9).

1.3 THE STAGE OF SUPERHEAVY PARTICLE DOMINANCE IN THE EARLY UNIVERSE

On the freezing of superheavy particles with mass m in the course of the expansion of the Universe, the ratio of the density of such particles $\varepsilon_m = m\nu n_r$ to the density of relativistic particles $\varepsilon_r \sim Tn_r$, grows as m/T. At the moment

$$t_o \sim m_{pl}/(\nu m)^2 \tag{1.11}$$

when the temperature drops down to

$$T_o \approx \nu m \tag{1.12}$$

the stage of superheavy particle dominance begins. At this stage the effective equation of state of the Universe is close to $p = 0$. Thus, we come to the conclusion that one of the cosmological consequences of GUT models, which predict the existence of superheavy metastable particles, is an early dust-like stage of cosmological expansion.

It would be possible to analyze in detail the transition from the stage with the equation of state $p = \varepsilon/3$ to the dust-like stages of expansion, with the use of results of Doroshkevich & Khlopov (1981), where a similar transition to massive neutrino dominance in a later period of cosmological expansion was studied. It would be possible to take into account non-zero pressure due to relativistic particles on the dust-like stage both in the case of superheavy particles, interacting with relativistic particles (monopoles or heavy quarks) and in the case of noninteracting particles (right-handed neutrinos, gravitino). But since we aim to show the principal relationship between GUT models and cosmology in the very early Universe we will consider a very approximate, but simple, temporal dependence of the equation of state:

$$p = \begin{cases} \varepsilon/3, & \text{at } t < t_o \\ 0 & \text{at } t_o < t < t_k \\ \varepsilon/3 & \text{at } t > t_k \end{cases}$$

where t_k is the moment of the end of early dust-like stage, determined either by decays of particles, or by their annihilation in the inhomogeneities formed at $p = 0$ stage.

2. PRIMORDIAL BLACK HOLES AND INHOMOGENEITY OF EARLY UNIVERSE

2.1 THE SPECTRUM OF INITIAL PERTURBATIONS

One of the main problems of modern cosmology is to explain the observed structure of the Universe (Zeldovich & Novikov 1975). In the big bang theory this problem is closely related to the question of the physical nature and spectrum of initial density fluctuations. It is known from galaxy formation theory what such fluctuations must be at the period of recombination of hydrogen to form the modern structure.

Ways of relating these perturbations to thermal or quantum fluctuations in the very early Universe (near the singularity) caused by the existence of a DeSitter stage of exponential cosmological expansion have appeared recently (Starobinsky 1980). Such a stage may arise due to vacuum polarization near the singularity or due to strong first-order phase transitions predicted by GUTs (Guth 1981). The exponential inflation of the Universe may lead to: i) exponential growth of fluctuation scales, ii) to their parametric enhancement (Lukash 1980; Komponeetz *et al* 1982). Besides that, the physical processes (bubble formation (Starobinsky 1982), and collisions (Hawking 1982) etc.) at the transition (Grishuk 1974; Starobinsky 1979) from a De Sitter to a Friedman stage of expansion may lead to additional generation of small scale perturbations. Thus, at the end of the exponential stage a certain spectrum of perturbations may arise, and the amplitude of large scale perturbations is large enough (Chibisov & Mukhanov 1981; Lukash 1980; Komaneetz *et al* 1982; Starobinsky 1982) to explain the large scale structure of the Universe (Zeldovich & Novikov 1975).

The study of the astrophysical impact of perturbations at small scales provides additional tests of theories which predict exponential inflation

2.2 FORMATION AND EVAPORATION OF PBH

Important information on small scale perturbations can be obtained from the restrictions on the spectrum of primordial black holes (Zeldovich & Novikov 1966; Hawking 1971)(PBH), which arise from initial perturbations. Even if the mean squared fluctuations are small, PBH are formed from fluctuations corresponding to the exponential tail of a Gaussian distribution of the amplitude (Carr mechanism) (Carr 1975, 1978). PBH formation is possible (Hawking 1982) in the course of bubble wall collisions at phase transitions.

Afterwards PBH may evaporate (Hawking 1974, 1975), thereby influencing a number of physical processes (Novikov *et al* 1979). If the mass M of PBH is smaller than 10^{15}g, such PBH evaporate by the present time. If their mass is larger than 10^{15}g, PBH remain up to the present time and contribute to the modern cosmological density. So a question arises concerning the relationship between the fraction of the matter $\beta(M)$ contained in PBH's of mass M at the moment of their formation, and the relative contribution $\alpha(M)$ of such PBH's to the cosmological density at the period of their evaporation (if $M < 10^{15}$g) or to the modern cosmological density (if $M > 10^{15}$g).

At the stage $p = \varepsilon/3$ (RD stage) for PBH to form, metric perturbations ~1 are needed. The moment of PBH formation is unambiguously related to its mass M (Zeldovich & Novikov 1966)

$$t_f \sim t_{pl}(M/m_{pl}) \tag{2.1}$$

Here $t_{pl} = (G\hbar/c^5)^{1/2}$ and $m_{pl} = (\hbar c/G)^{1/2}$ are the Planck time and the Planck mass. From the moment t_f up to the moment of evaporation (Hawking 1975)

$$t_e \approx t_{pl}(M/m_{pl})^3, \tag{2.2}$$

if PBH of mass M evaporate at the stage $p = \varepsilon/3$ or up to the end, t_*, of the RD stage, the PBH contribution to the density of the Universe grows, since PBH density decreases as a^{-3} and the density of ultrarelativistic matter outside PBH decreases more rapidly, as a^{-4}, where a is the scale factor. During the RD stage $a(t) \propto t^{1/2}$ (Zeldovich & Novikov 1975). Thus the relationship between α and β is

$$\alpha/\beta \approx [\min\{t_e, t_*\}/t_f]^{1/2} \tag{2.3}$$

Eqs. (2.1)-(2.3) lead to a direct dependence of the ratio α/β on the PBH mass, M (see curve A in fig. 1)

$$\frac{\alpha(M)}{\beta(M)} \approx \begin{cases} \dfrac{M}{m_{pl}} \min\{1, (M_{**}/M)^{3/2}\}, & M < M_* \\ 1 & M > M_* \end{cases} \tag{2.4}$$

where M_{**} is the mass of PBH evaporating at the end, t_*, of the RD stage

$$M_{**} \sim m_{pl}(t_*/t_{pl})^{1/3} \sim 10^{13}g \tag{2.5}$$

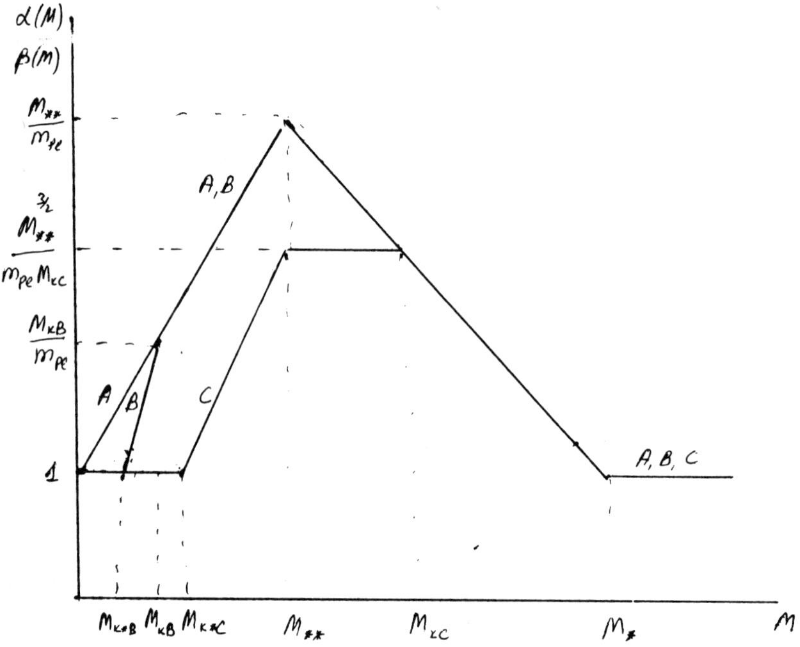

Fig. 1. The relationship between $\alpha(M)$ and $\beta(M)$: A - without early dust-like stages; B - $t_k < 10^{-25}s$; C - $t_k > 10^{-25}s$. $M_* \sim 10^{15}M_O$, $M_{**} \sim 10^{13}g$.

It is obvious that for $M > M_* \approx m_{pl}(t_*/t_{pl}) \sim 10^{15}M_\odot$, i.e. for PBH formed after the end of the RD stage, their relative contribution to the cosmological density does not grow, i.e. $\alpha/\beta = 1$.

Eq. (2.4) has to be modified if early dust-like stages are present. Such stages (see Sec. 1.3) may arise due to the dominance of superheavy metastable particles in the Universe. In principle, dust-like stages may also arise due to the dominance of low mass PBH (Zabbotin *et al* 1980; Barrow 1980) being formed earlier .

Within the frame of GUT (see Sec. 1.2 and 1.3) such dust-like stages are realized later than the stages of exponential expansion, so that physical processes in the inflational stage do not influence the frozen concentration of superheavy particles (excluding those generated before or immediately after inflation - c.f. gravitinos,,monopoles). On the other hand, these dust-like stages take place sufficiently early, so that the small scale inhomogenities that are of interest to us still do not dissipate. Thus, study of the behaviour of small scale perturbations left after the inflationary stage, during the dust-like stage provides relationships between the parameters of these perturbations (determined by the parameters of inflation) and the parameters of the dust-like stage (determined by the parameters of GUTs). To establish such relationships, formation of PBH and their evolution with early dust-like stages taken into account will be considered in some detail.

2.3 FORMATION OF PBH AND EARLY DUST LIKE STAGES

Fig. 2 illustrates qualitatively the relationship between t_f and PBH mass M in the presence of a $p = 0$ stage. To each mass M there

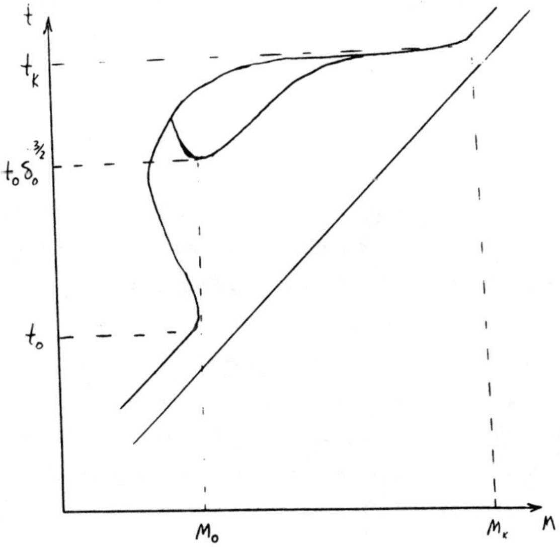

Fig. 2. The relationship between the time of PBH formation and mass.

corresponds a moment $t_h(M)$ when the mass M_h contained within the cosmological horizon is equal to M (line 1 on the fig. 2). This time $t_h(M)$ limits t_f from below: $t_f \geq t_h(M)$, since the notion of a "black hole" is senseless for $M > M_h$. On the other hand, PBH mass at the moment of PBH formation must exceed the Jeans mass M_J at that moment. If the $p = 0$ stage was absent, M_J would be smaller, but of the order of magnitude of M_h (line 2 of fig. 2) and PBH formation would be possible in the small region I, bounded by the lines 1 and 2, so that in this case it is possible to use Eq. (2.1) for order of magnitude estimates.

Note, that PBH formation automatically suggests that the metric perturbation ξ in the configuration from which PBH are formed, is of order 1. At $t < t_h(M)$ metric perturbations on the scale M are known to be (Lifshitz 1946) constant. Just these points lead to the conclusion (Carr 1975) that PBH formation at the $p = \varepsilon/3$ stage (region I in fig. 2) takes place in the exponential tail of the Gaussian distribution of the amplitude ξ of metric perturbations. If in the inflationary stage metric perturbations with spectrum $\delta(M) < 1$ are formed, the probability of the configuration with $\xi \sim 1$ being realized is

$$P(\xi) \sim \delta^{-1}(M) \exp\left[-\xi^2/\delta^2(M)\right] \tag{2.7}$$

For detailed discussion see Carr (1975). (The hydrodynamics of PBH formation with $\xi \sim 1$ in the $p = \varepsilon/3$ stage are discussed by Nadyozhin *et al* (1978).)

At the $p = 0$ stage the situation is different (Khlopov & Polnarev 1980; Polnarev & Khlopov 1981; Polnarev & Khlopov 1982a). The Jeans mass M_J from the moment t_o up to the moment t_k is much smaller than the mass within the cosmological horizon (curve 2 in the fig. 2). At $t > t_o$, M_J decreases according to the law determined by the properties of the particles dominating in the stage with $p = 0$. Thus the possibility of PBH formation from configurations with $\xi < 1$ (regions II and II^1 in fig. 2) arises. Indeed, a configuration with initial metric perturbations $\xi < 1$ at the moment when its size equals the horizon has the density contrast $\delta\rho/\rho \sim \xi$ as it was in the case $p = \varepsilon/3$. But subsequently $\delta\rho/\rho$ grows as $(t/t_o)^{2/3}$. At the moment

$$t_1 = t_o\ \xi^{-3/2} \tag{2.8}$$

when $\delta\rho/\rho \sim 1$, a nonlinear stage of the evolution of inhomogenity starts and the configuration separates from the cosmological expansion. During the subsequent contraction such configurations form PBH with a certain probability at the time $t_f \gtrsim t_1(M,\xi)$.

Following Khlopov & Polnarev 1980, Polnarev & Khlopov 1981 we can estimate the minimal probability of PBH formation in the nonlinear stage of evolution of inhomogenities at the stage of dominance of superheavy particles.

Just at the beginning of the contraction t_1 the density distribution of the configuration and its shape may be described by the following quantities: a) the mean density ρ_1 equal to within an order of magnitude to the mean density of the Universe at the moment t_1: b) the lengthscale of the configuration r_1: $m = (4\pi/3\rho) r_1^3$; c) the degree of deviation of the configuration from the spherical symmetry: $s = \max\{|\gamma_1-\gamma_2|, |\gamma_1-\gamma_3|, |\gamma_2-\gamma_3|\}$ where the parameters γ_1, γ_2 and γ_3 characterize the deformation along the three orthogonal main axes of the deformation tensor of the configuration considered; d) the inhomogenity u of the density distribution within the configuration:

$u \approx \dfrac{\rho_{c_1} - \rho_1}{\rho_1}$ where ρ_{c_1} is the central density of the configuration at the moment t_1.

When PBH is formed, i.e. the configuration approaches its gravitational radius r_g, the mean density of the configuration ρ_{PBH} at this moment t_{PBH} is equal to

$$\rho_{PBH} \sim \frac{M}{(4\pi/3) r_g^3} \sim \rho_1 x^{-3} \qquad (2.9)$$

where

$$x = r_g/r_1 \qquad (2.10)$$

On the other hand, the maximal density reached as a result of the contraction with retention of a nearly spherical form cannot exceed

$$\rho_{max} \sim \rho_1 . s^{-3} \qquad (2.11)$$

The estimate (2.10) follows from the fact that the minimal size of the configuration, $r_{min} > \min (t_{con}\Delta\nu, \Delta r)$, where t_{con} is the contraction time scale, $\Delta\nu \lesssim s\, r_1/t_{con}$ is the characteristic variation of initial velocities along the different axis, and $\Delta r \lesssim s\, r_1$ is the characteristic initial "flatness" of the configuration.

From Eqs. (2.10) and (2.11) we may conclude that PBH are formed only from configurations which are nearly spherically symmetric:

$$s \lesssim x \lesssim 1 \tag{2.12}$$

For $s \gtrsim x$, i.e. for most configurations, not PBH, but non-spherical structures with small gravitational potentials (c.f. "blini" or pancakes (Zeldovich & Novikov 1975)) are formed.

At the time t_{PBH} the equation of state in the centre of the configuration may once again change to $p = \varepsilon/3$. If the inhomogeneity of the configuration at the moment t_{PBH} is small enough

$$\frac{\delta\rho_{PBH}}{\rho_{PBH}} < 1 \tag{2.13}$$

the resulting pressure gradients do not exceed gravitational forces and thus surely do not prevent the formation of PBH. For particles not interacting with each other and with the radiation, pressure gradients are not essential. In this case PBH are surely formed if the configuration approaches its r_g before the moment of caustic formation, t_{caus}, in its centre:

$$t_{PBH} \lesssim t_{caus} \tag{2.14}$$

Otherwise, particles passing the caustic may escape from the central region of the configuration thus preventing PBH formation.

If by the moment t_{PBH} the configuration has been fragmented into weakly interacting grains with masses $\Delta M \ll M$, then allowing for the velocity dispersion of such grains gives just the same sufficient condition for PBH formation as in Eq. (2.14).

The contraction of a nearly spherical dust $(p = 0)$ configuration can be approximately described by the Tolman solution (see cf. Landau & Lifshitz 1973; Misner *et al* 1973)). It is shown in the Appendix that both conditions (2.13) and (2.14) imply the same restriction on the inhomogeneity of the configuration at time t_1

$$u \lesssim x^{3/2} \tag{2.15}$$

We now estimate the probability W_u, that the configuration has the degree of inhomogeneity u corresponding to the inequality (2.15). According to Zeldovich & Novikov (1975) the magnitude r_*, characterizing in Eq. (A.5) of the Appendix the inhomogeneity of the density distribution within the configuration, is unambiguously determined by the characteristic scale of the perturbation spectrum at the stage $p = 0$ and within a numerical factor of order unity is $\sqrt{r_*^2} \sim r_1$. Then, assuming the normal law of the density perturbation distribution, one obtains for the probability W_u (taking $\sigma^2 \simeq 1$):

$$W_u \sim \frac{1}{\sigma} \int_0^{x^{3/2}} e^{-u^2/\sigma^2} du \sim x^{3/2} \tag{2.16}$$

One can also estimate the probability W_s of the configuration having a sufficiently high degree of sphericity to satisfy the inequality (2.12). This rather cumbersome question was investigated by Doroshkevich (1970). If the random components of the deformation tensor are distributed according to the random law, since the operation of the reduction to the main axes is essentially non-linear, the distribution law of the introduced above main values $\gamma_1, \gamma_2, \gamma_3$ differs from the normal law by the factor $(\gamma_1-\gamma_2)(\gamma_1-\gamma_3)\ (\gamma_2-\gamma_3)$. Allowing for numerical factors, W_s is equal to

$$W_s \simeq \frac{27}{8\sqrt{5\pi}} \int_0^\infty d\gamma_1 e^{-\frac{9}{10}\gamma_1^2} \int_{\gamma_1(1-x)}^\infty (\gamma_1-\gamma_2)d\gamma_2 \int_{\gamma_1(1-x)}^\infty (\gamma_1-\gamma_3)(\gamma_2-\gamma_3)d\gamma_3 \tag{2.17}$$

$$\simeq 2.10^{-2}.x^5$$

It follows thus from (2.16) and (2.17) that PBH are surely formed with probability no less than

$$W_{PBH} \geq 2.10^{-2}\ x^{13/2} \tag{2.18}$$

Eq. (2.18) determines the minimal fraction of matter collapsed into the PBH with mass M at the time of PBH formation.

The magnitude x in the Eq. (2.19) may be related to the magnitude ξ. $x \sim GM/c^2(\rho_1/M)^{1/3} \sim \frac{G}{c^2} M^{2/3} \rho_o^{1/3} (to/t_1)^{2/3}$. But according to (2.8) $x \sim \frac{GM_o^{2/3} \rho_o^{1/3}}{c^2} . \xi . (M/M_o)^{2/3}$, where

$$M_o = \rho_o(ct_o)^3 \qquad (2.19)$$

is the mass contained within the cosmological horizon at the moment t_o. Since $\rho_o \sim 1/Gt_o^2$, one obtains finally

$$x(M) \sim \xi(M) \ (M/M_o)^{2/3} \qquad (2.20)$$

So Eq. (2.18) may be rewritten as follows

$$W_{PBH} \geq W(M,\xi) \approx 2\ 10^{-2}\ \xi^{13/2}\ (M/M_o)^{13/3} \qquad (2.21)$$

Using (2.7), one finds that the fraction of matter contained in PBH formed from configurations of mass M and initial amplitude ξ is given by

$$\beta(M,\xi) \gtrsim P(M,\xi)\ W(M,\xi) \propto \xi^{13/2}\ e^{-\xi^2/\delta^2} \qquad (2.22)$$

If $t_1(M, \delta(M))$ (curve 3 on fig. 2) is smaller than t_k, most configurations with mass M reach a nonlinear stage of their evolution. In this case, as it is easily seen from (2.22), in the region II the maximum of $\beta(M,\xi)$ is reached at $\xi \sim \delta$, so that there is no exponential suppression of the probability of PBH formation, as was the case in the RD stage. The region II^1 is bounded in the low mass region, $M < M_o$, by the curve 4, taking into account partial or complete dissipation of small scale perturbations, which will have left the cosmological horizon already at the RD stage.

Note that at the $p = 0$ stage, PBH with mass M may be formed from nonlinear evolution of more massive configurations (i.e. as a result of their fragmentation) as well as from coagulation of PBH or nonlinear configurations of lower mass.

2.4 EFFECT OF DUST-LIKE STAGES ON THE RELATIONSHIP BETWEEN $\alpha(M)$ AND $\beta(M)$

If in the early Universe at $t < t_*$, a stage with $p = 0$ existed, starting at the moment t_o and finishing at t_k, the contribution of PBH to the density of the Universe may increase only when $t_{pl} < t < t_o$ or $t_k < t < t_*$. In the presence of such a stage the dependence of t_f on M is not in general so simple as in Eq. (2.1). However for an arbitrary, unambiguous relationship between t_f and M, Eq. (2.3) may be generalized as follows

$$\alpha(M)/\beta(M) = [t_o/\min\{t_f,t_o\}]^{1/2} \cdot \begin{cases} 1, & t_e < t_k \\ [\min\{t_e,t_*\}/\max\{t_k,t_f\}]^{1/2}, & t_e > t_k \end{cases} \tag{2.23}$$

Eq. (2.23) shows that for PBH with $t_f < t_k$ an early stage with $p = 0$ influences essentially the relationship between α and β. For $t_f > t_k$ Eq. (2.23) goes into Eq. (2.3).

We have seen in Sec. 2.3 that the magnitude $\beta(M)$, defined earlier as the fraction of matter contained in PBH with mass M at the moment of their formation, loses its meaning in the presence of an early dust-like stage, since unlike the case of continuous RD-stage, $\beta(M)$ depends now not only on the mass of the PBH, but also on the initial properties of the configurations from which the PBH are formed, as well as on the very mechanism of their formation. However, it follows from (2.23) that the relationship between $\alpha(M)$ and $\beta(M)$ does not depend on t_f for PBH formed in the early dust-like stage. It is reasonable thus to redefine $\beta(M)$ as follows: for $M < M_k$ (M_k is the mass within the horizon at the moment t_k), $\beta(M)$ is the fraction of matter contained in PBH with mass M at the end of the early dust-like stage t_k, and for $M > M_k$, $\beta(M)$ has the original meaning, denoting the fraction of the matter contained in PBH with mass M at the moment of their formation $t_f > t_k$.

Now, we may explicitly write down the relationship between $\alpha(M)$ and the redefined magnitude $\beta(M)$ as a function of M and of the only parameter of the dust-like stage, t_k (curve B or C in fig.1). Note immediately that $\alpha(M)/\beta(M) \approx 1$ for PBH with $t_e < t_k$, i.e. with $M < M_* \approx M_{pl}(t_k/t_{pl})^{1/3}$, since even if such PBH had been present at the moment t_k, they would have evaporated so rapidly that their fractional contribution to the density of the Universe could not have increased at $t > t_k$. For more massive PBH it follows from (2.23) that

$$\frac{\alpha(M)}{\beta(M)} = \begin{cases} 1, & M < M_{k*} \\ \frac{M}{M_{pl}} \min\left\{1, \left(\frac{M_{**}}{M}\right)^{3/2}\right\} . \min\left\{1, \left(\frac{M}{M_k}\right)^{1/2}\right\}, & M_{k*} < M < M_* \\ 1, & M > M_* \end{cases} \quad (2.24)$$

For $M > M_*$, obviously $\alpha(M)/\beta(M) = 1$ as it was earlier.

2.5 ASTROPHYSICAL RESTRICTIONS ON THE SPECTRUM OF PBH

PBH formed in the very early Universe remain during subsequent stages of expansion. Their contribution to the cosmological density grows according to (2.24) and their effect on physical processes in the Universe is enhanced accordingly. As was mentioned earlier, according to Hawking (1974, 1975), PBH with masses less than 10^{15} g, must evaporate by the present time because of quantum processes. PBH evaporation can be consistently analyzed only for PBH masses exceeding $m_{pl} = 10^{-5}$ g, because otherwise their width due to evaporation is comparable with their mass. (PBH with masses smaller than 10^{-5} g have been studied by Zabbotin *et al* (1980)). PBH of mass 10^9 g evaporate before 1 s and restrictions on such PBH are obtained by analyzing the effects of their evaporation on the magnitude of the entropy of the Universe (Novikov *et al* 1979). The analysis must take into account both baryon charge generation (Zeldovich 1976) and additional entropy generation during the processes of PBH evaporation (Novikov *et al* 1979). Evaporation of PBH with mass 10^9 g takes place after 1 s. If PBH masses are in the interval 10^9-10^{10} g, evaporation takes place from 1 s to 10^3 s and therefore influences the ratio of neutron and proton concentrations, and hence the primordial helium abundance. So the observed helium abundance provides restrictions (Wainner & Nasselsky 1977; Zeldovich *et al* 1977) on the possible magnitude of $\alpha(M)$ in this mass interval. The observed deuterium abundance may put restrictions (Zeldovich *et al* 1977) on the magnitude of $\alpha(M)$ for PBH with masses from 10^{10}-10^{13} g which evaporate in the period 10^3 s-10^{12} s. Analysis of the dynamics of recombination provide restrictions on $\alpha(M)$ for $M = 10^{13}$-10^{14} g (Nasselsky 1978), since these evaporate at the stage of separation of the matter from radiation. The most stringent limits on $\alpha(M)$ are obtained for PBH with mass $M = 10^{14}$-10^{15} (Page & Hawking 1975; Carr

1976) from observations of the γ-background and γ-bursts, as well as from the analysis of synchrotron radioemission (Rees 1977; Blandford 1977).

PBH with $M > 10^{15}$ g remain at the present time and limits on their present concentration may be obtained by the contribution to the cosmological density (Zeldovich & Novikov 1966) or by their effect on supercluster dynamics (Novikov *et al* 1979)(for $M > 10^{16}\ M_{\odot}$). For a review of the observational limits on $\alpha(M)$ see Novikov *et al* (1979).

The limits on $\alpha(M)$ obtained from astrophysical considerations put upper limits on the magnitude $\beta(M)$. In the case of a continuous RD stage from the Planck time up to the period close to recombination of the hydrogen, the relationship between α and β is given by (2.4), and observational restrictions on the magnitude of $\alpha(M)$ result in the upper limits on $\beta(M)$, given in fig. 3 (curve 2). In the presence of early dust-like stages the

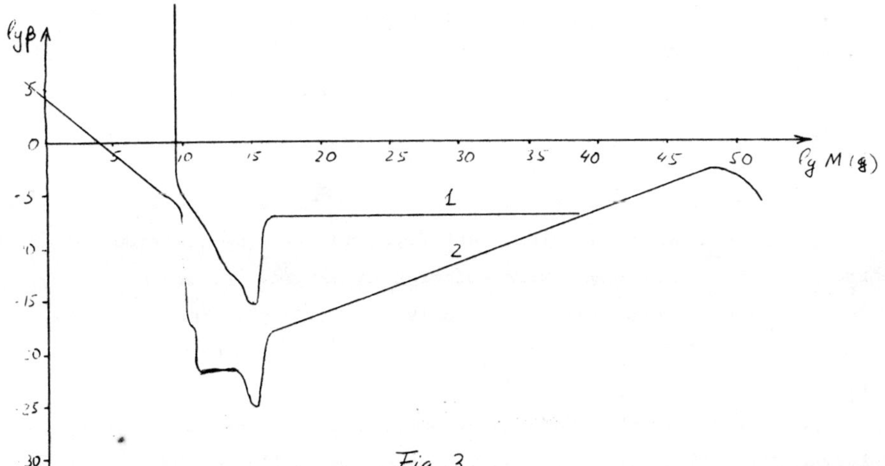

Fig 3

Fig. 3. Constraints on the magnitude $\beta(M)$: 1 - in the presence of an early dust-like stage with $t_k \lesssim 1$ s; 2 - in the absence of an early dust-like stage.

limits on the magnitude of $\beta(M)$, redefined in the preceding sections, depend essentially on t_k, the moment of the end of early dust-like stage.

The magnitude of t_k may depend on the parameters of specific models of grand unification, as well as on a number of details of the evolution of nonlinear structures formed in the dust-like stage. However, as it was shown by Polnarev & Khlopov (1982b), we may give a crude, but strongly reliable, upper limit for t_k:

$$t_k \lesssim 1 \text{ s} \tag{2.25}$$

Early dust-like stages must finish before 1 s so as not to contradict the observed abundance of light elements and the observed spectrum of relic radiation.

Restrictions on the PBH spectrum (curve 1 in the fig. 3), obtained on the basis of (2.25) and (2.24), though much weaker than those obtained earlier (see e.g. Novikov *et al* (1979), curve 2 in fig. 3) are more reliable, since they do not use the indirect assumption of the absence of early dust-like stages. (Relatively short dust-like stages in the period 1-10^4 s (Polnarev & Khlopov 1982a; Doroshkevich & Khlopov 1982) are not excluded in principle. However, because of their relatively short duration, such stages do not result in any essential softening of the restrictions on $\beta(M)$.)

Besides that, we do not exclude the possibility that at the end of the early dust-like stage additional baryon number as well as additional entropy may be generated. Initial entropy, and also the contribution to the entropy due to PBH of small mass evaporating before 1 s, are so to say, forgotten in this case. This practically removes any astrophysical restrictions on the magnitude of $\alpha(M)$, and consequently, on the magnitude $\beta(M)$ for $M < 10^9$ g.

Thus, in the presence of early dust-like stages expansion of the Universe which is arbitrarily inhomogeneous on scales $M < 10^9$ g cannot in principle be excluded.

Within the frame of specific models of grand unification which predict an early dust-like stage with $t_k \ll 1$ s, restrictions on the spectrum of PBH are noticeably enhanced as compared to the restrictions given by the curve 1 in fig. 3; but they remain weaker than those given by the curve 2 in fig. 3. Such models maintain the relationship between the CP-violating parameters, the possible amount of baryon number generated in the inequilibrium processes in the early Universe, and the parameters of superheavy particles. This puts additional limits on the duration of dust-like stages.

Indeed, if such a stage was induced by the dominance of particles with mass m and frozen concentration $\nu = (n_m/n_\gamma)_o$ beginning at t_o and ending at t_k, and if after t_k a baryon excess ΔB was added to the initial baryon excess $(\Delta n_B/n_\gamma)_o = \nu \frac{n_B}{n_m}$, the final baryon to photon ratio is determined by

$$(n_B/n_\gamma)_k = [(n_B/n_\gamma)_o + \nu\Delta B]\ (t_o/t_k)^{1/2} \qquad (2.26)$$

To obtain (2.26) it is sufficient to take into account that $(n_B)_k = (n_B)_o + \Delta n_B$, $(n_\gamma)_k = (n_\gamma)_o + m\nu(n_\gamma)_o/T_k$, $T_k \sim T_o(t_o/t_k)^{1/2}$, and $T_o \sim \nu m$. The quantities on the righthand side of Eq. (2.26) are determined by the parameters of concrete models. It is seen that unless a concrete model is given, it is impossible to put any constraint on the dust-like stage based only on the observed entropy of the Universe.

In the presence of a number of early dust-like stages, which alternate with RD stages in some way which depends on the details of the model, we must take t_k which appears in the relationship (2.24) between $\alpha(M)$ and $\beta(M)$ to be the moment of the end of the last early $p = 0$ stage.

Note that allowing for possible dust-like stages may lead to a softening of restrictions, obtained without such allowance, on the concentrations not only of PBH, but of other particles (Doroshkevich & Khlopov 1982) (e.g. gravitinos (Deser & Zumino 1977)) as well.

3. RESTRICTIONS ON GUTS AND ON THE INHOMOGENEITY OF THE EARLY UNIVERSE

3.1 THE RELATIONSHIP BETWEEN THE SPECTRUM OF PBH AND THE GUT PARAMETERS

In the preceding chapter the minimal probability $W_{PBH}(M)$ of the formation of PBH with mass M in an early dust-like stage was estimated. The magnitude $W_{PBH}(M)$ depends essentially on the form of the spectrum of initial perturbations $\delta(M)$. Suppose at the end of the inflationary stage the spectrum of metric perturbation is given by

$$\delta(M) = \delta_o (M/M_o)^{-n} \tag{3.1}$$

where M_o is the mass contained within the cosmological horizon at the beginning of the early dust-like stage. The corresponding density perturbations at the beginning of the dust-like stage are

$$\frac{\delta\rho}{\rho}(M,t_o) = \delta(M) \begin{cases} 1, & M_{min} < M < M_o \\ (M/M_o)^{-2/3}, & M > M_o \end{cases} \tag{3.2}$$

where M_{min} is the scale of dissipation of density perturbations. To make the subsequent restrictions on the GUT parameters independent of the concrete mechanisms of dissipation of the perturbations during the RD stage when $t < t_o$ (in particular independent of the magnitude M_{min}, which is evidently smaller than M_o), we shall restrict ourselves by considering scales which exceed M. Then from (2.21) one obtains

$$W_{PBH}(M) > 2\ 10^{-2}\ \delta_o^{13/2}\ (M/M_o)^{-13n/2} \tag{3.3}$$

This minimal estimate is valid for $M < M_{max}$, where M_{max} is the maximal mass of configurations which succeed in separating from the general cosmological expansion before the end of the early dust-like stage, i.e. the mass for which the moment of the beginning of the nonlinear stage, t_1, is determined by

$$\delta\rho/\rho\ (M_{max}, t_o)\ (t_1/t_o)^{2/3} \approx 1 \tag{3.4}$$

and is close to the moment of the end of early dust-like stage, t_k. From (3.1), (3.2), (3.4), (2.19), and (1.11) one obtains

$$M_{max} = m_{pl}(\tau/t_{pl})^{2/2+3n}(m_{pl}/\nu m)^{6n/2+3n}\delta_o^{\ 3/2+n} \tag{3.5}$$

The essence of the limits on τ and νm, which are the GUT parameters, is in the following: the GUT model contradicts the astrophysical restrictions on the PBH spectrum, if

$$W_{PBH}(M) > \beta(M) \tag{3.6}$$

for any value of M in the interval (M_o, M_{max}) (see qualitative fig. 4).

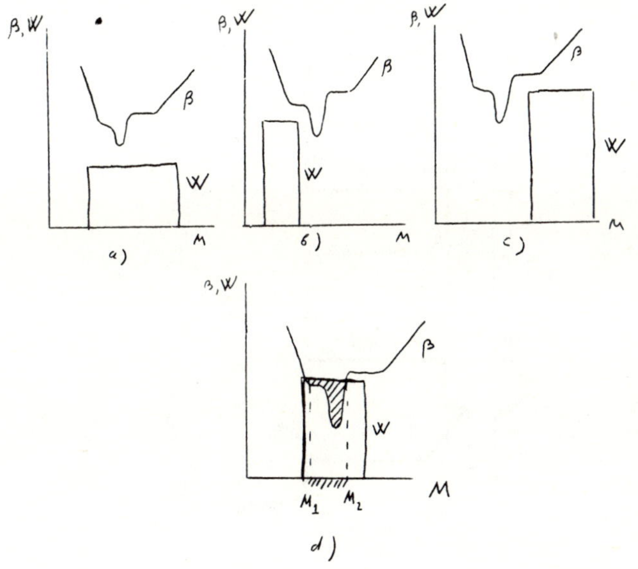

Fig. 4. Comparison of β and W: a,b,c - no restrictions; d - forbidden situation. (M_1, M_2) - forbidden region of PBH masses.

Let us illustrate this method in detail for the case of a flat spectrum of initial metric perturbations, i.e. putting $n = 0$. Then, the condition (3.6) reduces to a system of inequalities for a certain interval of PBH masses (M_1, M_2)

$$\frac{M}{m_{pl}}\,\delta_0^{-3/2} < \tau/t_{pl} < \frac{M}{m_{pl}}\left[W_{PBH}(M)/\beta_0(M)\right]^2 \tag{3.7}$$

and

$$\frac{\nu m}{m_{pl}} > (M/m_{pl})^{-1/2} \tag{3.8}$$

where $\beta_0(M)$ is determined from the observational restrictions on $\alpha(M)$ without allowing for early dust-like stages (curve 2 in fig. 3).

The restrictions obtained on τ and νm are given in fig. 5.

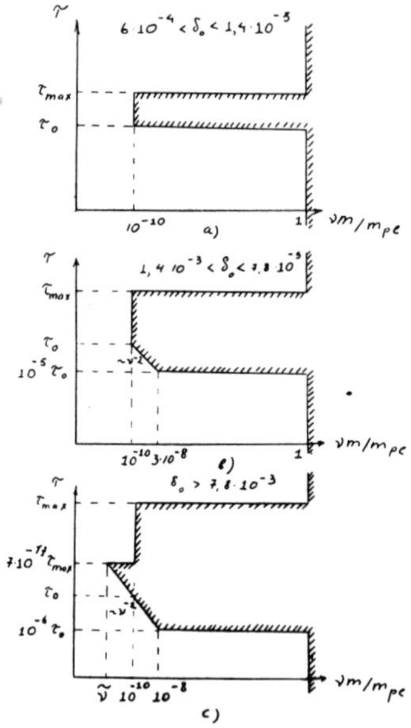

Fig. 5. Constraints on τ and νm.
$\tau_{max} = 6.10^{-19}(\delta_0/6.10^{-4})^{13}$s, $\tau_0 = 6.10^{-19}(\delta_0/6.10^{-4})^{-3/2}$s,
$\tilde{\nu} = 10^{-10}(\delta_0/7.8.10^{-3})^{-29/4}$

Note, that there are no restrictions on τ and νm for sufficiently small δ_o. As can be seen from (3.7) and curve 2 in fig. 3 this occurs for

$$\delta_o^{-3/4} > \frac{\max\left\{W_{PBH}(M)\right\}}{\min\left\{\beta_o(M)\right\}} \tag{3.9}$$

i.e. for

$$\delta_o \lesssim 6.10^{-4} \tag{3.10}$$

For $6.10^{-4} \lesssim \delta_o \lesssim 1{\cdot}4\ \ 10^{-3}$ the interval (M_1, M_2) is reduced to a narrow region around $M \sim 10^{15}$g $(M_1 \sim M_2 \sim 10^{15}$g) and the forbidden area $(\tau, \nu m)$ is shown in fig. 5a.

For $1{\cdot}4\ .10^{-3} \lesssim \delta_o \lesssim 7{\cdot}8\ .10^{-3}$, $M_1 \sim 10^{10}$g and $M_2 \sim 10^{15}$g, see fig. 5b.

And, finally, for $7{\cdot}8\ .10^{-3} \lesssim \delta_o \lesssim 0{\cdot}5$, $M_1 \sim 10^9$g and $M_2 \sim 10^{15}{\cdot}(\delta_o/7{,}8\ 10^{-3})^{29/2}$g. The corresponding region of forbidden values of the parameters $(\tau, \nu m)$ is shown in fig. 5c.

General restrictions on νm and τ shown in fig. 5 are valid for any type of particle predicted by GUTs. So the obtained restrictions not only demonstrate the principal possibility of establishing a relationship between GUT parameters and astrophysical limits on PBH spectrum, but also serve as a specific recipe for comparison of GUTs with astrophysical observations. Indeed, for any type of metastable particle, it is possible to estimate τ and νm within the frame of a GUT, as was demonstrated in Chapter 1 for the examples of magnetic monopoles, heavy quarks and leptons. Comparison of such estimates of τ and νm with the restrictions given in fig. 5, provide nontrivial restrictions on the properties of such particles, e.g. on their mass.

In the remainder of this Chapter we shall illustrate the effects of such an approach on the particles considered in Chapter 1.

3.2 RESTRICTIONS ON THE PROPERTIES OF VARIOUS PARTICLES

In Chapter 1 examples of superheavy metastable particles predicted by GUTs were given, and their lifetime relative to decay and frozen concentration ν were estimated.

Magnetic monopoles. In this case ν is given by (1.7) so that

$$\nu m \approx 10^7 (m/10^{16}\ \text{GeV})^2\ \text{GeV} \tag{3.11}$$

Monopoles are absolutely stable relative to decay and the end of the early dust-like stage of their dominancy may be related to their annihilation with antimonopoles in configurations which are separated at this stage. The annihilation timescale is

$$\tau \approx (n_m \langle\sigma v\rangle)^{-1} \tag{3.12}$$

where n_m is the monopole density within such configurations and $\langle\sigma v\rangle$ is the average annihilation rate. n_m to within an order of magnitude corresponds to the cosmological density at the beginning of the nonlinear stage, t_1, for configurations with mass M_o. Using (1.7) and (3.11) gives:

$$n_m \sim \rho_1/m \sim \delta_o^3\, 10^{53}\, (M/10^{16}\,\mathrm{GeV})^7\; 1/\mathrm{cm}^3 \tag{3.13}$$

Assume that the characteristic velocity of monopoles inside the configuration is of order of the virial velocity. Then, allowing for (2.20)

$$v \sim \sqrt{GM/r_1} \sim c x^{\frac{1}{2}} \sim c\, \delta_o^{\frac{1}{2}}\ . \tag{3.14}$$

The annihilation rate of monopoles, estimated in Chapter 1,is of order (Zeldovich & Khlopov 1978)

$$\langle\sigma v\rangle \sim \frac{g^4}{m^2}\, v^{-9/5} \sim \frac{g^4}{m^2}\, \delta_o^{-9/10} \tag{3.15}$$

where g is magnetic charge of the monopole. One obtains from (3.13)-(3.15)

$$\tau \sim 10^{-8}\ \mathrm{s}\ (10^{16}\,\mathrm{GeV}/\ m)^5\ \delta_o^{-21/10} \tag{3.16}$$

The restrictions on the monopole mass obtained from fig. 5 and using (3.11) and (3.16) at different values δ_o exclude the possibility of the existence of monopoles with $m > 10^{17}$ GeV at $\delta_o > 2.5\ 10^{-3}$ (see fig. 6). This restriction is weaker than the one obtained by Khlopov & Polnarev (1980) and Polnarev & Khlopov (1981), since they did not take into account the softening of restrictions on $\beta(M)$ due to the presence of early dust-like stages. It should be noted, however, that this restriction may be enhanced

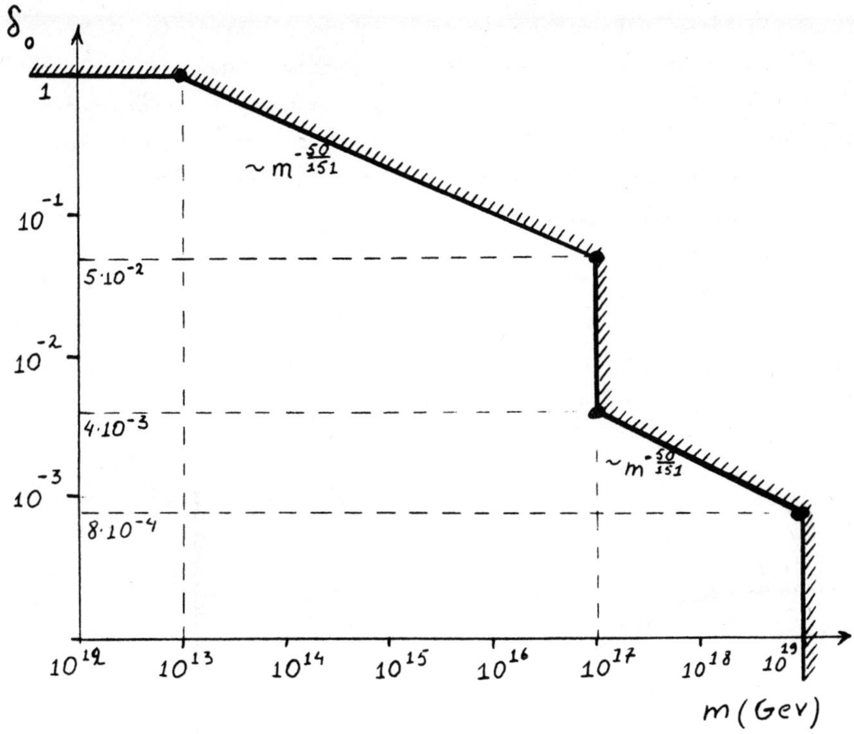

Fig. 6. Limits on the monopole mass at given δ_0.

if the fragmentation of separated configurations and PBH formation with masses smaller than M_0 is taken into account. The latter has not been done yet.

Heavy quarks. In the case of the heavy exotic quarks of the asymptotically free SU(5) theory (Kalashnikov 1980), τ and νm are related to the mass according to (1.2) and (1.10). The mass of such quarks may be expressed through the X-boson mass m_X, related to the proton lifetime, and the free parameter of the theory (Kalashnikov 1980) k_D:

$$m = r\, k_D\, m_X \tag{3.17}$$

There is a unique possibility to restrict the parameter k_D in the framework of our approach. In the theory described by Kalashnikov (1980), a very small mixing angle, determining the decay of heavy quarks, is predicted: $k_1 \sim 10^{-26}$, so that their lifetime is sufficiently large, and at $\delta_o > 10^{-7}$ the duration of the dust-like stage of their dominancy is determined by their annihilation in configurations separated from cosmological expansion. As in the case of monopoles, estimation of the annihilation time-scale for annihilation rate (1.9) gives

$$\tau = 10^{-27} k_D^{-5} \delta_o^{-3} \text{ s} \tag{3.18}$$

By using (1.10), (3.17) and (3.18) one obtains from fig. 5 restrictions on k_D for different δ_o (see fig. 7).

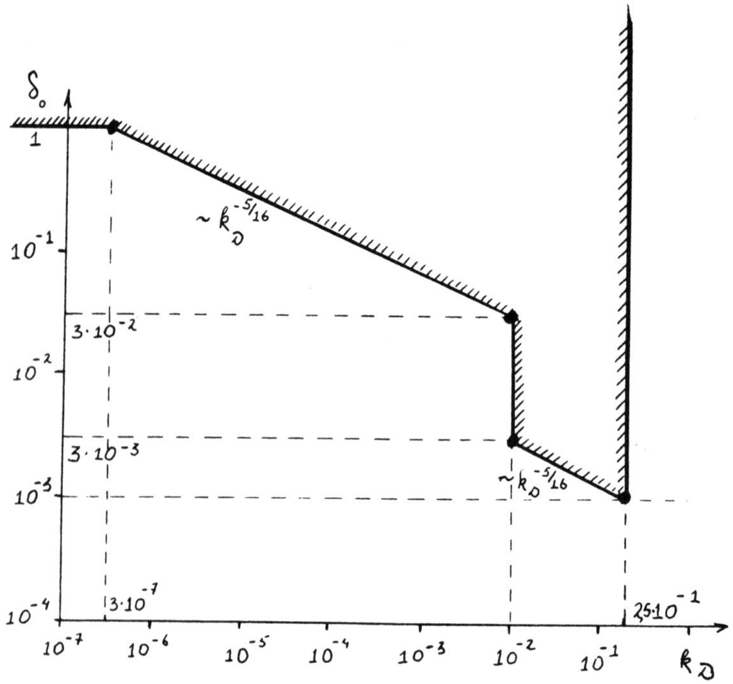

Fig. 7. Limits on the parameter k_D at given δ_o.

Heavy neutrinos (and gravitinos). In contrast with monopoles and heavy quarks, the lifetime of heavy neutrinos (as well as gravitinos (Deser & Zumino 1977; Fayet 1982)) which determines the duration of the dust-like stages of their dominance is given by the number κ of species of relativistic particles. The possibility of an inflationary stage after the freezing of gravitinos which would reduce their concentration is not considered below; see Weinberg (1982) (see Chapter 1).

Taking the mass m of these particles as a free parameter of the theory one obtains from fig. 5 restrictions on m at different δ_0 given in fig. 8 (heavy neutrinos) and fig. 9 (gravitinos; upper scale) for lifetimes of these particles given by (1.2) and (1.3), respectively.

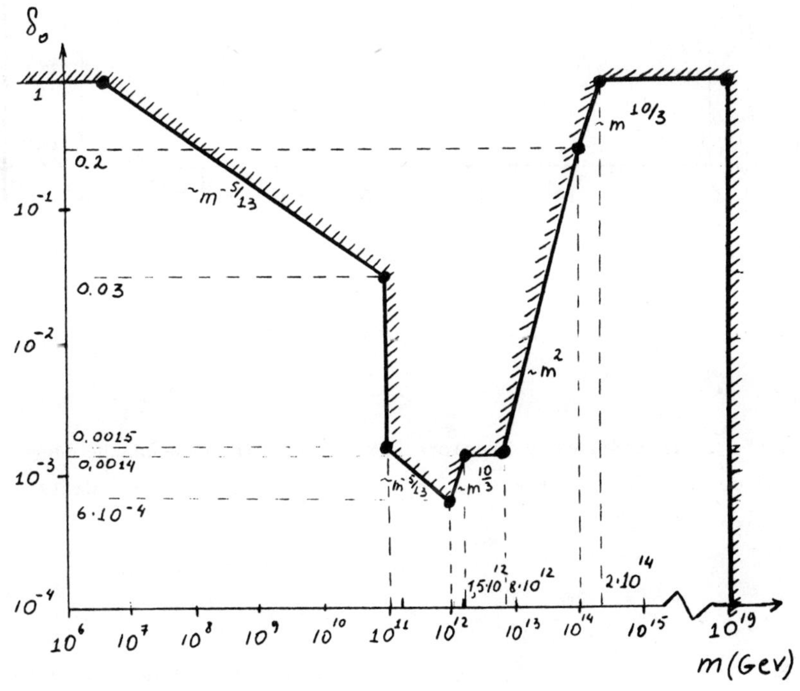

Fig. 8. Limits on the mass of superheavy neutrinos at given δ_0.

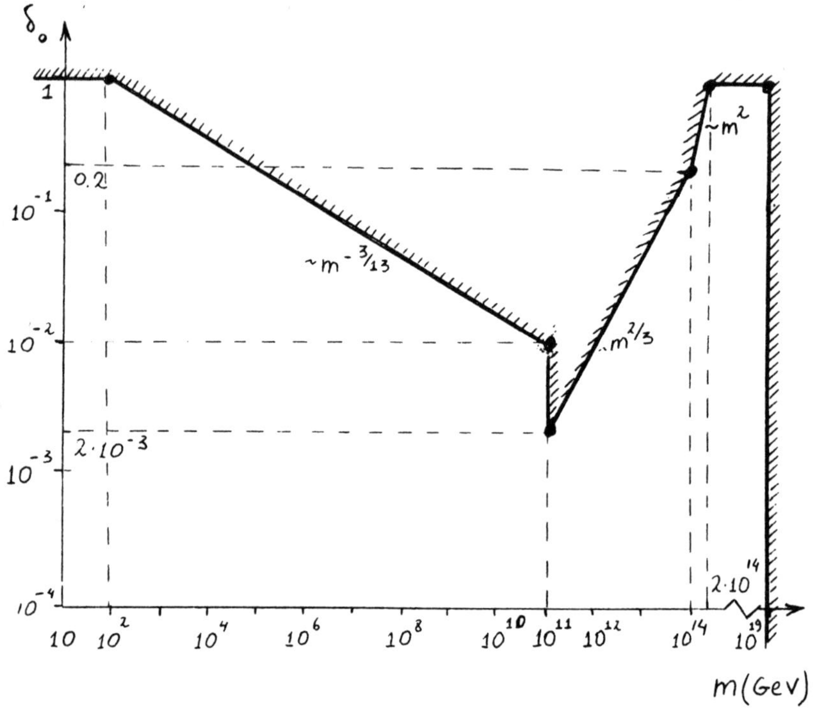

Fig. 9. Limits on the gravitino mass (at given δ_0).

If the theory predicts a relationship between the mass m of the heavy neutrino and the mass m_ν of the light neutrino (Witten 1979; Wetterich 1981)

$$m_\nu = \lambda/m \tag{3.19}$$

where $\lambda \sim m_W^2 \sim (100\ \text{GeV})^2$, our method in principle provides restrictions (depending on δ_0) on m_ν as well. It should be noted, however, that in the existing models (Witten 1979; Wetterich 1981), Eq. (3.19) implies an interaction of superheavy neutrinos with relatively light Higgs bosons, resulting in too rapid (before the establishment of an early dust-like stage) decay of heavy neutrinos into light fermions and Higgs bosons (in Eq. (1.1) $k_0 \gtrsim m/\kappa^2 m_{pl}$).

In the supersymmetric theories (Deser & Zumino 1977; Fayet 1982; Weinberg 1982), the gravitino mass m is related to the scale d of supersymmetry breaking

$$m = d/m_{pl}, \tag{3.20}$$

so that the restrictions on m provide restrictions on d.

CONCLUSION

GUTs together with the big bang theory may predict the existence of stages of superheavy particle dominance in the early Universe. The formation of PBH from small initial metric perturbations is an inevitable consequence of such stages. The relation between GUT parameters and the initial inhomogeneity of the Universe, connected (within the frame of the considered mechanism) with the PBH spectrum, provides a number of astrophysical observational restrictions on GUT parameters, depending on the initial inhomogeneity of the Universe. All the above estimates can be strengthened by detailed calculations, and in this sense all the quantitative results given above are by no means final. But already now, these results prove the principal possibility of obtaining information on GUT parameters from analysis of the observational consequences of the existence of early dust-like stages.

The detailed picture of evolution of such stages is to be elaborated. More detailed study of small initial fluctuations is to be done taking into account the smooth beginning and end of the stage as well as the final pressure of relativistic particles. Study of definite mechanisms of entropy losses at the non-linear stage of evolution of inhomogeneities may result in the growth of the lower bound of the probability of PBH formation, thus giving more stringent restrictions on GUT parameters. An attempt to give a more correct mathematical description of PBH formation from initial inhomogeneities, making it possible to refine the probability of PBH formation, is now under way.

Calculations within the frame of definite GUT models may increase the basis of astrophysical observational data, giving information on the properties of early dust-like stage, and, thus on GUT.

We are grateful to A.G. Doroshkevich, Ya.B. Zeldovich, D.A. Kompaneetz, I.D. Novikov, L.B. Okun, B.M. Pontecorvo and S.F. Shandarin for stimulating discussions.

APPENDIX

THE ACCOUNT FOR INHOMOGENEITY OF THE CONFIGURATION IN THE FRAMEWORK OF TOLMAN SOLUTION

In both the synchronized and comoving systems, the spherically symmetrical line element describing the behaviour of a dust-like matter has the form (Landau & Lifshitz 1973)

$$ds^2 = d\tau^2 - \exp \lambda(\tau,R)dR^2 - r^2(\tau,R)(d\theta^2 + \sin^2\theta \, d\psi^2) \qquad (A.1)$$

where τ is the proper time, R is the Lagrange radius and r is the Euler radius taken so that $2\pi r$ is the length of a circle with the centre at the origin.

Assume for simplicity (though it does not influence our previous estimates) that the expansion is changed by contraction simultaneously along the whole configuration at the moment t_1. Then, integrating the Einstein equations under such initial conditions one obtains the function $r(\tau,R)$ in the following parametrized form:

$$r = \tfrac{1}{2} r_1(R)(1 + \cos \eta) \qquad (A.2)$$

$$\tau = t_1 + \tfrac{1}{2} \frac{r_1^{3/2}(R)}{r_g^{1/2}(R)} (\eta + \sin \eta) \qquad (A.3)$$

Here $0 \le \eta \le \pi$, $r_g(R)$ is the gravitational radius for the mass contained within the Lagrange radius R and $r_1(R)$ is determined by the density distribution $\rho_1(R)$ at the moment t_1. The further evolution of the density determined by the Einstein equations follows the relationship:

$$\rho(\tau,R) = \rho_1(R) \frac{r_1^2(R) \, r_1'(R)}{r^2(R) \, r'(R)} \qquad (A.4)$$

Here prime denotes the partial derivative with respect to R. Let the density $\rho_1(R)$ of the distribution at the moment t_1 be as follows

$$\rho_1(R) = \rho_{c1}\left(1 - \frac{r_1^2(R)}{r_*^2}\right) \tag{A.5}$$

Here r_* characterizes the degree of inhomogeneity of the configuration, being determined by the second derivative of the density distribution with respect to the Euler radius at the centre of the configuration. The degree of inhomogeneity of the configuration defined earlier is $u = (r_1/r_*)^2$, where r_1 is the lengthscale of the configuration introduced above. It follows immediately from (A.5) that

$$r_g(R) = \frac{8\pi G\rho_{c1}}{3} r_1^3(R) \left[1 - \frac{3r_1^2(R)}{5\, r_*^2}\right] \tag{A.6}$$

The moment of caustic formation in the centre of the configuration occurs according to (A.2), (A.3) and (A.6) at t_{caus} given by

$$t_{caus} = t_1 + \left(\frac{3}{32\pi G\rho_{c1}}\right)^{\frac{1}{2}} \tag{A.7}$$

The moment t_{PBH}, when the whole configuration contracts under its gravitational radius, is

$$t_{PBH} \approx t_1 + \left(\frac{3}{32\pi G\rho_{c1}}\right)^{\frac{1}{2}}\left(1 - \frac{4}{3}x^{3/2}\right) \Big/ \left(1 - \frac{3}{5}u\right)^{3/2} \tag{A.8}$$

It follows from the Eqs. (A.7), (A.8) that the condition reduces to the inequality (2.15)

One obtains from the Eqs. (A.2)-(A.6)

$$\rho(\tau,R) = \rho_1(R)\left(\frac{2}{1+\cos\eta}\right)^3 \left\{1 + \frac{3}{5}\frac{\sin\eta(\eta+\sin\eta)}{(1+\cos\eta)^2}\frac{r_1^2(R)}{r_*^2}\right\}^{-1} \tag{A.9}$$

At the moment t_{PBH} the parameter η at the centre of the configuration is equal to

$$\eta_c = \pi - 2\,x^{\frac{1}{2}}\,(1 - \xi)^{1/3} \qquad (A.10)$$

where

$$\xi = \frac{9\pi}{40}\,\frac{u}{x^{3/2}} \qquad (A.11)$$

At the same moment of time the parameter η of the outer boundary of the configuration is equal to

$$\eta_B = \pi - 2\,x^{\frac{1}{2}} \qquad (A.12)$$

Then, with account for (A.9)-(A.12) the condition is fulfilled for $\xi \lesssim 1$, once again reducing to the inequality (2.15).

REFERENCES

Akhiezier, A.I. & V.B. Berestetsky (1965) *Quantum Electrodynamics* New York: Wiley, Interscience.

Barrow, J.D. (1980) *Mon. Not. RAS 192*, 427.

Bisnovatyi-Kogan, G.S. & I.G. Novikov (1980) *Astron. Zh. 57*, 899.

Blinnikov, S.I. & M.Yu. Khlopov (1982) *Yadernaya Fyzika 36*, 213.

Blandford, R.D. (1977) *Mon. Not. RAS 181*, 489.

Bond, J.R., G. Efstathiou & J. Silk (1980) *Phys. Rev. Lett. 45*, 1980.

Buras, A.J., J. Ellis, M.K. Gaillard & D.V. Nanopoulos (1978) *Nucl. Phys. B135*, 66.

Carr, B.J. (1975) *Astrophys. J. 201*, 1.

Carr, B.J. (1976) *Astrophys. J. 206*, 8.

Carr, B.J. (1978) *Comm. Astrophys. 7*, 161.

Cherry, M.L., Deakyne, M., Lande, K., Lee, C.K., Steinberg, R.I. & Cleveland, B. (1981) *Phys. Rev. Lett. 47*, 1507.

Chibisov, G.V. & V.F. Mukhanov (1981) *Pis'ma ZhETF 33*, 549.

Deser, S. & B. Zumino (1977) *Phys. Rev. Lett. 38*, 1433.

Dicus, D.A., D.N. Page & V.L. Teplitz (1982) *Phys. Rev. D26*, 1306.

Dolgov, A.D. & Ya. B. Zeldovich (1981) *Rev. Mod. Phys. 53*, 1.

Doroshkevich, A.G. (1970) *Astrofizika 6*, 581.

Doroshkevich, A.G. & M. Yu. Khlopov (1981) *Astron. Zh. 58*, 907.

Doroshkevich, A.G. & M.Yu. Khlopov (1982) *Pis'ma ZhETF 37*, 58.

Doroshkevich, A.G., M.Yu. Khlopov, R.A. Sunyaev, A.S. Szalay & Ya.B.Zeldovich (1981) Proc. X. Texas Symp. *Ann. N.Y. Acad. Sci. 375*.

Fayet, P. (1982) Preprint *LPTENS 82/10*.

Fidecaro, G. (1981) Preprint *CERN-EP/81*-136

Fiorini, E. (1982) *Proc. Int. Conf. High En. Phys. in Paris* (to be published).

Gerstein, S.S. & Ya.B. Zeldovich (1966) *JETP Letters 4*, 120.

Grishuk, L.P. (1974) *ZhETF 67*, 825

Gross, D.J. & F. Wilczek (1973) *Phys. Rev. Lett. 30*, 1343.

Guth, A.H. (1981) *Phys. Rev. D23*, 347.

Hawking, S.W. (1971) *Mon. Not. RAS 152*, 75.

Hawking, S.W. (1974) *Nature 248*, 30.

Hawking, S.W. (1975) *Commun. Math. Phys. 43*, 199.

Hawking, S.W. (1982) *Phys. Lett. B115*, 295.

Kalashnikov, O.K. (1980) *Lebedev Inst. Preprint - 79*.

Kibble, T.W.B. (1976) *J. Phys. A9*, 1387.

Khlopov, M.Yu. (1980a) *Yadernaya Fyzika 32*, 1890.

Khlopov, M.Yu. (1980b) *JETP Letters 33*, 162.

Khlopov, M.Yu. & A.G. Polnarev (1980) *Phys. Lett. B97*, 383.

Klinkhamer, F.R., Branco, G., Derendinger, J.P., Hut, P. & Masiero, A. (1981) *Astron. & Astroph. 94*, L19.

Kompaneetz, A.D., V.N. Lukash & I.D. Novikov (1982) *Astron. Zh. 59*, 424.

Kuzmin, V.A. (1970) *Pis'ma ZhETF 13*, 335.
Landau, L.D. & E.M. Lifshitz (1973) *Field Theory* Moscow: Nauka.
Learned, J., Reines, F. & Soni, A. (1979) *Phys. Rev. Lett. 43*, 907.
Lee, B.W. & Weinberg, S. (1977) *Phys. Rev. Lett. 39*, 165.
Linde, A.D. (1979) *Rep. Prog. Phys. 42*, 390.
Linde, A.D. (1982) *Phys. Lett. 108B*, 389.
Lifshitz, E.M. (1946) *ZhETF 16*, 587.
Lukash, V.N. (1980) *ZhETF 79*, 1601.
Lyubimov, V.A., Novikov, E.G., Nozik, V.Z., Tretyakov, E.F. & Kosik, V.S. (1980) *Phys. Lett. B94*, 266.
Misner, C.W., Thorne, K.S. & Wheeler, J.A. (1973) *Gravitation*. San Francisco: W.H. Freeman & Company.
Nadyozhin, D.K., Novikov, I.D. & Polnarev, A.G. (1978) *Astron. Zh. 55*, 216.
Nasselsky, P.D. (1978). *Pis'ma Astr. Zh. 4*, 387.
Novikov, I.D., Polnarev, A.G., Starobinsky, A.A. & Zeldovich, Ya.B. (1979) *Astron. & Astrophys. 80*, 104.
Page, D.N. & Hawking, S.W. (1975) *Astrophys. J. 206*, 1.
Okun, L.B. (1980) *Leptons and Quarks*. Moscow: Nauka.
Okun, L.B. (1981) *Sov. Phys. Usp. 24*, 341.
Peebles, P.J.E. (1980) *The Large-scale Structure of the Universe*. Princeton: Princeton University Press.
Polnarev, A.G. & Khlopov, M.Yu. (1981) *Astron. Zh. 58*, 706.
Polnarev, A.G. & Khlopov, M.Yu. (1982a) *Astron. Zh. 59*, 15.
Polnarev, A.G. & Khlopov, M.Yu. (1982b) *Astron. Zh. 59*, 800.
Polyakov, A.M. (1974) *JETP Letters 20*, 194.
Preskill, J.P. (1979) *Phys. Rev. Lett. 43*, 1365.
Rees, M.J. (1977) *Nature 268*, 333.
Salam, A. (1968) *Nobel Symp. 8*. ed. N. Svartholm. Stockholm.
Schvartzman, V.F. (1969) *JETP Letters 9*, 184.
Schwartz, A.S. (1976) *Nucl. Phys. 112*, 358.
Starobinsky, A.A. (1979) *Pis'ma ZhETF 30*, 719.
Starobinsky, A.A. (1980) *Phys. Lett. B91*, 100.
Starobinsky, A.A. (1982) Lecture at this Workshop.
Szalay, A.S. & Marx, G. (1976) *Astron. & Astrophys. 49*, 437.
t'Hooft, G. (1974) *Nucl. Phys. B79*, 276.
Vysotsky, M.I., Dolgov, A.D. & Zeldovich, Ya.B. (1977) *JETP Letters 26*, 200.
Wainer, B.V. & Nasselsky, P.D. (1977) *Pis'ma Astr. Zh. 3*, 147.
Weinberg, S. (1967) *Phys. Rev. Lett. 19*, 1264.
Weinberg, S. (1978) *The First Three Minutes*. Glasgow: Fontana.
Weinberg, S. (1982) *Phys. Rev. Lett. 48*, 1303.
Wetterich, C. (1981) *Nucl. Phys. B187*, 343.
Wilson, R. (1980) Harvard Preprint.
Witten, E. (1979) *Phys. Lett. 91*, 81.
Zabbotin, N.A., Marochnik, L.S. & Nasselsky, P.D. (1980). Preprint *ISR* 564.
Zeldovich, Ya.B. (1976) *Pis'ma ZhETF 24*, 29.
Zeldovich, Ya.B. (1981) *2nd International Seminar on Quantum Gravity, Moskow*. (To be published).
Zeldovich, Ya.B. & Khlopov, M.Yu. (1978) *Phys. Lett. 79B*, 239.
Zeldovich, Ya.B. & Khlopov, M.Yu. (1981) *Sov. Phys. Usp. 24*, 755.
Zeldovich, Ya.B., Klypin, A.A., Khlopov, M.Yu. & Chechetkin, V.M. (1980) *Yadernaya Fyzika 31*, 1286.
Zeldovich, Ya.B., Kobzarev, I.Yu. & Okun, L.B. (1974). *ZhETP 67*, 3.
Zeldovich, Ya.B. & Novikov, I.D. (1966) *Astron. Zh. 43*, 758.
Zeldovich, Ya.B. & Novikov, I.D. (1975) *Structure and Evolution of the Universe*. Moskow: Nauka.
Zeldovich, Ya.B., Starobinsky, A.A., Khlopov, M.Yu. & Chechetkin, V.M. (1977) *Pis'ma Astr. Zh. 3*, 208.

AN ATTEMPT TO GET RID OF THE COSMOLOGICAL CONSTANT

A.D. Dolgov
Institute of Theoretical and Experimental Physics,
Moscow 117259,
USSR.

Abstract. A model is proposed in which the cosmological constant is cancelled automatically because of the back reaction on it of a scalar field ϕ coupled to the curvature R. The cancellation takes place if two conditions are imposed on the theory, namely the vanishing of the mass and the selfinteraction of field ϕ. The vacuum energy does not disappear completely but only up to the order of $\rho_c \simeq M_p^2/t^2$. This leads to some modification of the standard expansion scenario.

As is well known, the Einstein equations can be generalized by adding a cosmological term:

$$R_{\mu\nu} - \frac{1}{2} R g_{\mu\nu} = -8\pi M_p^{-2} T_{\mu\nu} + \lambda g_{\mu\nu} \tag{1}$$

where $M_p \simeq 10^{19}$ GeV is the Planck mass, $T_{\mu\nu}$ is the energy-momentum tensor of matter and λ corresponds to the gravitational interaction of the vacuum. It is convenient to define the vacuum energy density which is connected with the last term in eq (1): $\rho_{vac} = -\lambda M_p^2/8\pi$. Modern astronomical data show that ρ_{vac}, if nonvanishing, is extremely small:

$$|\rho_{vac}| < 10^{-46}\ m_N^4 \tag{2}$$

where m_N is the nucleon mass. Any energy scale in elementary particle physics is enormous in comparison with this number.

The first question which arises is why does the vacuum not gravitate? Why does the zero point energy not gravitate? (Zeldovich, 1968). There are infinite contributions from any particle species to this energy. Probably the infinities in the total sum of these contributions are cancelled out as a consequence of supersymmetry. In reality however supersymmetry is broken so the finite part in ρ_{vac} should generally survive and be of the order of m^4, where m is a characteristic mass scale. For an unknown reason this energy does not influence the evolution of the universe.

The situation is even more weird in gauge theories with spontaneous symmetry breaking. The point is that a vacuum expectation value of a Higgs field χ necessarily leads to a nonzero cosmological term, $T_{\mu\nu}(\chi) \sim g_{\mu\nu}\langle\chi\rangle^2 m_\chi^2$. The corresponding vacuum energy is about $10^8\ m_N^4$ for the Weinberg-Salam model and is of the order of $10^{60}\ m_N^4$ for Grand Unification Theories. As condensates of the Higgs fields develop after sufficient expansion and cooling of the universe (Kirznits 1972, Kirznits and Linde 1972, Linde 1974), the initial vacuum energy should be nonzero and its value has to coincide with that of the later developed condensates with an accuracy which is better than a hundred orders of magnitude. One can hardly believe in such superfine tuning and some attempts (Polyakov 1982, Hawking 1982) have been made to find a more natural explanation of the smallness of the cosmological constant.

This paper presents an attempt to find a model of selfcancellation of cosmological terms. It is assumed that there exists a field ϕ which interacts with the vacuum energy in such a way that the amplitude of the field rises so that vacuum condensate of ϕ is produced. The energy of this condensate should cancel the vacuum energy which initially was the source of the condensate development. Naturally the rate of this process should be of the order of H^{-1} , where H is the Hubble constant, and the accuracy of this compensation is about H^2 , which is close to the existing bound.

To get insight into the essential features of the mechanism proposed we consider a simple, but as it turns out, not absolutely satisfactory, model consisting of a massless scalar field coupled to gravity and free of all other interactions. Later we shall see that coupling to other fields has to be introduced. Assume for simplicity that the universe is spatially flat (k = 0):

$$ds^2 = dt^2 - a^2(t)d\sigma^2 \qquad (3)$$

In this metric ϕ satisfies the equation

$$\phi^{;\lambda}_{\ ;\lambda} + \xi R\phi = \partial_t^2\phi + 3H\partial_t\phi - \frac{1}{a^2}\partial_i^2\phi + \xi R\phi = 0 \qquad (4)$$

where $H = \dot{a}/a$ is the Hubble constant and ξ is a numerical constant.

The scalar curvature R can be expressed, using the Einstein equations, as

$$R = 8\pi M^{-2}(4\rho_{vac} + T^{\mu}_{\ \mu}) \tag{5}$$

where M does not coincide with M_p because of the corrections due to nonzero ϕ (see below). For conformal fields $T^{\mu}_{\ \mu} = 0$ if quantum corrections are neglected (i.e. in the tree approximation).

If $\xi R < 0$, long-wave fluctuations of ϕ are unstable and there exist solutions of eq (4) which rise with time. For example, if the energy-momentum tensor is dominated by the vacuum term $g_{\mu\nu}\rho_{vac}$ ($\rho_{vac} > 0$) then $H = (8\pi\rho_{vac}/3M^2)^{1/2}$, $R = 12H^2$ and eq (4) has spatially homogeneous solutions of the type

$$\phi = \exp\{(-\frac{3}{2} \pm \sqrt{\frac{9}{4} - 12\xi})Ht\} \tag{6}$$

If $\xi < 0$ one of these solutions increases infinitely with t.

There is no known reason for ξ being negative. Two specific values of ξ have some justification. One is $\xi = 1/6$ which leads to conformal invariance of the theory and the other is $\xi = 0$ which should be true for Goldstone bosons (Dolgov and Valoshin 1982). However if vacuum energy is positive the assumption of negative ξ is a necessary condition for rising ϕ. Accordingly the constribution of ϕ into energy-momentum tensor should be taken into account:

$$T_{\mu\nu}(\phi) = \phi_{;\mu}\phi_{;\nu} - \frac{1}{2}g_{\mu\nu}\phi_{;\lambda}\phi^{;\lambda} - \xi\phi^2(R_{\mu\nu} - \frac{1}{2}g_{\mu\nu}R)$$
$$- \xi(\phi^2)_{;\mu;\nu} + \xi g_{\mu\nu}(\phi^2)^{;\lambda}_{\ ;\lambda} \tag{7}$$

The corresponding scalar curvature is

$$R = \frac{8\pi[4\rho_{vac} + (6\xi - 1)\dot{\phi}^2]}{M^2 + 8\pi\xi(6\xi - 1)\phi^2} \tag{8}$$

and H is given by

$$H^2 = \frac{8\pi}{3}\ \frac{\rho_{vac} + \frac{1}{2}\dot{\phi}^2 + 6H\xi\phi\dot{\phi}}{M^2 - 8\pi\xi\phi^2} \tag{9}$$

The back-reaction of ϕ on gravity slows down the explosive rise shown in eq (6) but no real stabilization is achieved. Indeed ϕ

does not tend to a constant as $t \to \infty$; instead $\dot{\phi} \to \text{const}$, so that $\phi \to At$ where $A = O(\rho_{vac}^{1/2})$. Anyway the de Sitter expansion law is destroyed by the classical field $\phi(t)$. Consequently $H \sim t^{-1}$ and $R \sim t^{-2}$ as $t \to \infty$. Unfortunately it is not only R than vanishes but also gravity as a whole because the effective gravitational constant tends to zero:

$$G_{eff} = [M^2 - 8\pi\xi\phi^2]^{-1} \sim t^{-2} \tag{10}$$

Before discussing a better model let us comment on the spatial dependence of the unstable solution of eq (4). If the scalar curvature is not spatially constant it will drive the formation of the condensate of ϕ only in the case of rather large characteristic scale of the matter inhomogeneity

$$\ell > M[32\pi\rho|\xi|]^{-1/2} \tag{11}$$

For example, for $\rho = 1 \text{ g cm}^{-3}$ the ϕ-condensate would be formed if the matter distribution is homogeneous up to distances $10^{13}(8\pi|\xi|)^{-1/2}$ cm .

Apart from killing the gravitation, the model discussed above does not lead to compensation of the cosmological term because $T_{\mu\nu}(\phi) \not\to \lambda g_{\mu\nu}$, the scalar curvature being compensated however. Moreover, an infinitely rising field $\phi(t) \sim t$ does not look very appealing. It is desirable to find a model where $\phi \to \phi_o = \text{const}$ as $t \to \infty$ and $T_{\mu\nu}(\phi_o) \sim g_{\mu\nu}$.

The simplest way to realize this idea, by giving the field ϕ a nonzero mass m or a selfcoupling $\lambda\phi^4$, immediately encounters difficulties. In this case extra terms appear in eq (4):

$$\ddot{\phi} + 3H\dot{\phi} + m^2\phi + \lambda\phi^3 + \xi R\phi = 0 \tag{12}$$

As before R can be expressed in terms of the trace of $T_{\mu\nu}$ in which new terms now appear which do not vanish for $\phi = \text{const}$:

$$\delta T^{\mu}{}_{\mu} = m^2\phi^2 + \beta_{\phi}\phi^4 \tag{13}$$

The last term on the r.h.s. of this equation is the quantum conformal anomaly (Crewther 1972, Chanowitz and Ellis 1972) produced by loops of self-interacting ϕ . One should remember that in matrix elements of T^{μ}_{μ} the anomalous term survives only for momenta exceeding m .

With the appropriate choice of the parameter signs eq (12) has solutions which tend to a nonzero constant ($\phi \to \text{const} \neq 0$) as $t \to \infty$. This solution however leads to the condition $m^2 + \lambda\phi^2 + \xi R = 0$ but not to $R = 0$. If we put $m = \lambda = 0$ then the anomalous term in eq (13) also disappears and we return to the first model.

From this consideration it follows that coupling of ϕ with other fields should be introduced with the following conditions being imposed (in flat space-time):

$$m_{\phi} = 0 \tag{14a}$$

$$U_{eff}(\phi) = 0 \tag{14b}$$

$$T^{\mu}_{\mu}(\phi)\Big|_{\phi=\text{const}} \neq 0 \tag{14c}$$

Whereas m_{ϕ} is a free parameter of the theory conditions (14b) and (14c) taken together look rather surprising. Of course it would be very desirable to find a model where conditions (14) appear naturally but I did not succeed. As a poor man's substitution it should be at least checked up whether conditions (14b) and (14c) are not self-contradictory. At one loop level this can be explicitly done. Let ϕ have the following couplings:

$$L_{int} = \lambda\phi^2\chi^2 + g\phi\bar{\psi}\psi + e^2A_{\mu}^2\phi^2 + ieA_{\mu}\phi^* \overleftrightarrow{\partial}_{\mu}\phi + \ldots \tag{15}$$

where χ , ψ , and A_{μ} are respectively a scalar, a spinor, and a vector (gauge) field.

This interaction being taken into account, T^{μ}_{μ} becomes nonvanishing:

$$T^{\mu}_{\mu} = \phi^4(\beta_o + \beta_{\frac{1}{2}} + \beta_1 + \ldots) + \ldots \tag{16}$$

where β_j comes from the loops formed by virtual particles with spin j . There are some other terms which are omitted in eq (16) (e.g. proportional to $\phi^2\chi^2$, χ^4 , $(\partial_{\mu}A_{\nu} - \partial_{\nu}A_{\mu})^2$ etc.) but, assuming that only ϕ is unstable with respect to nonzero space-time curvature, we neglect all other contributions.

The signs of β_j depend on whether the underlying quantum field theory is asymptotically free or not, so $\beta_o < 0$, $\beta_{\frac{1}{2}} < 0$, and $\beta_1 > 0$ for a non-abelian gauge field and $\beta_1 < 0$ for an abelian field.

In the one loop approximation the effective potential of ϕ is known to have the following form (Coleman and Weinberg 1973)

$$U_{eff}(\phi) = m^2\phi^2 + \lambda_\phi \phi^4 + \phi^4 \ln(\phi/\Lambda)\ (\gamma_o + \gamma_{\frac{1}{2}} + \gamma_1 + \ldots) \tag{17}$$

where Λ is an ultraviolet cutoff and γ_j comes from the contribution of particles with spin j. Constants γ_j are almost the same as β_j except that their signs are the following: $\gamma_o > 0$, $\gamma_{\frac{1}{2}} < 0$, and $\gamma_1 > 0$. So in principle, the parameters of the model can be fixed in such a way that the contributions of different particle species into $U_{eff}(\phi)$ cancel each other whereas the analogous contributions to T^μ_μ remain nonvanishing. Simultaneously the normalisation conditions to be imposed on the theory are the following:

$$m_\phi = 0\ , \quad \lambda_\phi = 0$$

Clearly this is the weakest point of the construction. At this stage it looks as if fine tuning in the initial value of vacuum energy has been replaced by fine tuning of the parameters of the model. But the important difference is that there may be some symmetry arguments, not found yet, which lead to conditions (14). One possibility is a supersymmetric model and another is a model invariant under the transformation $\phi \rightarrow \phi + \text{const}$ except for coupling to gravity. However in the latter case it is difficult to reach a definite conclusion because the model is non-renormalizable.

So at the moment no natural foundation for conditions (14) has been found and they are imposed by hand. Thus we assume that the classical homogeneous field ϕ satisfies the equation

$$\ddot{\phi} + 3H\dot{\phi} + \xi R\phi = \lambda\chi^2\phi + g\bar{\psi}\psi + e^2A_\mu^2\phi + \ldots \tag{18}$$

and that for $\phi \rightarrow \text{const}$

$$R \rightarrow \frac{8\pi[4\rho_{vac} + \beta\phi^4]}{M^2 + 8\pi\xi(6\xi - 1)\phi^2} \tag{19}$$

One can see that $R \to 0$ if $\phi = \phi_o = (4\rho_{vac}/|\beta|)^{1/4}$. In what follows we argue that ϕ does indeed tend to ϕ_o and that this classical field indeed eats up the cosmological term. In this model the gravitational constant is changing with time from its initial value $G_i = M^{-2}$ to its asymptotic value $G_o^{-1} = M^2 + 8\pi\xi\phi_o^2 = M_p^2$. This variation is not in contradiction with astronomy because ϕ changes predominantly in a very early epoch. The analogous changing of the gravitational constant due to the classical scalar field has been discussed by Smolin (1979), Englert et al. (1976), Zee (1979), and Linde (1980).

If $\xi R < 0$, eq (12) has unstable solutions leading to vacuum condensates of ϕ . As a result the spinor field becomes massive, $m_\psi = g\phi_o$. It is noteworthy that despite the obtained mass of ψ the energy-momentum tensor remains traceless in the tree approximation if $\xi = 1/6$. (If one considers a spinor field which is massive from the beginning interacting with a scalar field, the condition $T^\mu_\mu = 0$ can be also fulfilled by adding a total derivative $(1/6)(g_{\mu\nu}\partial^2_\alpha - \partial_\mu\partial_\nu)(\phi + g^{-1}m_\psi)^2$ to $T_{\mu\nu}$ in flat space-time.) If $\xi \neq 1/6$ extra terms proportional to $g\phi\bar{\psi}\psi$ etc. appear in T^μ_μ. We will not take them into account however because they prove to be small. The main contribution to T^μ_μ , which compensates ρ_{vac} , is given by the anomalous term $\beta\phi^4$.

Thus, with all the weak points kept in mind, the proposed feed-back mechanism forces the scalar curvature to vanish when ϕ tends to a constant value. We will argue in what follows that not only R but also the entire cosmological term $\rho_{vac}g_{\mu\nu}$ tends to zero. In other words, the energy-momentum tensor of the classical field ϕ_o is proportional to $g_{\mu\nu}$. First of all note that $T_{\mu\nu}(\phi)$ does not vanish when $\phi = \text{const}$ because $T^\mu_\mu \sim \phi^4$. $T_{\mu\nu}(\phi)$ has a rather complicated nonlocal structure but when $\phi = \text{const}$ only terms proportional to $g_{\mu\nu}$ should survive. To see this consider the equation

$$\frac{\dot{\rho}}{\rho + p} = -3H \tag{20}$$

where ρ and p are respectively the energy and pressure density of ϕ and ψ in the spatially homogeneous case (i.e. $\phi = \phi(t)$ and $\psi = \psi(t)$). This equation is equivalent to the Klein-Gordon equation for ϕ . In the tree approximation $\dot{\rho} = \dot{\phi}\ddot{\phi}$ + (terms not containing $\ddot{\phi}$) and $p + \rho = \dot{\phi}^2 + \ldots$ If higher orders of perturbation theory (loops) led to $p + \rho \neq 0$ for $\phi = \text{const}$, it would drastically change the wave equation for ϕ :

$$\ddot{\phi} + 3H[\dot{\phi} + \frac{A(\phi)}{\dot{\phi}}] + \ldots = 0 \tag{21}$$

where $A(\phi) = p + \rho$ at $\rho = \text{const}$. This seems improbable and one would expect $p + \rho \to 0$ and consequently $T_{\mu\nu} \sim g_{\mu\nu}$ as $\phi \to \text{const}$.

Returning to eq (20) let us note that the condition $\dot{\rho} = 0$ in flat space-time is fulfilled separately in each order of perturbation theory.

A detailed investigation of solution of eq (18) is impossible because the dependence of R on derivatives of ϕ is unknown. Moreover the production of quanta of the ϕ-field by the varying classical field $\phi(t)$ is not taken into account. The latter is proportional to $\dot{\phi}(t)$ to some power and is equivalent to a friction force.

When $\phi(t)$ is close to its asymptotic value ϕ_o one can expand eq (18) in powers of $\phi_1(t) = \phi_o - \phi(t)$ and obtain in linear approximation the following coupled equations governing the behaviour of $\phi_1(t)$ for large t:

$$\begin{aligned} &\ddot{\phi}_1 + 3H\dot{\phi}_1 + \Gamma(\phi_1)\dot{\phi}_1 + \mu^2\phi_1 = g\bar{\psi}\psi + \phi_o e^2 A_\mu^2 + \ldots \\ &H^2 = \frac{8\pi}{3M_p^2}(\rho_{matter} + 4\rho_{vac}\frac{\phi_1}{\phi_o}) \\ &\dot{\rho}_{matter} = -kH\rho_{matter} \end{aligned} \tag{22}$$

where $\Gamma(\phi_1)\dot{\phi}_1$ appears because of particle production by the varying field $\phi_1(t)$, k is a numerical factor ($k = 4$ for radiation and $k = 3$ for nonrelativistic matter) and $\mu^2 = -2^7\pi\xi\rho_{vac}M_p^{-2}$ is the mass of ϕ_1.

If damping due to particle production is large, $\Gamma > H$, then there exists solution which goes asymptotically as: $\phi_1 \sim t^{-2}$, $\rho_{matter} \sim t^{-2}$, $H \sim t^{-1}$, $\bar{\psi}\psi \sim t^{-2}$. This corresponds more or less to the usual picture. Large damping also prevents oscillations of ϕ which could make H^2 negative (see eq (22).

Such behaviour can be expected if at the 'beginning' vacuum energy was positive and large, $\rho_{vac} > \rho_{vac}^{GUT} \simeq m_{GUT}^4$. Moreover ϕ should tend to its asymptotic value slowly enough so that a fast change of ρ_{vac} due to the phase transitions mentioned above leaves the sum $\rho_{vac} + \delta\rho$ positive (note that the change of vacuum energy due to the phase transitions is negative, $\delta\rho < 0$).

Another problem which has to be solved is whether the energy density of ψ, A_μ, etc is too large. There are ways of altering the model, if necessary, for example by introducing an extra massless vector field interacting with ψ and A_μ. At any rate it is premature to form a definite conclusion prior to finding a model in which conditions (14) appear naturally.

It is noteworthy that a model of the type described above can also operate if initially $\rho_{vac} < 0$. In this case of course ξ should be positive. In a sense this possibility seems to be more natural. First, there is no necessity for choosing a large initial value of ρ_{vac}; it could be zero. Second, large damping Γ (which is not very easy to get) is not needed. Maybe however in this case a contradiction between the age of the universe and the value of the Hubble constant (see below) becomes stronger. Quantitative estimates have not been made.

Thus with some effort the cosmological term might be made small. The standard scenario of universe evolution is changed in this case. First of all the expression for the age of the universe

$$t_u = \frac{1}{H_o} \frac{1}{1 + \frac{1}{2}\sqrt{\Omega}} \tag{23}$$

becomes invalid. If $\rho_{vac} > 0$ the contradiction between the modern value of the Hubble constant $H_o = 100$ km/sec Mps and $t_u = (15 - 17)\times 10^9$y could be avoided. The bound on the number of neutrino species (Schwarzman 1969, Steigman et al. 1977, Dolgov and Zeldovich 1981), which follows from the data on cosmological abundances of He^4 and H^2, becomes less restrictive. If the cosmological constant always satisfies inequality (2), it would not influence the primordial nucleosynthesis. In the discussed model, however, ρ_{vac} (like the Hubble constant) is not constant in time but goes like M_p^2/t^2. That is why its role could be noticeable during all the history of the universe.

It is interesting also to consider the spatial dependence of the classical field ϕ. As was noted above its variation is possible on scales restricted by condition (11). For objects larger than ℓ a drastic change of gravitational forces could be possible. A detailed discussion seems premature until a more realistic and more definite model is found. One could hope nevertheless that the principal features of this approach remain the same in a better model.

I would like to thank A.D. Linde and Ya.B. Zeldovich for discussions and criticisms, and S. Siklos, G. Gibbons and S. Hawking for hospitality during this workshop.

during this workshop.

REFERENCES

Coleman S. and E. Weinberg (1973) Phys.Rev. D7, 1888
Chanowitz and J. Ellis (1972) Phys.Lett. 40B, 397
Crewther R.J. (1972) Phys.Rev.Lett. 28, 142
Dolgov A.D. and M.B. Voloshin (1982) Yadernaya Fizika 35, No.1
Dolgov A.D. and Ya.B. Zeldovich (1981) Rev.Mod.Phys. 53, 1
Englert F., C. Truffin and R. Gastmans (1976) Nucl.Phys. B117, 407
Hawking S (1982) Cambridge University Preprint
Kirznits D.A. (1972) JETP Lett. 15, 529
Kirznits D.A. and A.D. Linde (1972) Phys.Lett. 42B, 471
Linde A.D. (1974) JETP Lett. 19, 183
Linde A.D. (1980) Phys.Lett. 93B, 394
Polyakov A.M. (1982) Uspekhi Fiz. Nauk 136, No.3
Schwartzman V.F. (1969) JETP Lett. 9, 315
Smolin L. (1979) Nucl.Phys. B160, 253
Steigman G., D.N. Schramm and J.E. Gunn (1977) Phys.Lett. 66B, 202
Zeldovich Ya.B. (1968) Uspekhi Fiz. Nauk 95, 209
Zee A. (1979) Phys.Rev.Lett. 42, 417

AN INSTABILITY OF DE SITTER SPACE

M. J. Perry
Department of Physics, Princeton University, New Jersey 08544
U.S.A.

Abstract. The Euclidean formulation of quantum gravity is used to show that de Sitter space is unstable to the formation of black holes. This process is similar to the formation of black holes in hot Minkowski spacetime.

Minkowski spacetime at zero temperature appears to be absolutely stable. One can think of this as being a physical consequence of the positive energy theorem. The only asymptotically flat spacetime that has zero energy is Minkowski space, thus there is nothing that it could possibly decay into. If however we consider flat space at finite temperature, there will be energy present in the form of thermal gravitons. In this case, there are two distinct forms of instability. The first, the Jeans instability, can be seen in perturbation theory (Jeans 1902). It corresponds to there being a exponential enhancement of density fluctuations on distance scales larger than some critical length, the Jeans length. The second instability is due to the possibility of tunneling into a black hole state (Gross, et al. 1982). The rate for this process per unit volume Γ is given by

$$\Gamma \sim \exp\left(-\frac{1}{16\pi T^2}\right) \qquad (1)$$

where T is the temperature of the heat bath. (Units such that $G = c = \hbar = k = 1$ are used throughout this article).

In some respects, de Sitter spacetime can be thought of as at finite temperature. To be precise, an observer moving along a geodesic will see a heat bath at a temperature of $T = (\Lambda/12\pi^2)^{\frac{1}{2}}$ (Gibbons & Hawking 1977). This can be thought of as arising from the presence of a cosmological event horizon. This is made most transparent by writing the metric for de Sitter space in spherical polar co-ordinates. It is

$$ds^2 = -(1-\tfrac{1}{3}\Lambda r^2)dt^2 + (1-\tfrac{1}{3}\Lambda r^2)^{-1}dr^2 + r^2(d\theta^2+\sin^2\theta d\phi^2) \qquad (2)$$

The event horizon is located at $r = (3/\Lambda)^{1/2}$. This "temperature" does not manifest itself as a collection of thermal quanta however. This is seen by computing the expectation value of the energy-momentum tensor in this state. For pure gravity (Christensen & Duff 1979) we find

$$\langle T_{ab} \rangle = -\frac{26}{135\pi^2} \Lambda^2 g_{ab} \tag{3}$$

This energy-momentum tensor does not break de Sitter invariance. (This is in contrast to the expectation value of the energy-momentum tensor for thermal gravitons in Minkowski space which does break Lorentz invariance.)

This particular quantum state has Green's functions which are obtained by analytic continuation from a Euclidean space. The Euclidean version of de Sitter spacetime is obtained by setting $t = i\tau$ in equation (2). The singularity at $r = (3/\Lambda)^{1/2}$ can only be removed by identifying τ with period $(12\pi^2/\Lambda)^{1/2}$.

If we adopt the Euclidean approach to quantum gravity, integration over metrics which are asymptotic to de Sitter space corresponds to integrating over Euclidean metrics which are compact with $\Lambda > 0$. The Euclidean version of de Sitter space has an action $I = -3\pi\Lambda^{-1}$. This is the least possible action for such solutions of the Einstein equations. The only other known configuration which is a critical point of the Einstein action, and admits a spin structure, is a metric on the product space $S^2 \times S^2$. It can be obtained by an analytic continuation of the Kottler (1918), or Schwarzschild-de Sitter solution. The metric takes the form

$$ds^2 = -(1-\frac{2m}{r}-\frac{1}{3}\Lambda r^2)dt^2 + (1-\frac{2m}{r}-\frac{1}{3}\Lambda r^2)^{-1}dr^2 + r^2(d\theta^2+\sin^2\theta d\phi^2) \tag{4}$$

m is an arbitrary parameter. This represents a black hole immersed in de Sitter space. To construct a Euclidean metric, again we continue by setting $t = i\tau$. If $9m^2\Lambda = 1$, the apparent singularities in the metric can be eliminated. However, since $9m^2\Lambda = 1$, the black hole event horizon and the cosmological event horizon have merged. In this special case, the Lorentzian metric (4) is sometimes referred to as the Nariai (1950,1951) metric. Regarded as a Euclidean solution, we find that the metric is

$$ds^2 = (1-\Lambda p^2)d\tau^2 + \frac{dp^2}{1-\Lambda p^2} + \frac{1}{\Lambda}(d\theta^2 + \sin^2\theta d\phi^2) \tag{5}$$

By performing a co-ordinate transformation of the form

$$P = \Lambda^{-\frac{1}{2}} \cos\chi \tag{6}$$

we see that the metric takes the form

$$ds^2 = \frac{1}{\Lambda}\left[d\chi^2 + \Lambda\sin^2\chi d\tau^2 + d\theta^2 + \sin^2\theta d\phi^2 \right] \tag{7}$$

This metric is clearly a product metric on a pair of S^2's, each of radius Λ^{-1}. This instanton has a Euclidean action of $I = -2\pi\Lambda^{-1}$.

Two other compact instantons with $\Lambda > 0$ are explicitly known. These are $\mathbb{C}P^2$ with the Fubini-Study metric (Gibbons & Pope 1978), and the $\mathbb{C}P^2\#\overline{\mathbb{C}P^2}$ with the Page (1978) metric.

Neither of these metrics possesses a spin structure, and so do not contribute to any functional integral which contains fermions obeying the usual charge-statistics relations. It is probable that there are other instantons with $\Lambda > 0$ which have actions greater than $-2\pi\Lambda^{-1}$. These are not included in the present discussion.

If we now attempt to look at quantum corrections, it is necessary to determine whether these solutions are minima of the Einstein action or merely saddle-points. A solution is a minimum if the operator G:

$$G = -\Box - 2\,(\text{Riemann}) \tag{8}$$

has a positive semi-definite spectrum when restricted to transverse tracefree symmetric tensors (Gibbons & Perry 1979). $G > 0$ for the de Sitter metric. However, on $S^2 \times S^2$ there is a single eigenmode of G which has negative eigenvalue (Friedan 1980). In the co-ordinate system of (7) when the metric is g_{ab}, it takes the form

$$\phi_{ab} = \begin{pmatrix} 1 & 0 & 0 & 0 \\ 0 & \sin^2\chi & 0 & 0 \\ 0 & 0 & -1 & 0 \\ 0 & 0 & 0 & -\sin^2\theta \end{pmatrix} \tag{9}$$

Regarded as a perturbation of $S^2 \times S^2$, it causes one of the two spheres to shrink, and the other to expand. (This is rather analogous to a similar instability when two soap bubbles of equal size are connected together.) It therefore follows that the quantum corrections to the functional integral develop an imaginary part in this sector.

The interpretation of this imaginary part is that it is

caused by the possibility of a topology change in the spacetime (Gross, et al. 1982). This type of process may well be classically forbidden. However, it can occur as a result of quantum mechanical processes. To see the classical Lorentzian spacetime formed as a result of this tunneling process, we take a maximal slice through the instanton, and use this 3-surface as an initial data set for the spacetime. Thus the initial data for the evolution after tunneling is given by equation (5) with $d\tau = 0$. In the absence of any further quantum effects, the spacetime is static. However, there will presumably be some evolution in the system which could be found by some form of self-consistent (Hartree-Fock) back-reaction computation, analogous to the usual black hole computations. We can regard the Nariai solution as being a black hole of mass $(9\Lambda)^{-\frac{1}{2}}$ immersed in a background de Sitter space.

The probability of nucleating the black hole can be estimated from the partition function, which is, approximately

$$Z \sim e^{+3\pi/\Lambda}[1 + ie^{-\pi/\Lambda} V] \times (\text{quantum corrections}) \tag{10}$$

V is a factor of the volume of 3-space and arises from zero-mode contributions to Z. No attempt has been made to include contributions from multi-instanton configurations. However, on the basis of this, the probability of nucleating a black hole per unit volume per unit time, Γ , is approximately given by

$$\Gamma \sim e^{-\pi/\Lambda} \tag{11}$$

This number is incredibly small given the current upper-bound on Λ , ($\sim 10^{-122}$). However, if the universe had an inflationary epoch in which $\Lambda \sim 1$ (see the article by Starabinsky in this volume for example), then such a process could give rise to a large number of black holes in the early universe, which themselves could have interesting cosmological consequences.

A more complete and coherent account of this work can be found in Ginsparg & Perry (1982). I wish to thank Gary Gibbons, Paul Ginsparg, and Don Page for useful comments, and the National Science Foundation for financial support under contract PHY80-19754.

References

Christensen, S.M. & Duff, M.J. (1979). Quantizing gravity with a cosmological constant. Nucl. Phys., B170, 480-506.

Friedan, D. (1980). Nonlinear models in 2+ε dimensions. LBL Report 11517.

Gibbons, G.W. & Hawking, S.W. (1977). Cosmological event horizons, thermodynamics, and particle creation. Phys. Rev., D15, 2738-51.

Gibbons, G.W. & Perry, M.J. (1978). Quantizing gravitational instantons. Nucl. Phys., B146, 90-108.

Gibbons, G.W. & Pope, C.N. (1978). $\mathbb{C}P^2$ as a gravitational instanton. Commun. Math. Phys., 61, 239-48.

Ginsparg, P. & Perry, M. J. (1982). Instabilities in de Sitter space. To be published.

Gross, D.J., Perry, M.J. & Yaffe, L.G. (1982). The instability of flat space at finite temperature. Phys. Rev. D25, 330-355.

Jeans, J.H. (1902). The stability of a spherical nebula. Philos. Trans. Roy. Soc. (London), A199, 1-53.

Kottler, F. (1918). Über die physikalischen Grundlagen der Einsteinchen Gravitationstheorie. Ann. Phys. (Leipzig), 56, 401-62.

Nariai, S. (1950). On some static solutions to Einstein's gravitational field equations in a spherically symmetric case. Science Reports of the Tohoku University, 34, 160-7.

Nariai, S. (1951). On a new cosmological solution of Einstein's field equations of gravitation. Science reports of the Tohoku University, 35, 62-7.

Page, D. N. (1978). A compact rotating gravitational instanton. Phys. Lett., 79B, 235-8.

THE COSMOLOGICAL POTENTIAL OF SUPERGRAVITY

C. M. Hull

Department of Applied Mathematics and Theoretical Physics, Silver Street, Cambridge, England.

In the very earliest stages of the universe, quantum gravity is likely to have played an important role. One possible way of avoiding the difficulties associated with the non-renormalizability of quantum theories of gravity is to seek to embed gravity in some finite theory. In theories that possess a supersymmetry transforming boson and fermion degrees of freedom into one another there may be a cancellation of ultra-violet divergences. They also offer the prospect of unifying all the fundamental matter and interactions in a single theory. Such a superunified theory might then, in some sense, present an ultimate theory of nature.

Candidates for supsymmetric theories of gravity are the supergravity and supersymmetric string models. (For reviews and further references, see van Nieuwenhuizen (1981) and Schwartz (1982)). In the low energy limit the string theories tend to supergravity models and here only supergravity will be considered.

The Higgs mechanism plays a crucial role in modern physics and the scalar potential V determines the phase transitions and the effective cosmological constant. This is rather unnatural, however, since the potential is usually put in by hand to get a desired answer, such as an inflationary universe. It is of interest to try instead working from a possible `ultimate' theory, using the potential determined uniquely by the symmetries of the model and investigate symmetry breaking.

Extended Supergravity

Extended supergravity has N independent local supersymmetries. If the maximum spin is to be two, then N must be less than 8. Then N = 8 supergravity (Cremmer & Julia 1979) is the largest model and is likely to be the `most finite' since it has the greatest symmetry.

The super-multiplet consists of a graviton e^{α}_{μ}, eight spin - 3/2 gravitini ψ^{i}_{μ}, twenty-eight Abelian gauge bosons $A_{\mu[ij]}$, fifty-six spin - 1/2 fermions $\lambda_{[ijk]}$ and seventy scalars $\varphi_{[ijkl]}$. The spinors are Majorana (real), each field is antisymmetric in the internal indices

$i,j,k,\ldots = 1,\ldots,N$ and the scalars φ_{ijkl} are complex and self-dual in these internal indices. The gravitini are the 'gauge particles' introduced in making the supersymmetry local and the theory possesses a rigid SO(N) symmetry, rotating the N supersymmetries into one another.

This SO(N) symmetry has been gauged for all $N \leqslant 8$ by de Wit & Nicolai (1981). The $A_{\mu ij}$ become SO(N) gauge fields in the adjoint representation which couple minimally with gauge coupling g. The scalars gain a potential $V(\varphi)$ giving rise to a cosmological term $V(\varphi_0)$, which is proportional to g^2 where φ_0 is at a stationary point of V. It has been shown by Christensen et al., (1980) and Stelle & Townsend (1982) that at one loop infinities cancel so that g receives only finite renormalization for N > 4.

The potential $V(\varphi)$ determined by the theory is, at first sight, problematic. In the N = 4 theory, which has one complex scalar φ, the potential is negative and unbounded below,

$$V(\varphi) = -2g^2(1+2(1-|\varphi|^2)^{-1}) \qquad (1)$$

where $|\varphi|<1$. On its own the scalar sector is unstable and since the N = 4 model is a sub-theory of the higher N models, this problem persists there also.

It has been shown, however, by Breitenlohner & Freedman (1982a,b) and Gibbons et al., (1982) that coupling to gravity renders the theory stable. In particular, it will be seen that anti-de Sitter (AdS) space, with the cosmological constant given by a local maximum of the potential, provides a classically stable ground state about which a quantum theory might be constructed and this vacuum will also be quantum mechanically stable if a quantum theory exists that is supersymmetric. The gravity-scalar sector of the model is also stable, showing that potentials that are negative and unbounded below can make sense.

Positive Mass and Stability

In Newtonian theory, the gravitational potential is negative, since gravity is attractive, so that shrinking a system can make the total energy arbitrarily negative. In general relativity, the total energy receives contributions from the positive local energy density and the gravitational potential energy, which is non-local and even more negative in the

Newtonian case. However, for spaces that tend asymptotically to Minkowski or AdS space the total mass can never become negative. It appears that, provided the 'cosmic censorship' hypothesis holds, if a system collapses beyond a certain size a horizon forms, resulting in a black hole which has positive mass, with a singularity hidden inside.

The positivity of mass in general relativity was proved for asymptotically flat spaces by Schoen & Yau (1979, 1981), with Minkowski space as the unique zero energy state, and proving a similar result in gauged supergravity theories for asymptotically AdS (AAdS) spaces is the key to the stability result. This is because if the zero-energy state is unique and all other configurations with appropriate boundary conditions have higher energy, then there is no system that the zero-energy state can decay to and so it must be stable to all processes preserving the boundary conditions.

Asymptotically Anti-de Sitter Spaces

Since one might expect the ground state of a theory to be maximally symmetric and, in particular, translationally invariant it is convenient to consider backgrounds that are only non-trivial in the gravitational and scalar sectors, with the scalar fields constant, $\varphi(x) = \varphi_0$. The scalar field equation then implies that φ_0 lies at a stationary point of the potential V giving rise to a cosmological constant

$$\Lambda = V(\varphi_0) \tag{2}$$

so that the background metric $\bar{g}_{\mu\nu}$ satisfies

$$\bar{R}_{\mu\nu}(\bar{g}_{\mu\nu}) = -V(\varphi_0)\bar{g}_{\mu\nu}. \tag{3}$$

An appropriate choice of solution $\bar{g}_{\mu\nu}$ to (3) is the AdS metric which can be written

$$ds^2 = (1-\Lambda r^2/3)dt^2 - (1-\Lambda r^2/3)^{-1}dr^2 - r^2(d\theta^2 + \sin^2\theta d\varphi^2), \tag{4}$$

where $0 \leqslant r < \infty$, $-\infty < t < \infty$ and θ,φ are the standard spherical polar angles. (The conventions are given in Gibbons, et al., (1982)). This background is invariant under the AdS group O(3,2) generated by the ten Killing vector fields $\xi^{AB}_{\mu}(x)$ of (4), $(A, B = 0, 1, \ldots 4,\ \xi^{AB}_{\mu} = -\xi^{BA}_{\mu})$. If $\Lambda = -6g^2$, the space is also invariant under N supersymmetries $\delta_Q(\epsilon^{(m)})$, labeled by $m = 1, \ldots N$, whose spinorial parameters $\epsilon^{(m)i}_{\alpha}(x)$ are known as Killing spinor fields. Here α labels spinor components and the index i is an internal space index, $i = 1, \ldots N$.

For example, the change of the gravitino field $\psi^{i}_{\mu}(x)$ under an infintesimal supersymmetry is given by

$$\delta_Q(\epsilon^{(m)})\psi^{i}_{\mu} = \hat{\nabla}_{\mu}{}^{i}{}_{j}\epsilon^{i(m)} \tag{5}$$

where $\hat{\nabla}$ is a differential operator which for N = 8 is of the form

$$\hat{\nabla}_{\mu}{}^{i}{}_{j}\epsilon^{j} = D_{\mu}{}^{i}{}_{j}\epsilon^{j} - g\, i\sqrt{2}\, A_1^{ij}\gamma_{\mu}\epsilon_{j} + \ldots \tag{6}$$

where terms dependent on the fields that vanish in this AdS background are dropped, $D^{i}_{\mu j}$, is a gauge-covariant derivative and A_1^{ij} is a function of the scalar fields given by de Wit & Nicolai (1982). Thus a necessary condition for $\epsilon^{(m)}$ to be a Killing spinor is that

$$\hat{\nabla}_{\mu}{}^{i}{}_{j}\epsilon^{(m)j} = 0. \tag{7}$$

Consider now systems that tend asymptotically to this background with metric $g_{\mu\nu}$

$$g_{\mu\nu} \to \bar{g}_{\mu\nu} + 0(1/r) \tag{8}$$

and scalar fields $\varphi(x)$

$$\varphi \to \varphi_0 + 0(1/r). \tag{9}$$

The approximate symmetry group of the asymptotic region is then that of AdS space, the supergroup $OSp(N,4)$, and corresponding to each of the Killing spinors ϵ^{i} and vectors ξ^{AB} there will be total charges (Abbott & Deser 1982). These are the total four-momentum $J^{\mu 4}$, the total angular momentum $J^{\mu\nu}$, $(\mu, \nu = 0, \ldots 3,\ J^{AB} = J^{\mu 4}, J^{\mu\nu})$ and the total supercharge $Q^{(m)}$ which is given by a surface integral over a two-sphere S at infinity:

$$Q^{(m)} = \int_S \epsilon^{\mu\nu\rho\sigma} \bar{\alpha}_i^m \gamma^5 \gamma_\mu \psi_\nu^i \, d\Sigma_{\rho\sigma} \tag{10}$$

where $\alpha^{(m)i}$ is the commuting spinor that gives the component of $\epsilon^{(m)i}$ with respect to a basis of the Grassman algebra.

The operators corresponding to these charges satisfy the $OSp(N,4)$ algebra, which includes the relation

$$\{Q^{(m)},\bar{Q}^{(n)}\} = \hbar\sigma_{A\,B}^{(m)(n)} J^{AB} \tag{11}$$

where $\sigma_{\mu\nu} = \frac{1}{4}[\gamma_\mu,\gamma_\nu]$, $\sigma_{\mu 4} = \frac{i}{2}\gamma_\mu$ and the central charges are assumed to vanish. Relation (11) implies, on tracing with γ^0, that the energy operator J^{04} is given by

$$J^{04} = \frac{1}{4N}\hbar \sum_{\alpha,(m)} (Q_\alpha^{(m)} Q_\alpha^{(m)+} + Q_\alpha^{(m)+} Q_\alpha^{(m)}) \tag{12}$$

and so is a positive operator, giving positive energy for any physical state (Deser & Teitelboim 1977); Breitenlohner & Freedman 1982a,b). Grisaru (1978) has argued that this result should still apply in the classical limit of vanishing $\hbar$ and supercharge. However, this formal argument relies on the existence of a consistent supersymmetric quantum theory.

The Positive Mass Theorem

A rigorous classical proof of the positivity result for gauged supergravity has been given by Gibbons et al., (1982). It generalises the argument given by Witten (1981) for asymptotically flat spaces. The proof is based on a classical version of the global superalgebra (11) expressing the energy in terms of the charge of the supercharge $Q^{(n)}$ under the supergravity transformation $\delta_Q(\epsilon^{(m)})$ (Hull 1983a)

$$\delta_Q(\epsilon^{(m)})\, Q^{(n)} = \sigma_{A\,B}^{(n)\,(m)} J^{AB} \tag{13}$$

Letting $\beta^{(m)}(x)$ be a commuting asymptotic Killing spinor, that is one that tends to $\alpha^{(m)}$, the commuting 'part' of $\epsilon^{(m)}$,

$$\beta^{(m)} \to \alpha^{(m)}, \quad \hat{\nabla}_\mu \beta^{(m)} \to O(\frac{1}{r^2}) \tag{14}$$

as $r \to \infty$ and using (5) and (10), (13) becomes

$$J^{AB}\,\sigma_A^{(m)}\,{}_B^{(m)} = \int_S \epsilon^{\mu\nu\rho\sigma}\overline{\beta_i^{(m)}}\,\gamma^5\gamma_\mu\hat{\nabla}_\nu\beta^{(m)i}\,d\Sigma_{\rho\sigma} \tag{15}$$

On using the divergence theorem and the field equations this becomes a volume integral over a space-like three-surface Σ with boundary S.

$$J^{AB}\,\sigma_A^{(m)}\,{}_B^{(m)} = \int_\Sigma \Big[\overline{\beta_i^{(m)}}\,\gamma^\mu\beta^{(m)i}T^{MAT}_{\mu\sigma}$$

$$+\frac{1}{4}\epsilon^{\mu\nu\rho\sigma}\overline{(\hat{\nabla}_\mu\beta^{(m)})_i}\,\gamma^5\gamma_\nu(\hat{\nabla}_\rho\beta^{(m)})^i \tag{16}$$

$$+\overline{\delta\lambda_{ijk}}\gamma_\sigma\delta\lambda^{ijk}\Big]\,d\Sigma^\rho.$$

where $\delta\lambda_{ijk} = \delta_Q(\beta^{(m)})\lambda_{ijk}$ and $T^{MAT}_{\mu\nu}$ is the stress energy of any fields or sources that are not part of the supermultiplet. If $T^{MAT}_{\mu\nu}$ satisfies

$$T^{MAT}_{\mu\nu}\,U^\mu\,V^\nu \geqslant 0 \tag{17}$$

for all non-space-like vectors U, V and $\beta^{(m)}(x)$ is chosen to satisfy

$$\sum_{\mu=1}^{3}\gamma^\mu\hat{\nabla}_\mu\beta = 0, \tag{18}$$

then the right hand side of (16) is non-negative and vanishes if and only if $T^{MAT}_{\mu\nu} = 0$ and $\beta^{(m)}$ is an exact Killing spinor with

$$\delta\psi_\mu = \hat{\nabla}_\mu\beta^{(m)} = 0, \quad \delta\lambda_{ijk} = 0 \tag{19}$$

which is only possible if the space is exactly AdS, $g_{\mu\nu} = \bar{g}_{\mu\nu}$ (the details are given by Gibbons et al., (1982)).

Thus an AdS space admitting Killing spinors, that is one with $V(\varphi_0) = -6g^2$, is a unique zero-energy state and any other configuration tending asymptotically to this space will have a greater total energy. Then such a supersymmetric AdS space will be stable, in spite of the scalar potential being unbounded below.

The positivity theorems can be extended in a number of ways. They have been generalised by Gibbons et al., (1982) to include space-times that are regular outside apparent horizons. A similar theorem can be derived for models with supergravity coupled to supermatter in which a rather general scalar potential can be introduced (Hull 1983b). A lower bound can be given for the energy in terms of central charges (Gibbons & Hull 1982) and the 'magnetic mass' or NUT charge (Hull 1983c).

Symmetry Breaking, Bubbles and Foam

In the N = 4 theory, the potential (1) has a unique stationary point, a global maximum at $\varphi = 0$. Since $V(0) = -6g^2$ the space is indeed supersymmetric and will be stable. In the N = 5 theory in addition to a stable supersymmetric local maximum at $\varphi = 0$ there is a set of saddle points $\varphi = \varphi_1$ where $V(\varphi_1) = -14g^2$. These solutions do not admit Killing spinors and the above argument does not apply. However, the analysis of Breitenlohner & Freedman (1982a,b) shows that these saddle points will be stable to small perturbations, although it seems likely that they will be unstable to large fluctuations or to tunnelling processes. Anti-de Sitter spaces with $\varphi = \varphi_1$ maintain the O(3,2) space-time symmetry but the SO(5) gauge symmetry is broken by this vacuum to SO(3) and all the supersymmetry is spontaneously broken.

The N = 8 theory has a stable local maximum at $\varphi = 0$ and a set of saddle points similar to those in the N = 5 model that break all supersymmetry and break the SO(8) gauge group to SO(4) but it is not known if other critical points of the N = 8 potential exist.

The asymmetric stationary points lead to locally stable vacua that spontaneously break supersymmetry and gauge symmetry and, since they arise in a completely natural way from the theory without any arbitrary choices of potential etc., merit further study.

The way gravity stabilses a scalar theory with a pathological potential can be understood physically in terms of bubble formation. The decay of an AdS space with cosmological constant $\Lambda = \Lambda_1$, into one with a lower value of Λ, $\Lambda = \Lambda_2$, would presumably proceed by the formation of bubbles of true vacuum. One would expect regions of AdS space with $\Lambda = \Lambda_2$ expanding into regions of AdS space with $\Lambda = \Lambda_1$ Such a bubble solution must have zero energy as measured at large distances in the old false vacuum since energy is conserved in semi-classical processes.

The positive energy theorems show that there can be no such bubble solution.

Coleman & De Luccia (1980) have suggested a way of understanding the means by which the geometry inhibits bubble formation. As a result of the highly non-Euclidean geometry of the surfaces of constant time in AdS space the ratio of three-volume to surface area of a sphere tends to a constant as the size of the sphere increases. This implies that the energy gained by reducing the energy density inside the bubble may not be sufficient to compensate for the energy needed for the bubble walls.

The cosmological constant in gauged supergravity theories is large and negative while observations suggest Λ should be extremely small or zero. It may be that through some non-perturbative mechanism the cosmological constant observed in astronomy may be very different from the 'bare' cosmological constant in the Lagrangian. One possibility is that space-time foam (Hawking 1978) may play a role, with space-time being highly curved at Planck scales while appearing flat at large scales. The preliminary results reviewed here suggest that, despite one's flat-space prejudices against unbounded negative potentials, gauged supergravity may be a physically interesting theory with naturally arising scalar potentials and symmetry breaking mechanisms.

Acknowledgements

Much of the original work reported here was done in collaboration with G.W. Gibbons and N.P. Warner. I would also like to thank S.W. Hawking and M. Rocek for useful discussions.

References

Abbott, L.F. & Deser, S. (1982). Stability of Gravity with a Cosmological Constant. Nucl. Phys. B195, 76-96.

Breitenlohner, P. & Freedman, D.Z., (1982a). Positive Energy in anti-deSitter Backgrounds and Gauged Extended Supergravity. Phys. Lett. 115B, 197-201

Breitenlohner, P. & Freedman, D.Z., (1982b) Stability in Gauged Extended Supergravity. M.I.T. Preprint.

Christensen, S.M., Duff, M.J., Gibbons, G.W. & Rocek, M. (1981). Vanishing One-Loop β Function in Gauged N > 4 Supergravity. Phys. Rev. Lett. 45, 161-165.

Coleman, S. & De Luccia, F. (1980). Gravitational Effects on and of Vacuum Decay. Phys. Rev. D21, 3305 -3315.

Cremmer, E. & Julia, B., (1979). The S0(8) Supergravity. Nucl. Phys. B159. 141-212.

Deser, S. & Teitelboim, C. (1977). Supergravity has Positive Energy. Phys. Rev. Lett. 39, 249-252.

de Wit, B. & Nicolai, M. (1981). N = 8 Supergravity with Local S0(8) x SU(8) Invariance. Phys. Lett. 108B, 285-290.

Gibbons, G.W., Hawking, S.W., Horowitz, G.W. & Perry, M.J. (1982). Positive Mass Theorems for Black Holes. Princeton University Preprint.

Gibbons, G.W. & Hull, C.M. (1982). A Bogomolny Bound for General Relativity and Solitons in N = 2 Supergravity. Phys. Lett. 109B, 190-194.

Gibbons, G.W., Hull, C.M. & Warner, N.P. (1982). The Stability of Gauged Supergravity. D.A.M.T.P. Preprint.

Grisaru, M.T., (1978). Positivity of the Energy in Einstein Theory. Phys. Lett. 73B, 207-208.

Hull, C.M. (1983). The Positivity of Gravitational Energy and Global Supersymmetry. D.A.M.T.P. Preprint.

Hull, C.M. (1983b). Spontaneous Breaking of Supersymmetry in Curved Space-Time. D.A.M.T.P. Preprint.

Hull, C.M. (1983c). In preparation.

Schoen, R. & Yau, S.T. (1979). On the Proof of the Positive Mass Conjecture in General Relativity. Comm. Math. Phys. 65, 45-76.

Schoen, R & Yau, S.T. (1981). Proof of the Positive Mass Theorem II. Comm. Math. Phys. 79, 231-260.

Schwartz, J. (1982). Superstring Theory. Physics Reports, 89C, 223.

Stelle, K.S. & Townsend, P.K. (1982). Vanishing β-Functions in Extended Supergravities. Phys. Lett. 113B, 25-30.

van Nieuwenhuizen, P. (1981). Supergravity. Physics Reports 68, 4, 192-398.

Witten, E. (1981). A New Proof of the Positive Energy Theorem. Comm. Math. Phys. 80, 381-402.

CONFERENCE SUMMARY AND CONCLUDING REMARKS

F. Wilczek
Institute for Theoretical Physics
University of California, Santa Barbara
Santa Barbara, California 93106 USA

It has been a very enjoyable and exciting three weeks, and now we all have a lot to take home and think about at leisure. In this summary I'd like to try not so much to summarize the individual contributions, which the authors have done ably themselves, but rather to step back and try to get some perspective on our current (rapidly changing!) understanding. Some observations which came up in discussions over lunch or tea are also mentioned.

INFLATIONARY UNIVERSE

By my count seventeen of the thirty-six lectures at the workshop were mainly about the idea of an inflationary Universe, and many of the other talks were heavily influenced by this idea. This count attests to the enormously seductive appeal of the idea. By supposing that the Universe inflates by an enormous factor, we explain at one stroke why it is observed to be so nearly flat, so nearly homogeneous, and perhaps why it is so nearly free of various topological monstrosities such as magnetic monopoles, axion domain walls, primordial black holes... The idea is so simple, and yet it provides a qualitative understanding of some of the deepest puzzles of cosmology!

All the same, we must recognize that both the theoretical framework and the observational foundations of inflationary cosmology are shaky indeed.

On the theoretical side, it must be admitted that our understanding of the cosmological term is thoroughly unsatisfactory. In a typical unified gauge theory, the symmetric vacuum differs in energy density from our current vacuum by

$(10^{15}\text{GeV.})^4$; on the other hand we know that gravity acts as if the current vacuum energy differs by no more than $(10^{-12}\text{GeV.})^4$ from zero. This is an extraordinary discrepancy, by 108 orders of magnitude! Even a factor $e^{-1/\alpha} \simeq 10^{-60}$ would not be quite enough here, though $e^{-\pi/\alpha}$ would do! Another possible form of the suppression factor which might arise is $e^{-M_{pl.}/\mu}$, where μ is the unification scale. What seems clear is that some sort of exponential suppression is required. At present there are only the most shadowy ideas about how to achieve such a suppression.

It is surely an act of cosmic *chutzpah* to use this dismal theoretical failure as a base for erecting theoretical superstructures, but of course this is exactly what is done in current inflationary models. The cosmological constant is renormalized to be zero in our vacuum, and extrapolated back naively to the symmetric vacuum as if our theories needed no modification. Of course this is the best one can do at present, but the provisional nature of these speculations cannot be overemphasized.

In a suspiciously similar way, current inflationary scenarios rely heavily on the Coleman-E. Weinberg mechanism -- vanishing of a certain complicated mixture of bare couplings and mass terms -- whose justification is just as elusive.

Finally, unified gauge theories have an internal problem -- the gauge hierarchy problem -- of a very similar sort, the *a priori* unexpected vanishing of a certain mass parameter.

For these reasons I think it is fair to say that while the general idea of an inflationary universe is extremely attractive, the specific models so far put forward do not inspire confidence in detail. Can we test the basic idea observationally?

The most dramatic qualitative prediction of the inflation idea is that the Universe should be essentially flat -- with well-known consequences for the mass density ($\Omega = 1$), the deceleration parameter, and the age of the Universe. All the observations marginally favor an open Universe, but probably they do not quite exclude flatness. The most interesting is probably the mass density. On scales of galaxies and small groups of galaxies, $\Omega \leqslant .3$ seems to be favored and $\Omega = 1$ pretty definitely excluded. If we are to have $\Omega = 1$ when averaged over the largest scales, there must be a form of non-luminous matter which does not cluster efficiently supplying the missing mass. Massive neutrinos, gravitinos, and now axion coherent field energy are popular candidates. Ordinary matter (baryons) will almost certainly not do, because the density of baryons needed to make $\Omega = 1$ now would drastically suppress production of deuterium and ^3He in the early Universe, contrary to observation. Massive neutrinos may also have problems -- neutrinos heavy enough to give $\Omega = 1$ probably cluster on small scales, which is not what is seen.

MORE ON FLUCTUATIONS

Beautiful work on the spectrum of fluctuations expected in detailed inflationary models was carried on by several groups at the workshop, and mutual agreement was obtained after some struggles. Although for the reasons mentioned above I think the models should be viewed with great scepticism, the results are of some methodological interest, and related techniques may be applicable to an eventual correct theory. It should be emphasized that the inflationary scenario is the first framework in which the zeroth-order result - essential isotropy and homogeneity of the Universe - is explained. Of course, without such an explanation no calculation of fluctuations can be complete.

The result of the calculations is that the fluctuations predicted in the canonical Coleman-E. Weinberg model are essentially scale-invariant, as speculated by Zeldovich, but too large in absolute magnitude by a factor 10^4.

STRINGS

The strings which arise from spontaneous breaking of a global symmetry are perhaps better called sausages, since their energy density falls off only as $1/r^2$ as we move away from the axis. Numerically, if the scale of symmetry breaking is 10^{14}GeV. then the density of string energy is greater than

ρ_c the critical density out to $r = 10^{21}$cm.! This sounds dramatic enough, but no one was able to suggest sensible means of detecting such objects; perhaps we missed something.

Another kind of string arises when a local gauge symmetry breaks down to a discrete subgroup. This typically happens when the Higgs field doing the symmetry breaking belongs to a non-fundamental representation of the group. The energy in this kind of string is much more (exponentially) localized. Vilenkin suggests that such strings might lead to observable light deflection, e.g. for quasars.

Strings with parameters not far from those suggested by unified theories might seed fluctuations suitable for initiating galaxy formation. For strings associated with broken global symmetries the scale of breaking is $\gtrsim 10^{14}$GeV., which is quite close to the unification scale but significantly larger than the scale suggested by axions. For strings associated with broken gauge symmetries the interesting scale of breaking is $\gtrsim 10^{16}$GeV., which is large for SU(5) but perhaps could be associated with an extended gauge symmetry. Whether a significant density of these strings would survive inflation is most doubtful, in particular if we are relying on inflation to dilute monopoles weighing 10^{16}GeV.

MONOPOLES AND AXIONS

You have my summary of the complicated situation regarding these subjects earlier in this volume.

A SHOPPING LIST OF QUESTIONS

I find a useful way both of gauging our progress and of keeping in touch with reality is to think about what significant questions we'd like (and cannot unreasonably expect) to see answered in the near future. So here is a short list suggested by this workshop:

1) Are the foundations secure?

All our speculation start from gauge theories: we must have more precise QCD tests, find the Z and W boson, and so forth. Specifically, by measuring the width of the Z one will be able to check that there are no more neutrinos (or perhaps one more?) as demanded by calculations of primordial nucleosynthesis. Measurement of the mass of the Z will give us a more precise value for the Weinberg angle, which will help us decide if SU(5) is adequate for unification.

Of course many speculations will be confirmed if nucleon decay is observed with a lifetime $\tau \sim 10^{30}$ years. The decay channels could tell us, among other things, whether supersymmetry is relevant to low-energy ($< 10^{15}$GeV.!) physics - if so we expect strange particles in the final state.

A crucial question, I think underestimated, is the nature and mass of the Higgs particle. If it exists with the mass suggested by Coleman and E. Weinberg ($\sim$ 10GeV.) it should soon be discovered in the decay of heavy vector mesons (toponium or Z).

2. Is $\Omega = 1$?

If not, we must give up on inflation. If $\Omega \to 1$ on the largest scales, a non-luminous, non-dissipative form of matter must dominate the mass of the Universe!

3. Is the Zeldovich blini (Russian pancake) picture of galaxy formation correct?

Simple pictures of baryonsynthesis lead to adiabatic fluctuations, and thereby to the Zeldovich picture of galaxy formation. In this picture, fluctuations begin on the scale of clusters of galaxies and steepen by a one-dimensional collapse into a shock wave leading to a pancake which fragments into galaxies and smaller objects. (Was there really enough time to create small, old objects like globular clusters by such an indirect means?)

One relevant fact of observation is the filamentary structure of galaxies which seems apparent to the naked eye in sky surveys -- certainly suggestive of pancakes. So far this structure has defied statistical characterization. It was suggested that filamentary structure would be more apparent if one considered correlation functions of holes (parts of the sky underpopulated by galaxies) rather than correlation functions of galaxies.

Finally, if galaxies developed from adiabatic fluctuations then the imprint of these fluctuations leaving inhomogeneities on the microwave background cannot be much below current experimental limits.

4. Do neutrinos have cosmologically interesting masses?

Neutrino oscillation experiments all but rule this out unless the mixing angles between neutrinos are extremely small; perhaps the limits on $\nu_\mu \to \nu_\tau$ can be pushed further.

5. What about monopoles?

Are they really observed? How efficiently do they catalyze baryon number violation?

6. What about axions and familons?

I'm certainly anxious to hear about rare K and μ decays!

7. What about the cosmological constant?

"Whereof one cannot speak, thereof one must be silent." - Wittgenstein.